BAYESIAN DECISION ANALYSIS

Bayesian decision analysis supports principled decision making in complex but structured domains. The focus of this textbook is on the faithful representation and conjugate analyses of discrete decision problems. It takes the reader from a formal analysis of simple decision problems to a careful analysis of the sometimes very complex and data rich structures confronted by practitioners. The book contains basic material on subjective probability theory and multiattribute utility theory, event and decision trees, Bayesian networks, influence diagrams and causal Bayesian networks. The author demonstrates when and how the theory can be successfully applied to a given decision problem, how data can be sampled and expert judgements elicited to support this analysis, and when and how an effective Bayesian decision analysis can be implemented.

Evolving from a third-year undergraduate course taught by the author over many years, all of the material in this book will be accessible to a student who has completed introductory courses in probability and mathematical statistics.

JIM Q. SMITH is a Professor of Statistics at the University of Warwick.

BAYESIAN DECISION ANALYSIS

Principles and Practice

JIM Q. SMITH
University of Warwick

CAMBRIDGE
UNIVERSITY PRESS

CAMBRIDGE UNIVERSITY PRESS
Cambridge, New York, Melbourne, Madrid, Cape Town, Singapore,
São Paulo, Delhi, Dubai, Tokyo, Mexico City

Cambridge University Press
The Edinburgh Building, Cambridge CB2 8RU, UK

Published in the United States of America by Cambridge University Press, New York

www.cambridge.org
Information on this title: www.cambridge.org/9780521764544

First published 2010

Printed in the United Kingdom at the University Press, Cambridge

A catalogue record for this publication is available from the British Library

Library of Congress Cataloguing in publication Data
Smith, J. Q., 1953–
Bayesian decision analysis : principles and practice / Jim Q. Smith.
p. cm.
Includes bibliographical references and index.
ISBN 978-0-521-76454-4
1. Bayesian statistical decision theory. I. Title.
QA279.5.S628 2010
519.5′42–dc22 2010031690

ISBN 978-0-521-76454-4 Hardback

Contents

Preface

This book introduces the principles of Bayesian Decision Analysis and describes how this theory can be applied to a wide range of decision problems. It is written in two parts. The first presents what I consider to be the most important principles and good practice in mostly simple settings. The second part shows how the established methodology can be extended so that it can address the sometimes very complex and data-rich structures a decision maker might face. It will serve as a course book for a 30-lecture course on Bayesian decision modelling given to final-year undergraduates with a mathematical core to their degree programme and statistics Master's students at Warwick University. Complementary material given in two parallel courses, one on Bayesian numerical methods and the other on Bayesian Time Series given subsequently at Warwick, is largely omitted although these subjects are motivated within the text. This book contains foundational material on the subjective probability theory and multiattribute utility theory – with a detailed discussion of efficacy of various assumptions underlying these constructs – and quite an extensive treatment of frameworks such as event and decision trees, Bayesian Networks, as well as Influence Diagrams and Causal Bayesian Networks. These graphical methods help draw different aspects of a decision problem together into a coherent whole and provide frameworks where data can be used to support a Bayesian decision analysis.

This is not just a text book; it also provides additional material to help the reader develop a more profound understanding of this fascinating and highly cross-disciplinary subject. First, it includes many more worked examples than can be given in a such a short programme. Second, I have supplemented this material with extensive practical tips gleaned from my own experiences which I hope will help equip the budding decision analyst. Third, there are supplementary technical discussions about when and why a Bayesian decision analysis is appropriate. Most of this supplementary material is drawn from various postgraduate and industrial training courses I have taught. However all the material in the book should be accessible and of interest to a final-year maths undergraduate student. I hope the addition of this supplementary material will make the book interesting to practitioners who have reasonable skills in mathematics and help them hone their decision analytic skills. An asterisk denotes that a section contains more advanced material and can be skipped without loss of continuity to the rest of the text.

The book contains an unusually large number of running examples which are drawn – albeit in a simplified form – from my experiences as an applied Bayesian modeller and used to illustrate theoretical and methodological issues presented in its core. There are many exercises throughout the book that enable the student to test her understanding. As far as possible I have tried to keep technical mathematical details in the background whilst respecting the intrinsic rigour behind the arguments I use. So the text does not require an advanced course in stochastic processes, measure theory or probability theory as a prerequisite.

Many of the illustrations are based on simple finite discrete decision problems. I hope in this way to have made the book accessible to a wider audience. Moreover, despite keeping the core of the text as nontechnical as possible, I have tried to leave enough hooks in the text so that the advanced mathematician can make these connections through pertinent references to more technical material. Over the last 20 years many excellent books have appeared about Bayesian Methodology and Decision Analysis. This has allowed me to move quickly over certain more technical material and concentrate more on how and when these techniques can be drawn together. Of course some important topics have been less fully addressed in these texts. When this has happened I have filled these gaps here.

Obviously many people have influenced the content of the book and I am able here only to thank a few. I learned much of this material from conversations with Jeff Harrison, Tom Leonard, Tony O'Hagan, Chris Zeeman, Dennis Lindley, Larry Phillips, Bob Oliver, Morris De Groot, Jay Kadane, Howard Raiffa, Phil Dawid, Michael Goldstein, Mike West, Simon French, Saul Jacka, Steffen Lauritzen and more recently with Roger Cooke, Tim Bedford, Joe Eaton, Glen Shafer, Milan Studeny, Henry Wynn, Eva Riccomagno, David Cox, Nanny Wermuth, Thomas Richardson, Michael Pearlman, Lorraine Dodd, Elke Thonnes, Mark Steel, Gareth Roberts, Jon Warren, Jim Griffin, Fabio Rigat and Bob Cowell. Postdoctoral fellows who were instrumental in jointly developing many of the techniques described in this book include Alvaro Faria, Raffaella Settimi, Nadia Papamichail, David Ranyard, Roberto Puch, Jon Croft, Paul Anderson and Peter Thwaites. Of course my university colleagues and especially my PhD students, Dick Gathercole, Simon Young, Duncan Atwell, Catriona Queen, Crispin Allard, Nick Bisson, Gwen Tanner, Ali Gargoum, Antonio Santos, Lilliana Figueroa, Ana Mari Madrigal, Ali Daneshkhah, John Arthur, Siliva Liverani, Guy Freeman and Piotr Zwirnick have all helped inform and hone this material. My thanks go out to these researchers and the countless others who have helped me directly and indirectly.

Part I

Foundations of Decision Modelling

1
Introduction

1.0.1 Prerequisites and notation

This book will assume that the reader has a familiarity with an undergraduate mathematical course covering discrete probability theory and a first statistics course including the study of inference for continuous random variables. I will also assume a knowledge of basic mathematical proof and notation.

All observable random variables, that is all random variables whose values could at some point in the future be discovered, will be denoted by an upper case Roman letter (e.g. X) and its corresponding value by a lower case letter (e.g. x). In Bayesian inference parameters – which are usually not directly observable – are also random variables. I will use the common abuse of notation here and denote both the random variable and its value by a lower case Greek letter (e.g. θ). This is not ideal but will allow me to reserve the upper case Greek symbols (e.g. Θ) for the range of values a parameter can take. All vectors will be row vectors and denoted by bold symbols and matrices by upper case Roman symbols. I will use $=$ to symbolise a deduced equality and denote that a new quantity or variable is being defined as equal to something via the symbol $\triangleq$.

1.0.2 Bayesian decision analysis and the scope of this book

This book is about Bayesian decision analysis. Bayesian decision analysis seriously intersects with Bayesian inference but the two disciplines are distinct. A Bayesian inferential model represents the structure of a domain and its uncertainties in terms of a single probability model. In a well-built Bayesian model logical argument, science, expert judgements and evidence – for example given in terms of well-designed experiments and surveys – are all used to support this probability distribution. In their most theoretical forms these probability models simply purport to *explain observed scientific phenomena or social behaviour*. In their more applied settings it is envisaged that the analysis can be structured as a probabilistic expert system for possible used in the support decision processes whose precise details are currently unknown to the experts designing the system.

In contrast a Bayesian decision analysis is focused on solving a *given* problem or class of problems. It is of course important for a decision maker (DM) to take due regard of the expert judgements, current science and respected theories and evidence that might be summarised within a probabilistic expert system. However she needs to *apply* such domain knowledge to the actual problem she faces. She will usually only need to use a small subset of the expert information available. She therefore needs not only to draw on that small subset of the expert information that is *relevant* to her problem at hand – augmenting and complementing this as necessary with other context-specific information but also to use this probabilistic information to help her make *the best decision she can* on the basis of the information available to her. When modelling for inference it is not unusual to conclude that there is not enough information to construct a model. But this will not usually be an option for a DM. She will normally have to make do with whatever information she does have and work with this in an intelligent way to make the best decision she can in the circumstances.

The Bayesian decision analyses described in this book provide a framework that:

(1) is based on a formalism accommodating beliefs and preferences as these impact on the decision making process in a *logical* way;
(2) *draws together* sometimes diverse sources of evidence, generally acknowledged facts, underlying best science and the different objectives relevant to the analysis into a single coherent description of her given problem;
(3) provides a description that *explains* to a third party the reasons behind the judgements about the efficacy and limitations of the candidate decisions available so that these judgements can be understood, discussed and appraised;
(4) provides a framework where *conflict* of evidence and conflict of objectives can be expressed and managed appropriately.

The extent to which the foundations of Bayesian decision analysis has been explained, examined and criticised is unparalleled amongst its competitors. As stated in Edwards (2000); French *et al.* (2009) there is an enormous literature on this topic and it would be simply impossible in a single text to do justice to this. However the level of scrutiny it has attracted over the last 90 years has not only refined its application but also defined its domain of applicability. In Chapters 3, 4 and 6 I will review and develop some of this background material justifying the encoding of problems so that uncertainties are coded probabilistically and decisions are chosen to maximise expected utility.

I have therefore severely limited the scope of this book and addressed only a subset of settings and problems. This will allow me not only to present what I consider to be core material in a logical way but also to outline some important technical material in which I have a particular interest. The scope is outlined below.

(1) I will only discuss the arguments for and against a *probabilistic* framework for decision modelling. Furthermore, for practical reasons I will argue throughout the book, for a decision analysis the probabilistic reasoning assumed here is necessarily *subjective*.
(2) I consider only classes of decision problem where a single or group decision maker (DM) must find a *single agreed rationale* for her stated beliefs and preferences and it is this DM who is *responsible* for and has the *authority* to enact the decisions made. The DM will often take advice from experts to inform her beliefs. However if she admits an expert's judgement she adopts it as her own and is responsible for the judgements expressed in her decision model. Similarly

whilst acknowledging, as appropriate, the needs and aspirations of other stakeholders in the expression of her preferences, the DM will take responsibility for the propriety of any such necessary accommodation.

(3) The DM has the time and will to engage in building the type of *logical* and coherent framework that gives an *honest* representation of her problem. The model will support decision making concerning the current problem at hand in the first instance. However there will often be the promise that many aspects of the architecture and some of the expert judgements embodied in the model will be *relevant to analogous future problems* she might face.

(4) The DM is responsible for explaining the rationale behind her choice of decision in a compelling way to an *auditor*. This auditor, for example, may be an external regulator, a line manager or strategy team, a stakeholder, the DM herself or some combination of these characters. In this book we will assume that the auditor's role is to judge the plausibility of the DM's reasoning in the light of the evidence and the propriety of the scale and scope of her objectives.

(5) It is acknowledged by all players that the decision model is likely to have a limited shelf life and is intrinsically provisional. The DM simply strives to present an honest representation of her problem as she sees it at the *current* time. All accept that in the future her judgements may change in the light of new science, new surprising information and new imperatives and may later be adjusted or even discarded for its current or future analogous application.

The limited scope of this book allows us to identify various players in this process. There is the DM herself whose role is given above. There is an analyst who will support her in developing a decision model that can fulfil the tasks above as adequately as possible. There are domain experts to help her evaluate the potential effects on the objects of their expertise an enacted decision might have. Different experts may advise on different aspects of the DM's problem, but for simplicity we will assume that there is just one expert informing each domain of expertise. Throughout we will assume that the advice given by an expert will be no less refined than a probability forecast of what he believes will happen as a result of particular actions the DM might take.

Recent advances in Bayesian methodology have made it possible to support decision making in complex but highly structured domains, rich in expert judgements and informative but diverse experimental and survey evidence; see for example Cowell *et al.* (1999); Pearl (2000). Explanations and illustrations of how these advances can be implemented are presented in the second half of the book. The practical implementation of such decision modelling has its challenges. The analyst needs to guide the DM to first structure her problem by decomposing it into smaller components. Each component in the decomposition can then be linked to possibly different sources of information. The Bayesian formalism can then be used to recompose these components into a coherent description of the problem at hand. This process will be explained and illustrated throughout this book.

There are now many such qualitative frameworks developed and currently being developed, each useful for addressing a certain specific genre of problems. Perforce in this book I have had to choose a small subset if these frameworks that I have found particularly practically useful in a wide set of domains I have faced. These are the event/decision tree – discussed in Chapter 2 – the Bayesian network – discussed in Chapter 7 – and the influence diagram and causal Bayesian network discussed in Chapter 8.

In most moderate or large-scale decision making, the DM not only needs to discover good decisions and policies but also has to be able to provide reasons for her choice. The more compelling she can make this explanation the more likely it will be that she will not be inhibited in making the choices she intends to make. If her foundational rationale is accepted – and for the Bayesian one expounded below this is increasingly the case – she usually still has to convince a third party that the judgements, beliefs and objectives articulated through her decision model are appropriate to the problem she faces.

The frameworks for the decomposition of a problem discussed above are helpful in this regard because – being qualitative in nature – the judgements they embody are more likely to be shared by others. Furthermore they enable the DM to draw on any available evidence from statistical experiment and sample surveys, commonly acknowledged as being well conducted, to support as many quantitative statements she makes and use this to embellish and improve her probabilistic judgements. This draws us into an exploration of where Bayesian inference and Bayesian decision analysis intersect. In Chapter 5 we review some simple Bayesian analyses that inform the types of decision modelling discussed in this book. In Chapter 9 we discuss this issue further with respect to larger problems where significant decomposition is necessary.

One difficulty the DM faces when trying to combine evidence from different sources is when these pieces of evidence seem to give very different pictures of what is happening. When should the DM simply act as if aggregating the information and when should she choose a decision more supported by one source than another? Conflict can also arise when a problem has two competing objectives where all decisions open to the DM score well in one objective but not the other or only score moderately in both. When should the DM choose the latter type of policy and compromise and when should she concentrate in attaining high scores in just one objective? The Bayesian paradigm embodies the answers to these questions. Throughout the book I will show how various types of conflict within a given framework are being automatically managed and explained within the Bayesian methodology in the classes of problem I address.

1.0.3 The development of statistics and decision analysis

It is useful to appreciate why there has been such a growth in Bayesian methods in recent years. Some 35 years ago data-rich structures were only just beginning to be analysed using Bayesian methods. At that time inference still focused on deductions from data from a single (often designed) experiments. The influence of the physical sciences on philosophical reasoning – often through the social sciences which were striving to become more "objective" – was dominant and the complexity of inferential techniques was bounded by computational constraints. Bayesian modelling was not fashionable for a number of reasons:

(1) If decision making was to be objective then the Bayesian paradigm – based on subjective prior distributions and preferences represented via a utility function – was a poor starting point.
(2) Many of the top theoretical statisticians focused on problem formulations based on the physical and health sciences. This naturally led to the study of distributions of estimators from single

experiments that were well designed, likelihood ratio theory, simple estimation, analysis of risk functions and asymptotic inference for large data sets where distributions could be well approximated. Many foundational statistics courses in the UK still have this emphasis. In such problems where data could often be plausibly assumed to be randomly drawn from a sample distribution lying in a know parametrised family it was natural to focus inference on the development of protocols which remotely instructed the experimenter about how to draw inference over different classes of independent and structurally similar experiments. Here the obvious framework for inference was one which built on the properties of different tests and estimators which gave outputs that could be shared by any auditor. The framework of Bayesian inference, with its reliance on contextual prior information, seemed overly complicated and not particularly suited to this task.

(3) The development of stochastic numerical techniques was in its infancy. So for most large-scale problems, asymptotics were necessary. The common claim was that even if you were convinced that a Bayesian analysis should be applied in an ideal world the computations you would need to make were impossible to enact. You would therefore need to rely on large sample asymptotics to actually perform inferences. But these were exactly the conditions where frequentist approaches usually worked as well and more simply than their Bayesian analogues.

The environment had changed radically by the 21st century. In a post-modern era it is much more acceptable to acknowledge the role of the observer in the study of real processes. This acknowledgement is not just common in universities. Many outside academia now accept that a decision model *needs* to have a subjective component to be a valid framework for an inference: at least in an operational setting. Therefore when implementing an inferential paradigm for decision modelling the argument is moving away from the question of *whether* subjective elements should be introduced into decision processes on to *how* it is most appropriate to perform this task. The fact that Bayesian decision theory has attempted to answer this question over the last 90 years has made it a much more established, tested and familiar framework than its competitors. Standard Bayesian inference and decision analysis is now an operational reality in a wide range of applications, whereas alternative theories – for example those based on belief functions or fuzzy logic – whilst often providing more flexible representations – are less well developed. When looking for a subjective methodology which can systematically incorporate expert judgements and preferences the obvious prime candidate to try out first is currently the Bayesian framework.

Secondly the dominant types of decision problems have begun to shift away from small-scale repeating processes to larger-scale one-off modelling and high-dimensional business and phenomenological applications. For example in one of the examples in this book we were required to develop a decision support system for emergency protocols after the accidental release of radioactivity from a nuclear power plant. Here models of the functionality and architecture of a given nuclear plant needed to be interfaced with physical models describing the atmospheric pollutant, the deposition of radioactive waste, its passage into the food chain and into the respiratory system of humans and models of the medical consequences of different types of human behaviour. The planning of countermeasures has to take account not only of health risks and costs but also of political implications. In this type of scenario, data is sparse and often observational and not from designed experiments. Furthermore direct data-based information about many important features of the problem is simply not available. So expert judgements *have* to be elicited for at least some components of the problem. Note that to address such decision problems using a framework which embeds the plant in a sample space of similar plants appears bizarre. In particular the DM is typically concerned about the probability and extent of a *given* population adversely affected by the incident at a *given* nuclear plant, not features of the distribution of sample space of similar such plants: often the given plant and the possible emergency scenario is unique! A Bayesian analysis directly addresses the obvious issue of concern.

Thirdly the culture in which inference is applied is changing. Concurrently it is not uncommon for policy and decision making to be driven by stakeholder meetings where preferences are actively elicited from the DM body and need to be accommodated into any protocol. The necessity for a statistical model to address issues contained in the subjectivity of stakeholder preferences embeds naturally into a subjective inferential framework. Moreover businesses – especially those private companies taking over previously publicly owned utilities – now need to produce documented inferences supporting future expenditure plans. The company needs to give rational arguments incorporating expert judgements and appropriate objectives that will appear plausible and acceptable to an inferential auditor or regulator. Here again subjectivity plays an important role. The most obvious way for a company to address this need is to produce a probabilistic model of their predictions of expenditure based as far as possible on physical, structural and economics certainties, but supplemented by annotated probabilistic expert judgements where no such certainty is possible. The auditor can then scrutinise this annotated probability model and make her own judgements as to whether she believes the explanations about the process and expert judgement are credible. Note here that the auditor cannot be expected to discover whether the company's presentation is precisely true in some objective sense, but only whether what she is shown appears to be a credible working hypothesis and consistent with known facts. In the jargon of frequentist statistics by following Bayesian methods the company tries to produce a single plausible (massive) probability distribution that forms a simple null hypothesis which an auditor can then test in any way she sees fit!

Fourthly computational developments for the implementation of Bayesian methodologies have been dramatic over the last 30 years. We are now at a stage where even for straightforward modelling problems the Bayesian can usually perform her calculations more easily than the non-Bayesian. Routine but flexible analyses can now be performed using free software such as Winbugs or R and Bayesian methodology is often now taught using Bayesian methods (see e.g. Gelman and Hill (2007); Lancaster (2004)). The analysis of high-dimensional problems has been led by Bayesians using sophisticated theory developed together with probabilists to enable the approximation of posterior distributions in an enormous variety of previously intractable scenarios, provided they have enough time. The environment is now capable of supporting models for many commonly occurring multifaceted contexts and for providing the tools for calculating approximate optimal policies. So the Bayesian modeller can now implement her trade to support decision analyses that really matter.

1.1 Getting started

A decision analysis of the type discussed in this book needs to be customised. A decision analysis often begins by finding provisional answers to the following questions:

(1) What is the broad specification of the problem faced and its context? How might a decision analysis help?
(2) Who is the DM – with the authority to enact and responsible for the efficacy of any chosen policy?
(3) Who will scrutinise the DM's performance? In particular who will audit her assessment of the structure and uncertain features of her problem (sometimes of course this might be the DM herself)?
(4) What are the viable options the DM can choose between?
(5) What are the agreed facts and the uncertain features that embody a plausible description of what is happening? In particular what is the science and what are the socially accepted theories that inform the decision process? Is expert advice required on these issues and if so who should be asked?
(6) What are the features associated with the process on which the decision or policy impinges that are uncertain? How and to what extent do these uncertainties impact on the assessed efficacy of a chosen policy? How compelling will these judgements be to the auditor? Who knows about this interface?
(7) How are the intrinsic and uncertain features that determine the efficacy of any given policy related to one another? Who can advise on this? Who judgements can be drawn on?
(8) Where are the sources of information and data that might help reduce uncertainty and support any assertions the DM wants to make to an auditor? How might these sources be supplemented by expedient search or experimentation?

A Bayesian analyst will support the DM by helping her to build her own subjective probability model capturing the nature of uncertainties about features of the model which might affect her decision, helping her to annotate with supporting evidence why she chose this particular model of the underlying process. The analyst will proceed to elicit her utility function which will take due regard of the needs of stakeholders. He will then help the DM in calculating her expected utility associated with each decision viable to her. The best decisions will then be identified as those having the highest expected utility score. These terms will all be formally defined below and the theoretical justification and practical efficacy of following this methodology explored throughout this book.

1.2 A simple framework for decision making

Bayesian decision analysis has developed and been refined over many decades into a powerful and practical tool. However to appreciate some of the main aspects of such analysis it is helpful to begin by discussing the simpler methodologies. So we start by discussing problems where the responsible DM receives a single reward – usually a financial one – as a result of her chosen act. We will later show that these earlier methods were simple special cases of the fully developed theory: the scope for the efficacious use of these simple methods is, from a practical perspective, just rather restrictive. Subsequently in the book these simple techniques will be refined and elaborated to produce a broad platform on which to base a decision analysis for collections of problems of increasing complexity.

Notation 1.1. Let D – called the *decision space* – denote the space of all possible decisions d that could be chosen by the DM and Θ the space of all possible outcomes or *states of nature* θ.

In this simple scenario there is a naive way for a DM to analyse a decision problem systematically to discover good and defensible ways of acting. Before she can identify a good decision she first needs to specify two model descriptors. The first quantifies the consequences of choosing each decision $d \in D$ for each possible outcome $\theta \in \Theta$. The second quantifies her subjective probability distribution over the possible outcomes that might occur.

More specifically the two descriptors needed are:

(1) A *loss function* $L(d,\theta)$ specifying (often in monetary terms) how much she will lose if she makes a decision $d \in D$ and the future outcome is $\theta \in \Theta$. We initially restrict our attention to problems where it is possible to choose Θ big enough so that the possible consequences θ are described in sufficient detail that $L(d,\theta)$ is known by DM for all $d \in D$ and $\theta \in \Theta$. Ideally the values of the function $L(d,\theta)$ for different choices of decision and outcome will be at least plausible to an informed auditor.

(2) A *probability mass function* $p(\theta)$ on $\theta \in \Theta$ giving the probabilities of the different outcomes θ or possible states of nature just before we pick our decision d. If we have based these probabilities on a rational analysis of available data we call this mass function a *posterior mass function*. This probability mass function represents the DM's current uncertainty about the future. This will be her judgement. But if she is not the auditor herself then it will need to be annotated plausibly using facts, science, expert judgements and data summaries.

Note that if the spaces D and Θ are finite of respective dimensions r and n then $p(\theta)$ is a vector of n probabilities, whilst $\{L(d,\theta) : d \in D, \theta \in \Theta\}$ can be specified as an $r \times n$ matrix all of whose components are real numbers. If both $D = \{d_1, d_2, \ldots, d_r\}$ and $\Theta = \{\theta_1, \theta_2, \ldots \theta_n\}$ are finite sets then the losses $\{L(d_i,\theta_j) = l_{ij} : i = 1,2,\ldots r, j = 1,2,\ldots n\}$ can be expressed as a table called a *decision table* and shown below.

$$
\begin{array}{llcccccc}
 & & & & \text{States} & \text{of} & \text{nature} & \\
 & & \theta_1 & \theta_2 & \cdots & \theta_j & \cdots & \theta_n \\
 & d_1 & l_{11} & l_{12} & \cdots & l_{1j} & \cdots & l_{1n} \\
 & d_2 & l_{21} & l_{22} & & l_{2j} & & l_{2n} \\
 & \vdots & & & \ddots & & & \vdots \\
\text{Decisions} & d_i & l_{i1} & l_{i2} & & l_{ij} & & l_{in} \\
 & \vdots & & & & & \ddots & \vdots \\
 & d_r & l_{r1} & l_{r2} & \cdots & l_{rj} & \cdots & l_{rn}
\end{array}
$$

Note that instead of providing a loss function the DM could equivalently provide a *pay-off* $R(d,\theta) = -L(d,\theta)$. In this book we will move freely between these two equivalent representations choosing the one with the most natural interpretation for the problem in question.

One plausible-looking strategy for choosing a good decision is to pick a decision whose associated expected loss to the DM is minimised. This strategy is the basis of one of the oldest methodologies of formal decision making. Because of its simplicity and its transparency to an auditor it is still widely used in some domains. It will be shown later that such a methodology is in fact a particular example of a full Bayesian one. It therefore provides a good starting point from which to discuss the more sophisticated approaches that are usually needed in practice.

Definition 1.2. The *expected monetary value* (EMV) strategy instructs the DM to pick that decision $d^* \in D$ minimising the expectation of her loss [or equivalently, maximising her expected payoff], this expectation being taken using DM's probability mass function over her outcome space Θ.

To follow such a strategy, the DM chooses $d \in D$ so as to minimise the function

$$\overline{L}(d) = \sum_{\theta \in \Theta} L(d,\theta)p(\theta)$$

where $\overline{L}(d)$ denotes her expected loss or, equivalently, maximises

$$\overline{R}(d) = \sum_{\theta \in \Theta} R(d,\theta)p(\theta)$$

where $\overline{R}(d)$ denotes her expected payoff.

Definition 1.3. A decision $d^* \in D$ which minimises $\overline{L}(d)$ (or equivalently maximises $\overline{R}(d)$) is called a *Bayes decision*.

Remark 1.4. As we will see later there are contexts when $p(\theta)$ may be a function of d as well as θ.

Consider first the simplest possible EMV analysis of a medical centre's treatment policies of a mild medical condition which is not painful and where the doctor – our DM – aims to treat patients so as to minimise the treatment cost. Here the centre (or her representative doctor) is the responsible DM. An auditor might be government health service officials. Note that this is a specific example where a cause of interest – here a disease – is observed indirectly through its effects – here a symptom.

Example 1.5. A patient can have one of two illnesses $I = 1, 2$ and is observed to exhibit symptom A or not, $\overline{A}$. Two treatments d_1 and d_2 are possible and the associated costs and the conditional probabilities of $I|A$ and the marginal probabilities $P(A)|P(\overline{A})$ are given below.

Costs	$I=1$	$I=2$
d_1	100	200
d_2	400	50

$I \mid A$	$I=1$	$I=2$	$P(A)\|P(\overline{A})$
A	p	$1-p$	π
$\overline{A}$	$(1-q)$	q	$1-\pi$

It follows that the expected costs of the two treatments given the existence or not of a symptom are given by

$$\begin{aligned}
\overline{L}(d_1|A) &= 100P(I=1|A) + 200P(I=2|A) \\
&= 100p + 200(1-p) = 50(4-2p), \\
\overline{L}(d_2|A) &= 400P(I=1|A) + 50P(I=2|A) \\
&= 400p + 50(1-p) = 50(1+7p), \\
\overline{L}(d_1|\overline{A}) &= 100P(I=1|\overline{A}) + 200P(I=2|\overline{A}) \\
&= 100(1-q) + 200q = 50(2+2q), \\
\overline{L}(d_2|\overline{A}) &= 400P(I=1|\overline{A}) + 50P(I=2|\overline{A}) \\
&= 400(1-q) + 50q = 50(8-7q).
\end{aligned}$$

So if using the EMV strategy DM will prefer d_1 to d_2 if and only if

$$\overline{L}(d_1|.) < \overline{L}(d_2|.).$$

Thus if a patient exhibits symptom A then the DM will prefer d_1 to d_2 if and only if

$$50(4-2p) < 50(1+7p) \Leftrightarrow p > \frac{1}{3}$$

and if symptom $\overline{A}$ is presented then DM will prefer d_1 to d_2 if and only if

$$50(2+2q) < 50(8-7q) \Leftrightarrow q < \frac{2}{3}.$$

Now suppose the DM believes that $p = q = \frac{3}{4}$, so if A is observed she will choose d_1 and if $\overline{A}$ is observed she will choose d_2. The DM might then be interested in how much she should expect to spend if she acts optimally. This is discovered simply by substituting the appropriate probabilities in the formulae above. Thus

$$\overline{L}(d_1|A) = 50\left(4 - 2 \times \frac{3}{4}\right) = 125,$$

$$\overline{L}(d_2|\overline{A}) = 50\left(8 - 7 \times \frac{3}{4}\right) = 187.5.$$

So, since $P(A) = \pi$ and $P(\overline{A}) = 1 - \pi$, under optimal action the amount we expect to spend is

$$\overline{L} = 125\pi + 187.5(1 - \pi) = 187.5 - 62.5\pi.$$

Obviously exactly the same technique can be extended to apply to a problem when, for each symptom that might be presented, there are many explanatory illnesses or conditions n and many treatments t. For each presented symptom, the expected loss associated with each of the t treatments d can be calculated. In this case each expectation will be the sum of n products of a probability of an illness and an associated cost. The DM can then find the Bayes decision for each possible presented symptom by simply choosing a treatment with the smallest expected cost. Furthermore, by taking the expectation over symptoms under these optimal costs we obtain the expected per patient cost over the given population, just as she did in the simple example above. So at least in this problem the EMV decision rule is easy to implement and its rationale is fairly transparent, once the losses and the probabilities (p, q, π) are given. But where do the probabilities come from?

The answer is they need to come from the DM or from an expert she trusts. She will usually need to be able to provide these as a function of the information she has and present her reasoning to an auditor in a compelling way. We return to this activity later in the book. However the first important point to emphasise about following the EMV strategy in this way is that, perhaps surprisingly, in discrete problems like the one illustrated above, optimal acts are often found to be robust to minor misspecification of the values of the parameters of the model (here (p, q, π)). Thus in the above the DM only needed to know whether or not $p > 1/3$ or $q < 2/3$ before she could determine how to act. It is not unusual for only coarse information to be provided – for example the probabilities of illness given symptoms – before the DM can determine how to act well. Of course, the coarser the information needed to justify a certain decision rule as a good one, the easier it is for a DM to convince an auditor that she has acted appropriately. In the example above note that the associated expected costs under optimal acts are linear in $p(q)$. In general this type of output of an analysis can be more sensitive to the decision maker's expert judgements, though this is often robust too.

Note that if the probabilities are as given above then the DM needs to calculate only functions that are linear in the probabilities provided. This makes the whole analysis easy to perform, depict and communicate to an auditor. As we scale up the problem so that the size of r and n are large this linearity remains in these more refined scenarios.

Finally note that the policy does not only apply to a single patient but to all those presenting. However the more general analysis is provisional and time limited. Changing environments will inevitably cause the various probabilities to drift in time as will the associated costs, so the values of the thresholds governing the optimal policy will change. Furthermore in the medium to long term, as alternative treatments become available new policies which incorporate these new treatments may well become optimal and the nature of the best policy may change. Moreover there may well be changes in stakeholder's needs forcing the decision to be driven not just by cost but also by other factors, for example the speed of recovery of the treated patient. This again may well change the evaluation of the efficacy of each treatment policy and provoke changes in the optimal decision.

1.2.1 Reversing conditioning

Even in the straightforward context given the problem described above is simplified in two ways that make it an unsatifactory template for inference in such scenarios. First a doctor will normally observe several symptoms, not just one. Second we have noted that probabilities need to be provided by the DM to make it work. Psychological studies, see Chapter 4, have demonstrated that probability statements are usually most reliably and robustly estimated or elicited when conditioned in an order that is consistent with when they happened. In this example the event of contracting a disease clearly precedes the manifestation of any symptom. Furthermore any symptom can be seen as the result (or effect) of its cause – the disease. So the analyst should encourage the DM to specify her joint distribution over diseases and symptoms via the marginal probability of the cause – here the disease – and the probability of the effect given a cause: here the probability of the symptom given each disease. But in the illustrative example above, to obtain simple expressions, we have specified the inputs to the decision analysis as the symptom margin and the probability of a disease given a symptom. This is not consistent with their causal order.

Notation 1.6. Let I denote the disease of the patient with sample space $\{1, 2, \ldots, n\}$, so there are n possible explanatory diseases and m symptoms are observed. Define the random variable $\{Y_k : 1 \leq k \leq m\}$ to be indicators on the m symptoms – i.e. $\{Y_k = 1\}$ when the kth symptom is present and $\{Y_k = 0\}$ when it is absent, $1 \leq k \leq m$ – and write the binary random vector $\boldsymbol{Y} = (Y_1, Y_2, \ldots, Y_m)$.

Noting the comment on causation above, the information the doctor would often employ either from hard data or elicited scientific judgements would usually be about:

- The relative prevalence of the different possible n diseases as reflected by the marginal probabilities of diseases $\{P(I = i) : 1 \leq i \leq n\}$. Supporting information about these probabilities could be obtained from relevant previous case histories of the population concerned, or failing that be derived from scientific judgements about the typical exposures of this population.
- Scientific knowledge about the ways in which a given disease might manifest itself through the m observed symptoms. This could be expressed through the set of conditional probabilities $\{P(\boldsymbol{Y} = \boldsymbol{y}|I = i) : \boldsymbol{y}$ a binary m string, $1 \leq i \leq n\}$. Again support for these probabilities could come from either case histories or scientific judgements associated with each possible disease considered: for example the probability that the patient exhibited a high temperature given they had a particular disease. These issues are addressed in detail in Chapters 4, 5 and 9.

Fortunately this type of causally consistent information is enough to give the DM what is needed to perform an EMV analysis. Thus, provided that $P(\boldsymbol{Y} = \boldsymbol{y}) > 0$, i.e. provided there is at least a small chance of seeing the combination of symptoms $\boldsymbol{y}$, when presented by the set of symptoms $\boldsymbol{x}$, the probabilities $\{P(I = i|\boldsymbol{Y} = \boldsymbol{y}) : 1 \leq i \leq n\}$, can be calculated by simply applying *Bayes rule*

$$P(I = i|\boldsymbol{Y} = \boldsymbol{y}) = \frac{P(\boldsymbol{Y} = \boldsymbol{y}|I = i)P(I = i)}{\sum_{i-1}^{n} P(\boldsymbol{Y} = \boldsymbol{y}|I = i)P(I = i)}. \quad (1.1)$$

Incidentally note that if $P(\boldsymbol{Y} = \boldsymbol{y}) = 0$ then the doctor would believe she would never see this observation. So in this case there is no need to calculate the corresponding posterior probabilities.

We have illustrated above that to assess the expected costs of this action the DM needs $\{P(\boldsymbol{Y} = \boldsymbol{y}) : \boldsymbol{y}$ a sequence of 0's and 1's of length $m\}$ in this population. But this is also easy to calculate from these inputs: we use the *Law of Total Probability* which tells us that

$$P(\boldsymbol{Y} = \boldsymbol{y}) = \sum_{i=1}^{n} P(\boldsymbol{Y} = \boldsymbol{y}|I = i)P(I = i) \quad (1.2)$$

Note that these standard equations, whilst familiar to anyone who has studied an introductory course in probability, are nonlinear functions of their inputs. The results of mapping from

$$\{P(I = i) : 1 \leq i \leq n\}; \{P(\boldsymbol{Y} = \boldsymbol{y}|I = i) : \boldsymbol{y} \text{ a binary } m \text{ string}, 1 \leq i \leq n\}$$

to the corresponding pair

$$\{P(\boldsymbol{Y} = \boldsymbol{y} : \boldsymbol{y} \text{ a binary } m \text{ string}\}, \{P(I = i|\boldsymbol{Y} = \boldsymbol{y} : \boldsymbol{y} \text{ a binary } m \text{ string}, 1 \leq i \leq n\}$$

can often surprise the DM. Later in this chapter we discuss how this formula can be explained. But note that because the consequences of these rules have been examined exhaustively by probabilists over the last century it is relatively easy to convince an auditor that these are the rules that should be used to map from one set of belief statements to another. If the DM decides to use an alternative way of expressing her uncertainty than through probability then the appropriate maps between belief statements have to be justifiable. In practice this is likely to be a challenge.

1.2.2 Naive Bayes and conditional independence

Although Bayes rule and the law of total probability can be used to solve the technical problem described above there remains a serious practical issue to resolve. Thus for each disease $I = i$ – because we know that all these probabilities must sum to one – we need to obtain the $2^m - 1$ probabilities

$$\{P(\boldsymbol{Y} = \boldsymbol{y}|I = i) : \boldsymbol{y} \text{ a binary vector of length } m\}.$$

Even for moderate values of m this elicitation will be a resource-expensive task. Furthermore, because all these probabilities must be non-negative and sum to one, at least some will be very small. Experience has shown that to accurately estimate or elicit probabilities of events which occur with very small probability is difficult: see Chapter 4. However there are various accepted formulae from probability theory – called credence decompositions – that can be used to address this practical difficulty. One of these is introduced below.

To avoid, as far as possible, having to make statements about very small probabilities like those above, recall that from the definition of a conditional probability if $\boldsymbol{Y} = (Y_1, Y_2)$, then

$$P(\boldsymbol{Y} = \boldsymbol{y}) = P(Y_2 = y_2 | Y_1 = y_1) P(Y_1 = y_1)$$

and more generally if $\boldsymbol{Y} = (Y_1, Y_2, \ldots, Y_m)$,

$$P(\boldsymbol{Y} = \boldsymbol{y}) = \left(\prod_{j=2}^{m} P(Y_j = y_j | Y_1 = y_1, \ldots, Y_{j-1} = y_{j-1}) \right) P(Y_1 = y_1).$$

Using this rule but conditioning on $\{I = i\}$ therefore gives

$$P(\boldsymbol{Y} = \boldsymbol{y} | I = i) = \left(\prod_{j=2}^{m} P(Y_j = y_j | Y_1 = y_1, \ldots, Y_{j-1} = y_{j-1}, I = i) \right) P(Y_1 = y_1 | I = i).$$

This is a useful formula because all probabilities in the product on the right-hand side of this equation are typically much larger than those on the left. This is because a product of numbers all of which lie between zero and one is smaller than any of its components. It follow that the elicitation of these conditional probabilities will in practice be more reliable. However, the original problem still remains. There are still the same large number of probabilities input into this formula before $P(\boldsymbol{Y} = \boldsymbol{y} | I = i)$ can be calculated. But because we have respected the causal order inherent in this problem, it is sometimes appropriate to make a further modelling assumption which helps to circumvent the explosion of the elicitation task.

Definition 1.7. Symptoms $\boldsymbol{Y} = (Y_1, Y_2, \ldots, Y_m)$ are said to be *conditionally independent* given the illness I – written $\amalg_{j=1}^{m} Y_j | I$ – if for each value of i, $1 \leq i \leq n$,

$$P(\boldsymbol{Y} = \boldsymbol{y} | I = i) = \prod_{j=1}^{m} P(Y_j = y_j | I = i). \tag{1.3}$$

Note that, under this assumption, unlike the general equation above, each probability on the right-hand side is a function of only two arguments.

Definition 1.8. The *naive Bayes model* assumes that all symptoms are independent given the illness class $\{I = i\}$ for all possible illness classes i, $1 \leq i \leq n$.

Although a naive Bayes model embodies strong assumptions, for a variety of reasons such simple models often work surprisingly well in many applications – including medical ones – and provide a benchmark from which to compare more sophisticated models, some of which are discussed later in Chapters 7 and 9. Essentially the model asserts that if the discasc status of a paticnt is known then the presence or absence of one symptom would not affect the probability of the presence or absence of a second. So if for example, for a given disease, whenever a patient exhibited the symptom of a high temperature he always also exhibited the symptom of nausea, but otherwise there was no connection between the two symptoms, then the naive Bayes model would not be valid.

Note that, when there are n binary symptoms and n diseases in the model above, the naive Bayes model needs only $mn+n-1$ probabilities to be input, whilst the general model has $2^m n-1$. So for example if the doctor observes 8 binary symptoms and 10 possible illnesses, to build this inference engine under the naive Bayes needs 89 probability inputs, perhaps an afternoon's elicitation, whereas the general model needs $2,559$. In general, naive Bayes models are therefore much less expensive to elicit than some of their more sophisticated or general competitors.

1.2.3 Bayes learning and log odds ratios

Why does Bayes rule take the form it does and how exactly does it work? One of the best ways of explaining how information from symptoms transforms beliefs is to express this transformation in terms of a function of the illness probabilities – *log odds ratios (or scores)*. Note that odds are commonly used in betting – for example in horse racing gambles – so many people are familiar with the numbers as an expression of uncertainty. In fact some (e.g. O'Hagan (1988); Speigelhalter and Knill-Jones (1984)) have advocated the elictation of odds – or their logarithm the log odds – instead of probability.

Under the naive Bayes assumption, provided that $P(\boldsymbol{Y}=\boldsymbol{y})>0$, i.e. provided that it is not impossible to observe any one of the combination of symptoms $\boldsymbol{x}$, then

$$P(I=i|\boldsymbol{Y}=\boldsymbol{y})=\frac{P(\boldsymbol{Y}=\boldsymbol{y}|I=i)P(I=i)}{P(\boldsymbol{Y}=\boldsymbol{y})}=\frac{\prod_{j=1}^{m}P(Y_j=y_j|I=i)P(I=i)}{P(\boldsymbol{Y}=\boldsymbol{y})}$$

and

$$P(I=k|\boldsymbol{Y}=\boldsymbol{y})=\frac{P((\boldsymbol{Y}=\boldsymbol{x}|I=k)P(I=k)}{P(\boldsymbol{Y}=\boldsymbol{x})}=\frac{\prod_{j=1}^{m}P(Y_j=x_j|I=k)P(I=k)}{P(\boldsymbol{Y}=\boldsymbol{y})}.$$

So, provided that $P(I=k|\boldsymbol{Y}=\boldsymbol{y})>0$, dividing these two equation gives

$$\frac{P(I=i|\boldsymbol{Y}=\boldsymbol{y})}{P(I=k|\boldsymbol{Y}=\boldsymbol{y})}=\prod_{j=1}^{m}\left\{\frac{P(Y_j=y_j|I=i)}{P(Y_j=y_j|I=k)}\right\}.\frac{P(I=i)}{P(I=k)}$$

which, on taking logs can be written in the linear form

$$O(i,k|\mathbf{y}) = \sum_{j=1}^{m} \lambda_j(i,k,y_j) + O(i,k) \tag{1.4}$$

where the *prior log odds of I* are defined by

$$O(i,k) = \log \left\{ \frac{P(I=i)}{P(I=k)} \right\}$$

and the *posterior log odds of I* are defined by

$$O(i,k|x) = \log \left\{ \frac{P(I=i|\mathbf{Y}=\mathbf{y})}{P(I=k|\mathbf{Y}=\mathbf{y})} \right\}$$

and the *log-likelihood ratio of the jth observed symptom* is given by

$$\lambda_j(i,k,y_j) = \log \left\{ \frac{P(Y_j=y_j|I=i)}{P(Y_j=y_j|I=k)} \right\}.$$

Thus the posterior log odds between i and k are the prior log odds between these quantities plus a score $\lambda_j(i,k,x_j)$ reflecting how much more probable it was to see the observed symptom y_j were $I=i$ rather than $I=k$. The linearity of the relationship between prior and posterior odds means that the DM can quickly come to a good appreciation about how and why what she has observed – the symptoms – has changed her odds between two diseases. Note that:

(1) The larger $O(i,k)$, the more probable the DM believes the disease i to be relative to disease k a priori. In particular when $O(i,k)=0$ before observing any symptoms DM believes the diseases i and k to be equally probable.

(2) If the probability of the observed symptoms under two illnesses are the same then this implies that

$$\sum_{j=1}^{m} \lambda_j(i,k,y_j) = 0.$$

The formula (1.4) therefore tells us that the relative probability of the illnesses a posteriori is the same as it was a priori: a very reasonable deduction! On the other hand an observed symptom y_j contributes to an increase in the probability of illness i relative to the probability of illness k if and only if $\lambda_j(i,k,y_j) > 0$. This inequality is equivalent to the statement

$$P(Y_j=y_j|I=i) > P(Y_j=y_j|I=k)$$

i.e. when the probability of what we have seen is greater under the hypothesis that $I=i$ than $I=k$ then the probability of disease i will increase relative to the probability of disease k. Again this appears eminently reasonable.

(3) More subtly note that $\lambda_j(i,k,y_j)$ will be very large and have a dominating effect on $O(i,k|\mathbf{y})$ whenever $P(Y_j=y_j|I=i)$ is much greater than $P(Y_j=yx_j|I=k)$, even when $P(Y_j=y_j|I=i)$

is very small, i.e. even when the symptom is unlikely for the more supported illness. So unless these small probabilities can be elicited accurately – and often this is difficult – the posterior odds calculated by the formula above may well mislead both the DM and the auditor. In general Bayes rule updating can be very sensitive to elicited or estimated probabilities when the data observed has a small probability under all explanatory causes.

It is straightforward – if rather tedious – to express the posterior probabilities as a function of some of the log odds ratios (see Exercise 1.7). Thus, after a little algebra, it can be shown that

$$P(I = i|Y = y) = \frac{\exp[O(i, 1|y]}{\left[1 + \sum_{k=2}^{n} \exp[O(k, 1|y)]\right]}. \tag{1.5}$$

So having calculated $O(k, 1|y)$, $2 \leq k \leq n$ using (1.4) – i.e. the posterior logodds of the first illness against the rest we can find the posterior illness probabilities using equation (1.5). This formula holds for any labelling of the illness, although it is sometimes useful for interpretative purposes to choose the first listed illness to be the simplest or most common one.

Example 1.9. Suppose the DM believes that there are three possible diseases $I = 1, 2, 3$, all with equal prior probability. The doctor observes four binary symptoms $\{S_k : 1 \leq k \leq 4\}$ in that order. An expert trusted by the DM has judged that the naive Bayes model is appropriate in this context and has given the probabilities of the ith symptom being present given the different illness in the table below.

I	S_1	S_2	S_3	S_4
1	0.8	0.2	0.6	0.995
2	0.4	0.2	0.6	0.98
3	0.4	0.2	0.3	0.9

The doctor now observes the first three symptoms as present whilst the last is absent. Calculating her posterior odds and hence her probabilities $p_j(i)$ after observing the first j symptoms using the formulae above it is easily checked that they are given in the table below.

	$p_1(i)$	$p_2(i)$	$p_3(i)$	$p_4(i)$
$I = 1$	0.5	0.5	0.57	0.12
$I = 2$	0.25	0.25	0.29	0.24
$I = 3$	0.25	0.25	0.15	0.64

Notice that after seeing the first symptom the probability of the first illness is 0.5, the highest. This remains the same after the second symptom since what the DM observes is equally probable given all explanations. Therefore this symptom has no explanatory power. The third symptom lends further support to the first illness being the right one, but the absence of the last symptom – an unlikely explanation of any observation for any of the illnesses – reverses the order of the probability, making the first illness very unlikely.

This example illustrates the three points above. Note in particular that the diagnosis depends heavily on the absence of the last symptom. This has a small probability for all the illnesses and a very small probability for the first. So the reason for the diagnosis can be fed back to the doctor: that "the unlikely absence of the 4th symptom and the relatively better explanation of this observation provided by illness 3 outweighs in strength all other evidence pointing in the other direction". This might be acceptable to the doctor or the external auditor. Alternatively she or he may question whether the elicited small probabilities of this symptom were possibly inaccurate or not apposite to this particular patient. A reasoned and supported argument for adjusting these probabilities could then lead to a documented revision of the diagnosis.

A second point is that the particular symptoms actually observed from a patient with illness 3 turn out to be the most unlikely to be observed, being observed on such a patient with only the small probability 0.0024. So the reason illness 3 has been chosen is that it provides the best of a poor set of explanations of the observed symptoms. An auditor of the doctor – perhaps the doctor herself! – may well question why, on this basis, she did not search for a different explanation. The inference is only stable if the doctor really does believe that the three illnesses she has considered are the only possibilities and the relative odds are really accurate.

Thus for example suppose that a 4th illness with a prior probability of only 0.05 as probable as the other alternatives was omitted from consideration in the original analysis for simplicity. Suppose however that under this hypothesis the presence of the first three symptoms actually observed was very likely – for example > 0.8 – and the absence of the last symptom was also likely – again with probability > 0.8. Then it is easy to calculate that this illness would have a posterior probability more than 8 times larger than illness 3 provides. Incidentally note that, unlike probabilities, posterior odds of new alternatives like this can be calculated and appended to the original analysis without changing the posterior odds between the earlier calculated diseases.

So when any explanation of the data is poor this is a cue to feedback this information to the DM. Good Bayesian analyses run with *diagnostics* that are designed to allow an auditor to check whether *every* model in a given class models might give poor explanations of the evidence. These diagnostics can be designed to inform either the plausibility of the analysis as it applies to the case in hand, or alternatively to the class of problems to which the analyses purport to apply. They prompt the DM to creatively reappraise her model.

1.3 Bayes rule in court

1.3.1 Introduction

Recently, spurred on by the proliferation of DNA evidence, various experts have given probabilistic judgements in court about the strength of evidence supporting a match between the suspect and the crime. In principle at least, the juror's task is relatively straightforward. This gives an interesting new and accessible context for which the type of Bayesian analysis is pertinent.

Jurors – our DMs in this example – need to assess the probability of guilt (G) or otherwise ($\overline{G}$) of the suspect given any background information (B) they have already been given and the new piece of evidence (E) delivered by the expert. By law, the only persons allowed to make an assessment of the guilt or innocence of the suspect is a juror, whether this is before the new evidence arrives ($P(G|B)$) or the probability ($P(G|B,E)$) of guilt in the light of the new information E.

Let us assume that the juror is reasoning logically along the lines we have described above. Then her posterior odds need to be the product of her prior odds and the likelihood ratio. Thus explicitly she should calculate the odds of guilt over innocence given the background information and the new evidence using the formula

$$\frac{P(G|B,E)}{P(\overline{G}|B,E)} = \frac{P(G|B)}{P(\overline{G}|B)} \times \frac{P(E|G,B)}{P(E|\overline{G},B)}. \tag{1.6}$$

The important implication here is that – to encourage a juror to be rational – the expert should *only* be allowed to provide jurors with information about the strength of evidence by communicating – either explicitly or implicitly – the value of this likelihood ratio. This is the probability of the actual evidence observed given the suspect is guilty *relative* to the probability of that evidence given the suspect is innocent – both conditional on the background information B provided to everyone. For example for the expert to present the probability of the evidence given guilt, or given innocence on their own is superfluous and might potentially confuse the jury.

Tables of this formula can help jurors come to an appropriate revision of their beliefs. For example the table below gives prior and posterior log odds of guilt and prior and posterior probabilities of guilt when a credible expert witness asserts that the evidence is $100\times$ more probable when the suspect is guilty than when she is innocent.

Prior prob.	Post. prob.	Prior ln. odds	ln. LR	Post. ln. odds
0.001	0.09	−6.91	4.61	2.30
0.01	0.50	−4.61	4.61	0
0.30	0.98	−0.85	4.61	3.76
0.50	0.99	0.00	4.61	4.61
0.70	0.996	0.85	4.61	5.56
0.90	0.999	2.20	4.61	7.81

1.3.2 A hypothetical case study

To demonstrate these principles we now give a hypothetical study about a typical scenario that might be met in court. Woman A's first child died of an unexplained cause (B). When her second child also died of an unexplained cause (E) she was arrested and she was tried for the murder of her two children (hypothesis G). This was the prosecution case. The defence maintained that both her children died of SIDS (sudden infant death syndrome). An expert witness asserted that only one in 8,500 children die of SIDS so

$$P(E,B|\overline{G}) = \left[\frac{1}{8,500}\right]^2 \simeq \frac{1}{73 \times 10^6}.$$

Apparently the jury treated this figure as the probability of A's innocence – i.e. as $P(\overline{G}|E,B)$. This spurious inversion is sometimes called the *prosecutor fallacy*. So jury members calculating her probability of guilt as

$$\begin{aligned} P(G|E,B) &= 1 - P(\overline{G}|E,B) \\ &= 1 - \frac{1}{73 \times 10^6} \end{aligned}$$

found guilt "beyond reasonable doubt". On the basis of this and other evidence the jury convicted her for murder and she was sent to prison.

1.3.2.1 *The first probabilistic/factual error*

It is well known that if a mother's first child dies of SIDS then (tragically) her second child is much more likely to die too. For example it is conclusively attested that a tendency to the condition is inherited. The conditional independence assumption implicit expert witnesses' calculation is therefore logically false. Suppose on the basis of an extensive survey of records of such cases that

$$P(E|B,\overline{G}) \simeq 0.1.$$

Assuming the figure above,

$$P(E,B|\overline{G}) = P(E|B,\overline{G}) \times P(B|\overline{G}) \simeq \frac{1}{85 \times 10^3}.$$

This is over 850 times larger than the probability quoted by the expert.

1.3.2.2 *The second error: the prosecutor fallacy*

From the rules of probability we know that, in general,

$$P(\overline{G}|E,B) \neq P(E,B|\overline{G}).$$

To obtain $P(\overline{G}|E,B)$ the juror must apply the Bayes rule formula: something very difficult to do in her head. Note that it is not unreasonable for a statistically naive but otherwise intelligent juror to assume that these two probabilities are the same when expressed in words (i.e. "the probability this SIDS event will happen to an innocent parent suspect"). However an expert witness who presents the probability $P(E,B|\overline{G})$ as if it is $P(\overline{G}|E,B)$ is either statistically incompetent or consciously trying to mislead the jury.

1.3.2.3 A rational analysis of this issue

This uses the formula (1.6) above. Here we need the juror's prior odds of guilt. These are of course dependent on everything the juror has heard in court. However suppose that one useful statistic, obtained by surveying death certificates in the UK over recent years, is the following. Of children who die in ways unexplained by medicine, less than $\frac{1}{11}$ are subsequently discovered to have been murdered. With no other information taken into account other than this, a typical juror might set

$$\frac{P(G|B)}{P(\overline{G}|B)} \leq \frac{1}{11}\left(\frac{10}{11}\right)^{-1} = 0.1.$$

So after learning of the first child's death a juror believing this statistic would conclude the probability that the child was murdered by her parents was at least 10 times less probable than that there was an innocent explanation of the death. Note that this is consistent with actual policy in the UK where someone like A who loses a first baby for unexplained reasons but for whom there are no other strong reasons to assume she had murdered her baby, is freely allowed to conceive and have a second child. Surely if it was thought that such a woman probably did kill her baby, then at least one would expect that the second child would be taken into care or into hospital where he could be monitored. So the numbers given above could be expected to pass a reasonable auditor as at least plausible.

Logically, guilt as it is defined implies that

$$P(E|G,B) = 1$$

and we are taking

$$P(E|\overline{G},B) = 0.1$$

so equation (1.6) gives us that

$$\frac{P(G|B,E)}{P(\overline{G}|B,E)} \leq 0.1 \times \frac{1}{0.1} = 1$$
$$\Leftrightarrow P(G|B,E) \leq 0.5.$$

In the face of this evidence, the suspect should therefore not be seen as guilty beyond reasonable doubt. Although the value of $P(G|B,E)$ might vary between jurors, most rational people substituting different inputs into the odds ratio formula should convince themselves that, on the basis simply of the deaths, any conviction would be unsafe. This activity of investigating the effect of different plausible values of inputs into a Bayesian analysis is sometimes called a *sensitivity analysis*.

The example provides us with a scenario where a DM legitimately adopts some of the conditional probabilities she needs for her inference from an expert: here the forensic statistician, in a way likely to be acceptable to any auditor. She then combines these probabilities that legitimately come from herself to arrive at a robust and defensible decision.

Why do expert witnesses mislead the jury by providing $P(E, B|\overline{G})$ and not provide the analysis above? One reason is that many of them really don't understand ideas of probability and independence well enough to understand the issues discussed above. Indeed applying the rules of probability appropriately to a given scenario is quite hard without help. They therefore mislead themselves as well as the jury. A second possible reason is that they tend to see the most horrible cases and disproportionately few innocent ones. This selection bias, discussed in Chapter 4, makes their own assessments of the prior odds of guilt unreasonably high. Their own posterior assessments of guilt are consequently inflated as well and they genuinely try and convey these inflated odds to the jury. But whilst explicable the communication of this personal false judgement is clearly counter to the principle that the jury should decide on the basis of the evidence, not the prejudices of the expert!

Further discussion of this and related problems can be found in Dawid (2002b) and Aitken and Taroni (2004).

1.4 Models with contingent decisions

The scenarios illustrated in the previous sections have a very straightforward structure. In many decision problems the structure of the model is less obvious. Finding an EMV strategy is then not quite such a transparent task.

Example 1.10. A laboratory has to test the blood of 2^n people for a rare disease having a probability p of appearing in any one individual. The laboratory can either test each person's blood separately [decision d_0] or randomly pool the blood of the subjects into 2^{n-r} groups of size $x = 2^r$, $r = 1, 2, \ldots, n$ [decision d_r] and test each pooled sample of blood. If a pool gives a negative result then this would mean that each member of the pool did not have the disease. If the pool gave a positive result then *at least one* member of the pool would have the disease and then all the members of that pool would then be rechecked individually. Assuming that any test, either pooled or individual, costs £1 to perform what is DM's EMV strategy for this problem?

Note that $\overline{L}(d_0) = 2^n$ and if the DM decides $[d_r]$ to pool into groups of $x = 2^r$ then the probability that this pooled sample is positive is

$$P(\text{group +ive}) = 1 - P(\text{group −ive}) = 1 - (1 - p)^x \triangleq \lambda.$$

Therefore since under d_r the number of groups $2^{n-r} = 2^n x^{-1}$ the expected number of pooled samples to recheck is $\frac{2^n \lambda}{x}$. So the expected number of rechecked individuals under d_r is

$$\frac{x.2^n}{x}\lambda = 2^n[1 - (1 - p)^x].$$

The expected total cost of using decision d_r is the number of tests on groups plus the expected number of rechecked individuals under that regime

$$\begin{aligned}\overline{L}(d_r) &= 2^n x^{-1} + 2^n(1 - (1 - p)^x) \\ &= 2^n[1 + x^{-1} - (1 - p)^x]\end{aligned}$$

where $x = 2^r$, $1 \leq r \leq n$. For any fixed value of p the expected losses associated with d_r can be compared.

There are three main points to take away from this example. The first is that if probability distributions of outcomes depend on what the DM decides to do then following an EMV strategy becomes less transparent unless tools are developed to guide her calculations.

The second illustrates the provisional nature of any decision analysis. Thus having completed this analysis we might reasonably question why we only consider pooling groups of size a power of 2. In fact we could extend the analysis above in a straightforward way to calculate the expected losses associated with other pools of arbitrary group size. If we do this we find that the formulae are less elegant than the ones above but look very similar and are simple to calculate. More interestingly, if the DM decides to pool into a large group and this turns out to be positive, instead of subsequently then testing all the individuals in the group separately she could consider whether to check the further possibility of testing subgroups of this group first and only subsequently test individuals in positive subgroups.

Exploring new possibilities of extending the decision space in ways like those suggested above is an intrinsic part of a decision analysis. It involves both the DM – e.g. "Is it scientifically possible to split the blood sample into more that two groups and if so how many?" – and the analyst – "Is there some technical reason why a suggested new decision rule must be suboptimal and therefore not worth investigating". Note that such embellishments of the decision problem do not destroy the original analysis. The original expected losses associated with other decisions – and thus their relative efficacy – remain the same no matter how many alternative decisions we investigate. The earlier analyses of the relative merits of decision rules hold fixed, we might just find a new and better one.

Finally note that developing a structure to address this specific problem brings with it an analogous methodology for addressing problems like it. For example the analysis above applies to the detection of other similar blood conditions across other similar populations, albeit with an appropriate change of the probability p. We will see in many of the problems addressed in this book, it is possible to carry forward the structure of parts of the problem as well as some of the probability assessments. This is one feature that can make a decision analysis so worthwhile: it not only provides support for the decision problem at hand, but also informs analogous decision analyses that might be performed in the future.

1.5 Summary

We have seen illustrated above how if a DM is encouraged to choose a decision minimising her expected loss then this provides her with a framework that allows her both to systematically explore her options and develop and examine her beliefs. This methodology not only helps her make a considered choice but develop arguments explaining why she chose the policy she did to an external auditor in a logical and consistent manner. These analyses will also usually inform her decision making about future scenarios whenever these share features with the problem at hand.

Even in the very simple examples in this chapter we have been able to demonstrate that the role of the analyst is to support the DM to make wise and defensible decisions and help her to explore as many scenarios and options as she needs to in order to have confidence in her decisions. The analyst's task is never to tell the DM what to do but to provide frameworks to help her creatively explore her problem and to come to a reasoned decision she herself owns as well as providing a template framework which she might adjust in a decision analysis of similar problems she might meet in the future.

1.6 Exercises

1.1 The effect of d kilograms of fertiliser, $0 < d < 1,000$ on the expected yield λ of a crop is given by $\lambda = 10(8 + \sqrt{d})$. The cost of a kilogram of fertiliser is £5 and the profit from a unit of crop is £10. Find the EMV decision rule.

1.2 Prove the assertion in the text concerning the change in probability following the introduction of a 4th explanation of the symptoms in the first example on medical diagnosis.

1.3 In the example above prove that if $p > 1 - 2^{-1/2}$ then you should simply apply d_0 whilst otherwise you should first pool the blood in some way.

1.4 In the example above show that d_2 is at least as good as d_1 for all values of p so that DM should never pool into groups of 2: groups of 4 being always better.

1.5 DM is in charge of manufacturing T-shirts in aid of a sponsored marathon race in 50 weeks' time. Leasing a small machine for manufacturing these (decision d_1) will cost £100,000 whilst leasing a large machine (decision d_2) will cost £300,000 over these 50 weeks. DM hopes to obtain free TV advertising with probability p. If this happens she expects to sell 1,800 items a week but if not she expects to sell only 400 a week. If DM makes £10 clear profit for each T-shirt sold, show that her Bayes decision is to to buy the smaller machine if $p < 4/13$.

1.6 Items you manufacture are independently flawed with probability π and otherwise perfect. If DM dispatches a flawed item she will lose a customer with an expected cost to him of £A. DM can dispatch the item immediately (d_1) or inspect it with a foolproof method, and keep replacing the item and checking until she finds a good one (decision d_2). The cost of making an item is £B and the cost of checking it £C

   i) Show that when $A = 10,000$, $B = 3,000$ and $C = 1,000$ under an EMV decision rule you should prefer d_2 to d_1 when $0.2 < \pi < 0.5$.

ii) Show that if $\frac{A}{B} < 4$ the DM should never inspect regardless of the value of π.

iii) Show that the DM should inspect for some value of π if

$$A^2 + B^2 > A(C + 2B).$$

iv) Show that if $B = 0$ the DM should inspect if and only if $\pi < 1 - \frac{C}{A}$ and if $C = 0$ if and only if

$$\left(\pi - \frac{1}{2}\right)^2 < \frac{1}{4} - \frac{B}{A}.$$

1.7 Prove the formula (1.5) above which expresses probabilities in terms of log odds.

2
Explanations of processes and trees

2.1 Introduction

Some simple decision problems can be transparently solved using only descriptors like a decision table and some supplementary simple belief structure like naive Bayes model. However for most moderately sized problems the analyst will often discover that the explanation of the underlying process, the consequences and the space of possible decisions in a problem has a rich and sometimes complex structure. Whilst it is possible to follow an EMV strategy in such domains, the elicitation of the description of the whole decision problem is more hazardous. The challenge is therefore to have ways of encapsulating the problem that are transparent enough for DM, domain experts and auditors to check the faithfulness of the description of a problem but which can also be used as a framework for the calculations the DM needs to make to discover good and defensible policies.

One of the most established encompassing frameworks is a picture called a decision tree depicting, in an unambiguous way, an explanation of how events might unfold. Over the years historic trees have been used to convey the sorts of causal relationships which populate many scientific and social theories and hypotheses. These hypotheses about what might happen – represented by the root to leaf paths of the tree – describe graphically how one situation might lead to another and are often intrinsic to a DM's understanding of how she might influence events advantageously. It is often possible for DM to use this tree to describe, quantify and then evaluate the consequences of following different policies. This process will support her when she compares the potential advantages and pitfalls of each policy and finally comes to a plausible and defensible choice of a particular decision rule. In Chapter 1 some of the advantages of ordering variables consistently with their causal history have already been pointed out. In this chapter the elicitation and evaluation tool – the historic tree – is introduced which provides a compelling framework for communicating the input of a decision analysis to an auditor.

We also saw in the last chapter that in order to calculate an optimal policy it is often expedient to use Bayes rule to reverse the directionality of conditioning so that it is consistent with the order in which the DM becomes aware of information: for example – in the health diagnosis scenario described there – to condition on symptoms before considering their causes, the diseases. This encourages the analyst to transform a historic tree so that it accommodates fast calculation and a transparent taxonomy of the space of decision rules rather than a transparent explanation. This transformation process to a rollback tree is explained and illustrated below.

Another tree especially useful to help the DM and her auditor to become aware of the sensitivity of the inputs of her analysis is the normal form tree. In this chapter we will discuss various examples of different levels of complexity that illustrate how these different tree structures can be used to represent a problem, be a framework for the calculation of optimal policies and form the basis of a sensitivity analysis.

2.2 Using trees to explain how situations might develop

2.2.1 Drawing historic trees

Throughout this book directed graphs are used as various frameworks for describing a model. So it is useful to begin this section with some general definitions. A *directed graph* $\mathcal{G} = (V(\mathcal{G}), E(\mathcal{G}))$ is defined by a set of vertices denoted by $V(\mathcal{G})$ and a set of directed edges denoted by $E(\mathcal{G})$ connecting the vertices of the graph to one another. If a vertex $v' \in V(\mathcal{G})$ is connected by an edge in $E(\mathcal{G})$ to a vertex $v \in V(\mathcal{G})$ then v' is called a *parent* of v and v is called a *child* of v'.

A *directed tree* $\mathcal{T} = (V(\mathcal{T}), E(\mathcal{T}))$ is a directed graph with two additional properties. First it has a unique vertex with no parent called its *root vertex* $v_0 \in V(\mathcal{T})$. Second all other vertices v have exactly one parent v'. The vertex set $V(\mathcal{T})$ of a direct tree partitions into the set of *leaves* $L(\mathcal{T})$. $L(\mathcal{T}) \subset V(\mathcal{T})$ which are the vertices $v \in V(\mathcal{T})$ with no children and the set of *situations* $S(\mathcal{T}) = V(\mathcal{T})\backslash L(\mathcal{T})$. Finally a *floret* $\mathcal{F}(v)$ of a situation $v \in S(\mathcal{T})$ of the tree $\mathcal{T}$ is the directed subtree $\mathcal{F}(v|\mathcal{T}) = (V(\mathcal{F}(v)), E(\mathcal{F}(v)))$ where $V(\mathcal{F}(v))$ consists of the situation v and all its children and $E(\mathcal{F}(v))$ consists of the set of directed edges from v to each of its children. Note that any directed tree $\mathcal{T}$ is fully defined by its set of florets $\{\mathcal{F}(v|\mathcal{T}) : v \in S(\mathcal{T})\}$.

Possibly the most descriptively powerful use of a directed tree to faithfully express a problem in this book will be called a *historic tree* (Shafer, 1996). This depicts directly the different ways the DM believes that situations might develop, both in response to events that happen and also to decisions that can be taken by her. Our first example is a simplification of a process associated with product safety.

Example 2.1. A company is interested in the possible allergenic properties of a shampoo with a new ingredient. For the ingredient to have a toxic effect it must first penetrate the epidermis – the outer layer of the skin. This will certainly happen if the user has a wound where the shampoo can penetrate the skin. If the shampoo does penetrate then it might or might not inflame the dermis layer below. If this happens it will cause an allergic rash to appear on the epidermis. There is also a second possibility that can only occur if the dermis becomes inflamed. Proteins may react so that messages may pass to the lymph nodes causing sensitisation to occur – i.e. the individual will come out in a rash later even when exposed to very small quantities of the new ingredient. The company would like to ensure that – under standard applications of the shampoo – with high probability no more than a certain very small proportion of the population will suffer an allergic reaction to the new ingredient and an even smaller proportion will be sensitised.

Conventionally the root vertex of a historic tree is drawn to the left of the paper with subsequent vertices drawn above, to the right and below it. The root vertex is the starting point of the story of the problem. Each directed path away from the root to a leaf depicts a possible way the DM believes situations might unfold. The edges along this path label the sequences of events describing this development. The leaves of this directed tree can be used to label the root to leaf paths of the tree and hence are associated with one possible sequence of events from their beginning to their end. On the other hand the situation $S(\mathcal{T})$ of the tree describe intermediate states in the development of the history of the process. The edges of a floret $\mathcal{F}(v|\mathcal{T})$ describes the set of possible immediate developments of the unit that can occur once it reaches the situation v.

To illustrate the construction of a historic tree consider the product safety example above. Here the DM owning the explanation is the company representatives. Following events in their chronological order, the user will either have a wound, W, or not $\overline{W}$ when she applies the shampoo. So we let the root vertex v_0 have two edges out of it labelled by these contingencies. After the application of the shampoo in its usual concentration the first turn of events if she is not wounded – the situation described by vertex v_1 is that either the ingredient penetrates the epidermis – represented by the outgoing edge labelled by H or it does not – represented by an outgoing edge $\overline{H}$. If it does not then no adverse effect can happen. If it does then a second situation v_2 happens with emanating edges labelled by the events that inflammation I of the dermis occurs or does not $\overline{I}$. If it inflames the dermis then we reach a final situation v_3 where sensitisation S occurs or not $\overline{S}$. On the other hand if the user is wounded then by definition penetration will occur leading to a situation labelled by v_4. This in turn may lead to inflammation – an edge leading to a situation v_5 or to a leaf vertex along an edge representing no inflammation. Finally edges are drawn from v_5 representing whether or not the wounded user becomes sensitised. Using the obvious labelling of the edges, the full historic tree is depicted below.

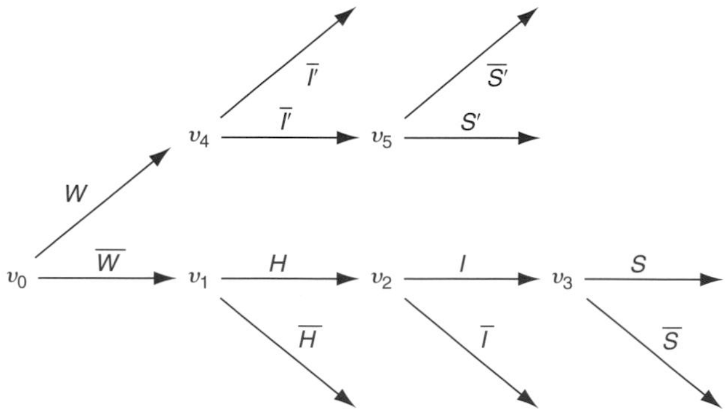

Note that the leaves of this tree $\overline{H}, \overline{I}, \overline{S}, S, \overline{I}', \overline{S}', S'$ label the possible out-turns of the process – by the final resolution – that might impinge on any decision making: namely, whether or not the user was wounded and in each of these contingencies whether the ingredient does not penetrate the skin and no adverse effects happen, that it penetrates the skin but causes no inflammation and so does not cause adverse effect, that it causes a rash but does not sensitise or that it causes a rash and also sensitises the customer.

This type of historic tree is called an *event tree* because the resolution of each of its situations are not in the control of DM but are determined by the nature of the unit described: here the person using the shampoo. In an event tree all its edges label certain important conditional events in the story. Here the initial edges labelled by W and $\overline{W}$ denote the event that a user is wounded. The edge $\overline{H}$ denotes the event that the ingredient does not penetrate the skin given she is not wounded whilst H denotes the event that it does. The event $\overline{I}, \overline{I}'$ denotes the event that given we reach situation 2 there has been penetration whilst the edge labelled I, I' denotes the event that this does happen conditional on penetration, respectively when the user has or does not have a wound. Finally edges $\overline{S}, \overline{S}'$ and S, S' denote respectively the event that sensitisation has not or has taken place given irritation has occurred in the respective cases of not wounding, wounding.

The edges of each floret of an event tree can be embellished with conditional probabilities. Here the probabilities $P(W)$ and $P(\overline{W})$ assigned to whether or not the user carries a wound can be associated to their respective edges. Similarly the probability $P(\overline{H}|\overline{W})$ can be associated to the edge $\overline{H}$ labelling no penetration given no wound, probability $P(H|\overline{W})$ to the edge H of the floret emanating from situation v_1, the conditional probability $P(I|H, \overline{W})$ to the edge I, the conditional probability $P(S|H, I, \overline{W}) = P(S|H, \overline{W})$ to the edge S of the floret emanating from situation v_2 and so on. When we embellish the edges of an event tree $\mathcal{T}$ with the appropriate conditional probabilities associated with that development of the story we call $\mathcal{T}$ a *probability tree*.

2.2.2 Parallel situations in historic trees

A simple example of a historic event tree as given above is where the DM has made no compromises about the chronological order used in advancing the story of the tree, nor have we conditioned on any event which we now know has happened but which might have been caused by other events depicted in the tree. Such trees are particularly useful descriptively. This is because it can often be agreed by all those involved in a decision process that they faithfully describe the possible ways in which the future might unfold. Moreover there will often be agreement about when the collections of the edges of florets rooted at two different situations should be assigned the same vector of probabilities.

Thus in the example above, both the auditor and the DM might agree that, for any individual drawn from a population $\Omega(\boldsymbol{x})$ of users without a wound described by a set of covariates $\boldsymbol{x}$ – for example indexing for their age and the amount of shampoo they use – the probability that this shampoo penetrates their skin would be the same: but simply unknown to both. The DM and auditor are often able to agree that the probabilities about the distribution associated with two florets on the tree describing the development of the *same* unit are the same. Thus in the example above they might agree that the conditional probabilities on the pair of edges $(I, \overline{I})$ emanating from v_2 should be given the same probabilities as on the pair $(I', \overline{I}')$ emanating from v_4 and edges $(S, \overline{S})$ emanating from v_3 the same probabilities as on the pair $(I', \overline{I}')$ from situation v_5 even if they may not agree as to what these probabilities should be.

Symmetries like these commonly occur and their identification is an essential feature of many tools that enable the DM to simplify and make sense of a complicated problem in a way that can be compelling to a third party. In particular they not only allow the DM to draw information about one unit in one situation and use that to make inferences about the probabilities in another – see Chapter 4 – but also lie behind other graphical descriptions of dependence that we will discuss in detail in Chapter 6.

Definition 2.2. Two chance situations $v_1 \in V(\mathcal{T}_1)$ and $v_2 \in V(\mathcal{T}_2)$ associated with their respective historic probability trees $\mathcal{T}_1$ and $\mathcal{T}_2$ (possibly the same) are said to be *parallel* if there is an invertible map of the edges of the floret $\mathcal{F}(v_1|\mathcal{T}_1) \to \mathcal{F}(v_2|\mathcal{T}_2)$ such that under this map each probability associated with an edge in $\mathcal{F}(v_2|\mathcal{T}_2)$ is the same as the probability associated with its corresponding edge in $\mathcal{F}(v_1|\mathcal{T}_1)$.

When the chronological order of the story has been faithfully followed in a tree – as in the one above – it is often plausible to conjecture that the interpretation of the meaning of all downstream edges in the adapted story is unchanged conditional on learning that certain upstream edges have happened. We will see that this type of stability is often an essential component of a decision analysis: sadly often neglected. When Kolmogorov elegantly axiomatised probability – and gave birth to modern probability theory – he provided a framework for a coherent theory structured around the event space which in a discrete problem corresponds to the leaves of the tree. However he threw away one of the most important bridges between theoretical probabilities and probabilities reasoned from *actual* belief systems: the topology of its historic tree.

For example note that the event tree

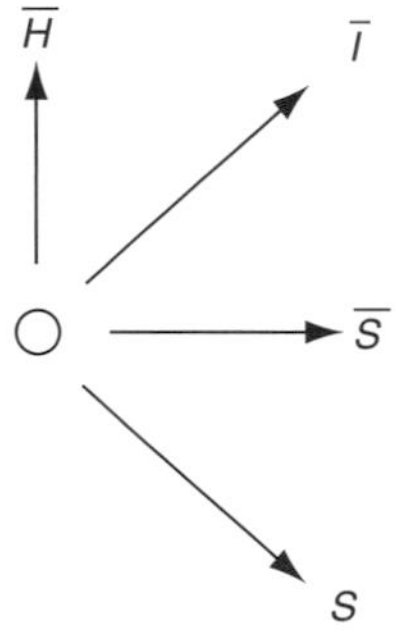

has the same associated probabilistic event space as the subtree of the tree above rooted at v_1. However when we condition on the event $P(\overline{H}) = 0$ the probabilities assigned to all the other edges in the probability tree of this event tree will change. In this tree the change happens to be a simple one. To condition on this event we scale up each of the remaining edge probabilities so they add to one and set $P(\overline{H}) = 0$. However for more complicated trees the necessary changes can be much less predictable. This illustrates how the tree is more expressive than the simple space of events: it embodies a generally agreeable understanding of the impact of important downstream conditioning events in terms of *local* changes in edge probabilities that simply reassign probabilities to either the value 0 – if impossible – or one if inevitable whilst keeping the remaining probabilities unchanged. These ideas strongly impact on how compellingly a Bayesian DM can argue her case to an external auditor.

2.2.3 Using parallel situations to predict the effects of controlling a system

In a decision analysis it is often important to try to predict the consequences of certain decisions that are enacted. If the *idle* probability tree – that is the tree representing the system when it is not subject to any control – is historic then it is quite often possible to produce compelling arguments for identifying some of the edge probabilities needed for the same system when it is subjected to various controls. Thus consider the historical tree below. The original tree has now been extended to include a new initial act, with edges labelled "control" and "idle". Under the "idle" development we simply allow shampoo to be applied to the user. However for those histories described by root to leaf paths starting with the edge labelled "control" we allow for the possibility that we first intervene in the system and cause a small wound in the scalp of the user, like one that might be naturally found in that population. The DM is interested in the effect this intervention/control/treatment might have on the subsequent development of the user.

The DM may well consider it reasonable in this context to assign probabilities to edges in such a way that the probabilities of the effects of a natural wound could be equated with the probabilities of the corresponding events of a wound created artificially in the way described above. She would then be able to assert that her probabilities associated to edges emanating from v_4 and v_5 were identical to the probabilities associated to the analogously labelled edges emanating from v_6 and v_7 respectively.

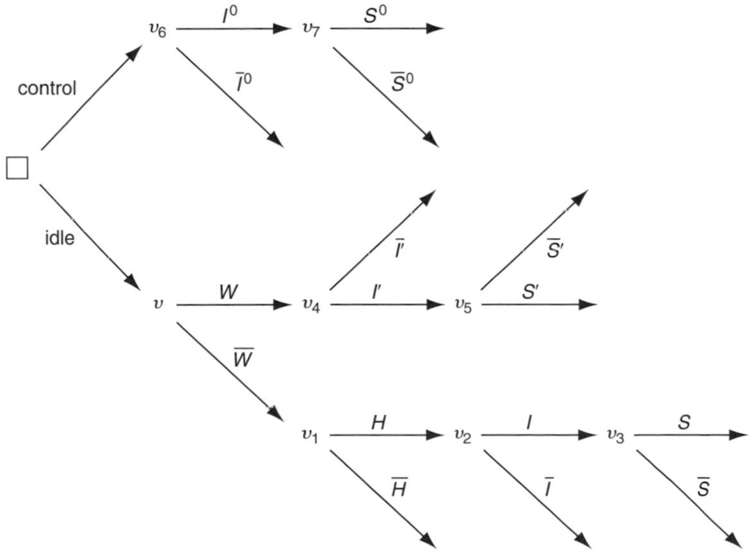

These types of parallel situation where the probabilities on the edges of a tree describe the development after a control or treatment have been called "causal" by some authors.

2.3 Decision trees

2.3.1 A more complicated example

Some of the most important advantages of a tree representation can only be fully appreciated when applied to less simple problems than the one above. The tree below is a simplification of the type of decision problem faced by forensic scientists when trying to balance evidence for and against a suspect (see e.g. Aitken and Taroni (2004)). Its more complicated underlying story will enable us to illustrate how such real historic trees contain decision situations where the DM can decide what to do as well as chance situations determined by her environment, how they often exhibit many symmetries and how different agents might be most informed about different subtrees and so act as the DM's trusted experts who adopts their subjective probabilities as her own. It also illustrates how a historic tree can become unwieldy and how to address the issue of making it as simple as possible. This serves as an introduction to why the toolkit of techniques we describe later in the book are necessary when addressing decision problems that are not simply textbook ones.

Example 2.3. A robber tortured an elderly householder in her front room until the victim told him the location of her savings which he then stole. A suspect was picked up an hour later for an unconnected driving offence and held in custody. The robbed woman was able to raise an alarm minutes after the robbery. A few moments before the crime was committed there were many witnesses attesting to the fact that the house was entered by a single man who was wearing a bright green pullover. This was accepted by all as an incontrovertible fact. A bright green pullover was later found in the suspect's wardrobe which he acknowledged was his.

If the case goes to court then the prosecution will assert that the suspect is the robber in the story line above. The defence will maintain that the suspect was not the robber and had never been to the house. Furthermore although the suspect agrees he wore a green pullover on the day of the crime he asserts that his brother had entered the house a day earlier to collect rent wearing this garment. If the police decide to prosecute then the suspect will be found either guilty or innocent.

The evidence found after a forensic search of the scene of the crime was a bloody fingerprint and a bright green fibre. The recovered mark of a finger left at the crime scene has already been discovered to give a partial match to the suspect's fingerprint. Because of other evidence both the defence and the prosecution agree that this mark was left by the culprit. Although not yet processed the forensic science department could be asked to match DNA from blood in the mark of the finger found at the crime scene to that of the suspect. A future analysis of the blood from this mark at the crime scene will give one of four results: no match, inconclusive, a partial match or a full match to the suspect's blood. The police must choose to either arrest the suspect and prosecute – when in concert with the prosecution they will have the further option of strengthening their case by testing for a match in the DNA or a match in the fibres found at the crime scene with those of the suspect's pullover – or to release the suspect without performing either of the additional tests.

Here assume that the decision analysis is performed on behalf of the prosecution in concert with the police. In the last section it was argued that the most compelling trees were historical ones which allow the chronological order of situations as they happened. In particular this helps us identify parallel situations. However trees of real-sized problems can get bushy very quickly and will be opaque to the DM if the analyst is not prepared to compromise over this. So it is sometimes expedient to violate this chronology for the sake of the simplicity of the tree. Here we know that an agreed part of the story is that whatever else has happened, the print has given a partial match. The historic tree would depict all developments of events from the past and include the developments if no match, a partial match or a full match had occurred. In the tree drawn below we have compromised the historic chronology of the edges and first condition on *a fact* – here the partial match – that can be accommodated into versions of the story. A tree like the one below where the historic chronology of root to leaf paths is only violated by introducing some agreed facts into the tree will be called *episodic*.

Draw the tree starting from the left of the page. The first event impacting on the case is whether or not the suspect's brother entered the house in the recent past wearing the suspect's pullover B or whether this did not happen $\overline{B}$. If he did enter then he either left the detected green fibre F_b or not $\overline{F}_b$. In all cases the next turn of events is whether or not the suspect entered the house and robbed the victim (event C) or whether the robber was someone else (event $\overline{C}$). At this point all agree that the partial print was found. If this print is assumed to be the culprit's then it is useful to record that both the events C and $\overline{C}$ are conditioned on this fact and whether the recovered fibre was the left by the suspect F_s or someone who was not the suspect or his brother $\overline{F}_{b,s}$. The beginning of the tree – henceforth called the *initial tree* – is given in the figure below.

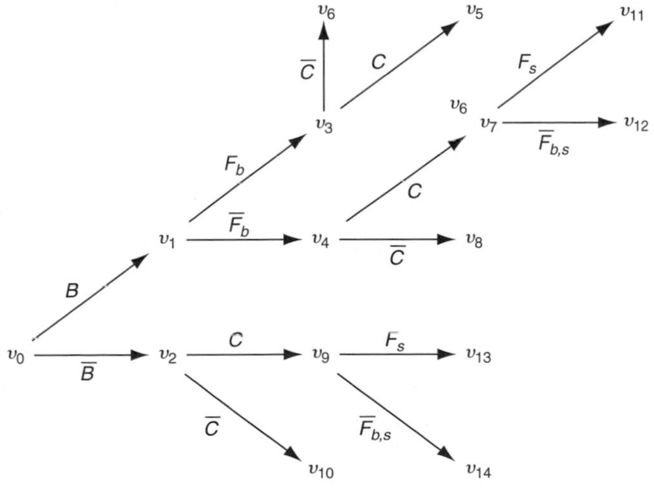

The first point this example illustrates is that in moderately large problems there is usually no *unique* historic or episodic tree for a given set of possible unfoldings of events. As a general principle the analyst should usually choose a tree with a minimum number of vertices that expresses all the DM's salient beliefs but no more. This will make it easier for the DM to understand the tree and take ownership of it. Thus in the example above note that if the brother had not gone to the house then he could not have left the green fibre. The simplest representation is to omit situations describing such impossible developments. So the relevant root to leaf paths in the tree above moved straight from the absence of the brother to the presence and perpetration or absence of the suspect.

Second the episodic and especially historic trees of real problems often exhibit many symmetries in both the nature of the events and their shape. The subsequent unfolding of events on reaching two particular situations – expressed by the subtrees rooted at each of these situations – can often be labelled identically to one another. In particular the directed subtree $\mathcal{T}_1'$ whose vertex and edge sets are respectively given by $V(\mathcal{T}_1')$ and $E(\mathcal{T}_1')$ and whose root is one situation – v_1(say) – may be isomorphic to a different directed subtree $\mathcal{T}_2'$ whose vertex set and edge sets are respectively $V(\mathcal{T}_2')$ and $E(\mathcal{T}_2')$ and whose root is a different situation – v_2(say). Trees $\mathcal{T}_1'$ and $\mathcal{T}_2'$ are called *isomorphic* if there is a bijective map from $\Psi : V(\mathcal{T}_1') \to V(\mathcal{T}_2')$ such that there is an edge $e_1 = (v_1', v_1'') \in E(\mathcal{T}_1')$ if and only if there is an edge $e_1 = (\Psi(v_1'), \Psi(v_1'')) \in E(\mathcal{T}_1')$.

The subtree from situation v_{11} to the end of the investigation, the *sampling subtree* from v_{11}, is given below. This depicts the unfolding of events after the brother actually came to collect the rent wearing the pullover but did not leave the recovered fibre and then the suspect robbed the woman and left the recovered fibre. Continuing the story from v_{11}, were the blood on the fingerprint to be analysed it will give no match [−], an inconclusive result [?] a partial match [+] or a complete match [++] to the suspect's blood. The prosecution and police could decide to arrest the suspect and prosecute P or let him go $\overline{P}$. If they arrest him they can choose to take a sample of fibre from his pullover and see if it matched that found at the crime scene and/or check for a match between his DNA and the blood at the crime scene. Let S_0 denote the decision not to do any further tests, S_f the decision to test the fibre match alone, S_b the blood match alone and $S_{f,b}$ the decision to test both.

Now certain symmetries in the subtrees of the full trees are apparent. For example it is easy to check that the possible unfolding of events after v_{13} until all sampling is completed – the *sampling subtree* from v_{13} – is, with the obvious identification of vertices and edges, topologically identical (or isomorphic) to the sampling subtree from v_{11}. In fact the sampling subtrees rooted at each of the situations $v_5, v_6, v_{12}, v_8, v_{10}, v_{11}, v_{13}, v_{14}$ are all isomorphic. So the topology of the sampling subtree above can be used as a template to represent all these developments: implicitly glued to each of the situations. This decomposition is helpful to the DM both in simplifying her depiction of the problem and encouraging her to focus on particular shared components of the problem separately.

The final outcome is whether or not the suspect is found guilty G or not $\overline{G}$ by the jury. Regardless of the unfolding of the past this can be represented by the *guilt subtree* given below. So again only one subtree needs to be drawn. The guilt subtree needs

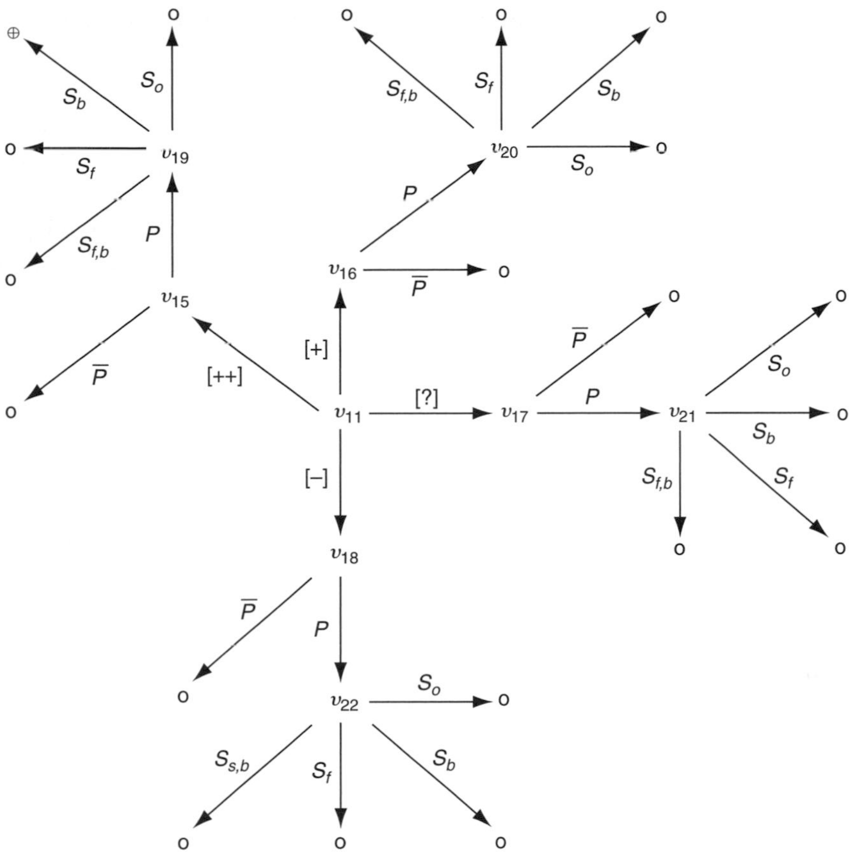

to be pasted on each of the leaves of all the sampling subtrees from the situations $v_5, v_6, v_{12}, v_8, v_{10}, v_{11}, v_{13}, v_{14}$.

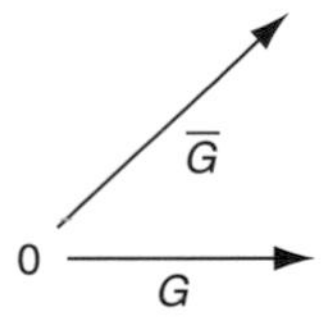

The full episodic tree is now obtained by pasting the sampling subtree on each of the leaves of the initial subtree and then the guilt subtrees on top of these. This gives us a tree with 320 leaves – the atoms of the sample space – and 265 situations. So the episodic tree of even a moderately complex decision problem like this one can be large. On the other hand the symmetries usually inherent in a problem allow the tree to be decomposed into much smaller and more manageable subtrees.

2.3.2 Chance and decision situations and consequences

The situations of trees representing a decision problem can usually be partitioned into those vertices whose emanating edges can be labelled by possible acts by the responsible agent – called *decision situations* – and those – called *chance situations* – associated with possible outcomes over which she has no direct control. Thus in the example above, where the responsible agent is the prosecution the decision situations are those deciding whether or not to prosecute and whether or not to sample the pullover for a match or the blood for a match. All the other situations depicted are chance situations. Traditionally decision situations are represented by □ vertices and chance situations by ○ vertices.

The second useful embellishment of a tree is to label the leaves – which represent the possible ways situations could next evolve – with the rewards determined by their consequences. We have argued in the last section that the analyst should have elicited a tree from a DM which is detailed enough for the rewards associated with the consequences of following a certain root to leaf path to be certain to the DM. In practice when these consequences are elicited it may well become apparent that the problem description embodied in the tree is simply not rich enough to allow the DM to specify these consequences unambiguously.

Thus turn to our first example concerning the allergic potential of the ingredient. Clearly the consequences are different if the company markets the product than if they shelve it. They may also consider trialling the product in a restricted market for a limited period. This would allow them to see whether there were allergenic effects not predicted by the lab experiment when the product was actually applied to real customers. On the basis of this pilot they could then decide whether or not to market the product. Note that our tree has already expanded into a decision tree below. The decision situations in this embellishment of the problem are vertices v_4, v_5, v_6, v_7: the decision whether not to market $\overline{M}$, market M or test the market T given the four different results from the lab together with decisions of whether or not to market after the test gave a good positive outcome $+$ or a poor one $-$. In this example we may well need to go further: for example including the types of customer that might be exposed and so on.

In some problems it is possible to express the impact of consequences that might arise from a sequence of decisions and consequent out-turn of events simply in terms of the financial reward. In this chapter we will focus on such problems. So in the example above the DM might argue that she wants to express any health consequences to potential customers purely in terms of the eventual financial damage that marketing an allergy-inducing product might cause.

However such circumstance are rather unusual. In the next chapter and Chapter 6 we develop techniques to address problems where the DM's rewards from measuring the consequences are not simply financial. For example, in the problem above the company may want to consider not only the short-term financial implications on the particular product line but also the legal consequences and the consequences on the reputation of the company of marketing an allergenic product. In Example 1.9 the consequences of whether or not to take the suspect to court and how much forensic evidence to gather to support the case are even

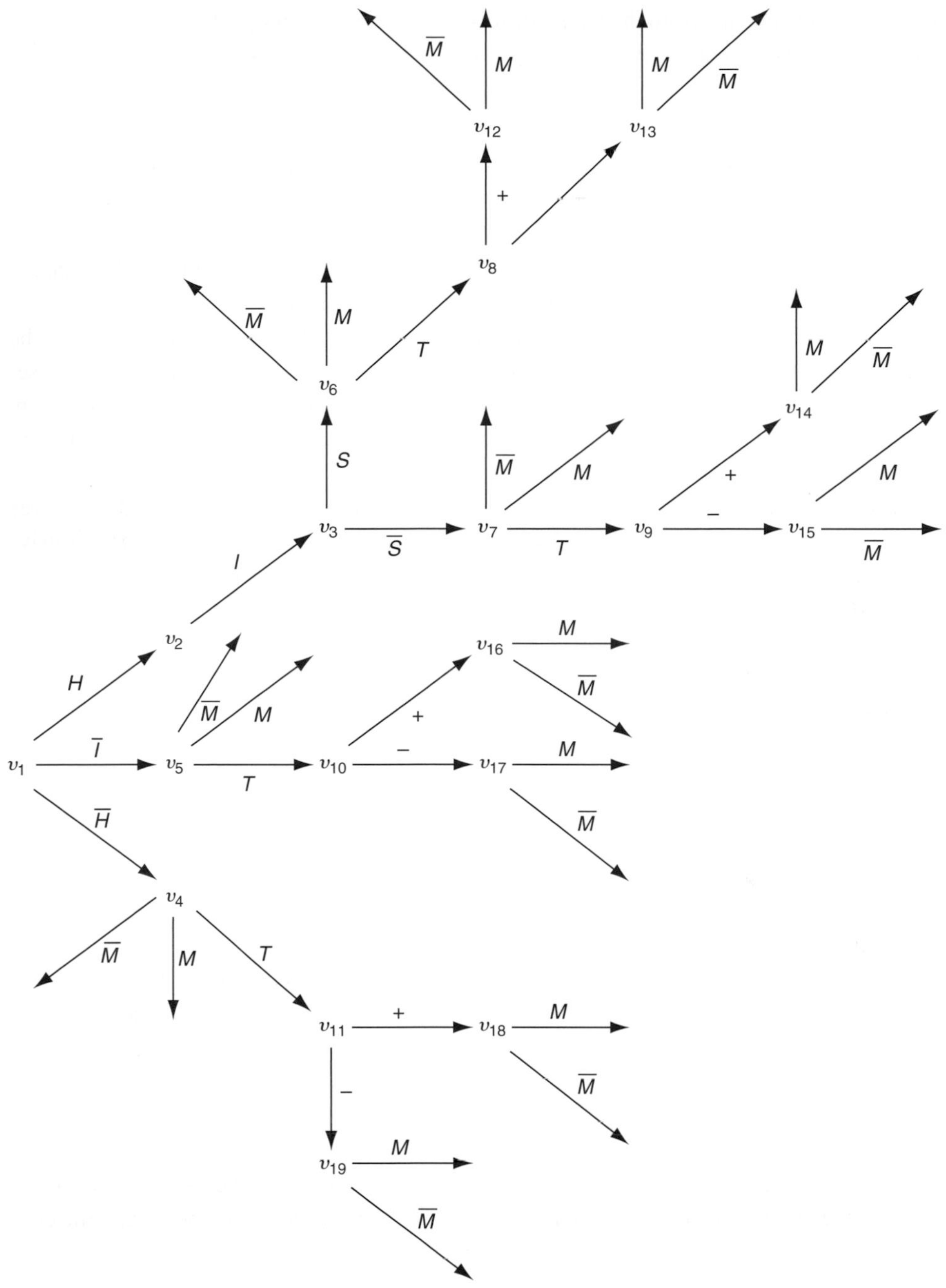

$\overline{M}$
M
M
$\overline{M}$
v_{12}
v_{13}
+
v_8
$\overline{M}$
M
T
M
$\overline{M}$
v_6
v_{14}
S
$\overline{M}$
M
+
M
v_3
v_7
v_9
–
v_{15}
$\overline{S}$
T
$\overline{M}$
I
v_2
v_{16}
M
H
$\overline{M}$
M
$\overline{M}$
+
v_1
$\overline{I}$
v_5
v_{10}
–
v_{17}
M
T
$\overline{H}$
$\overline{M}$
v_4
$\overline{M}$
M
T
v_{11}
+
v_{18}
M
–
$\overline{M}$
v_{19}
M
$\overline{M}$

more stark. On the one hand the DM is concerned to maximise the probability of obtaining a conviction of the suspect. But on the other this has to be set against resource constraints: both the financial cost of the forensic investigation and the resource cost of preparing this case against the possibility of deploying staff in a potentially more fruitful case.

2.3.3 Chance edge probabilities

Edges coming out from decision nodes cannot be labelled with probabilities at the start of the analysis, because these are chosen with certainty by the DM. However the edges out of chance nodes can. Consider the example above. The probabilities of B and $\overline{B}$ need to be chosen to reflect how plausible the prosecution/police find the presence of the brother to be. Moving through the events described by the initial tree there appear to be several plausibly parallel situations. For example the DM could well be happy to assign the same probability of the fibre being left at the brother's visit as the probability it was left at the suspect's visit. This would mean she could assert that v_1, v_2 and v_9 are parallel situations. In particular setting the conditional probabilities $P(F_b|.) = P(F_s|.)$ on the three edges labelling these events may all be equal. Similarly v_4, v_7 may also be considered parallel. These can be depicted below.

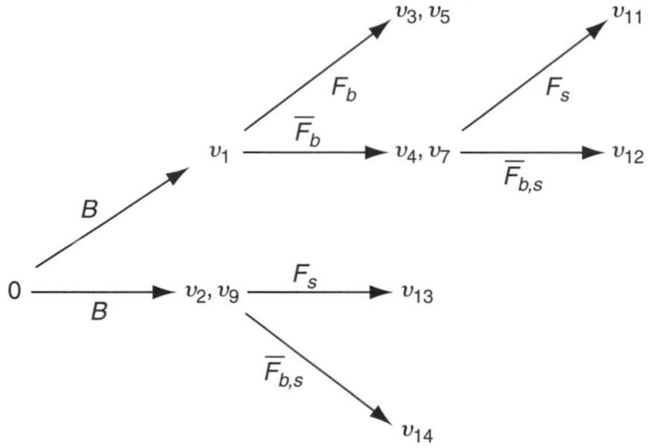

The probabilities associated with the four edges labelled $++, +, ?, -$ in the sampling subtrees are simply a function of the reliability of the fibre matching and DNA matching techniques. It would usually be accepted that these probabilities will not depend on the history of the case – and so be the same on all the isomorphic subtrees listed above. Thus again there are many parallel situations. Furthermore the judgement that these really are parallel – being linked to beliefs about the impartiality of the forensic scientists – is likely to be acceptable to all concerned: not only the DM but others – like an auditor – here the defence counsel, judge and jury. Note that the probabilities adopted by the DM are likely to be provided by the relevant experts – here the forensic scientists and their statisticians. The jury may well also take these judgements as their own. In fact these probabilities would typically be generic across many other cases, not just this one, with parallel situations in them and so sampling and experimental information is usually available and confident statements about the values of these probabilities can usually be made.

Finally the DM needs to provide probabilities for the edges labelled $\overline{G}$ or G – whether the jury find the suspect guilty – in the guilt subtree. The most straightforward way to do this is a statistical one. The DM simply embeds the event in question in a (sometimes hypothetical) population of past parallel situations represented by cases where in the judgement of the prosecution, the strength of evidence for or against the suspect is comparable.

For example consider the probabilities emanating from the situation labelled $\oplus$ correspond to the probability of the event a jury will find a suspect guilty given the suspect was the culprit and is prosecuted and a positive match found of his DNA to the DNA found in the blood at the scene, that no fibre match was searched for, and that the brother visited the house leaving no fibre whilst the culprit did. However the prosecution choose to assign this probability they might reasonably assume that the jury will ignore, or will be instructed to ignore, the issue of whether the brother visited the house or the culprit left the fibre since there is no evidence that the fibre found was from the culprit's pullover. So in particular it is quite irrelevant whether the brother had visited wearing the culprit's pullover or not: $P(G|\oplus)$ is just the probability that in a type of case like this one with a partial print and a full DNA match a suspect like the one prosecuted will be found guilty. Notice that because of arguments like the one above many of the situations in the tree will be associated with the jury's decision.

What these probabilities technically mean and how the analyst can try to measure them as accurately as possible will be deferred to Chapter 4 and how evidence can be used to support these judgements will be discussed and illustrated in Chapters 5 and 9. For the remainder of this chapter we will simply assume that these can be elicited accurately. The point we illustrate through this example is that in moderately sized problems the DM will often adopt as her own, probabilities provided by different experts. Here the probabilities for the initial subtree are likely to be provided by the investigating police officers, those associated with the sampling subtree by the forensic scientists and statisticians and the probabilities on the edges of the guilt subtree by statisticians working with prosecution counsel who have collated information about jury verdicts from past case histories.

2.4 Some practical issues*

2.4.1 How detailed should an episodic tree be?

Recall that the set of root to leaf paths on an episodic tree represent the possible outworkings of history as envisaged by the DM and hence the atoms of her event space. But when drawing a tree like the one above one practical question is how *refined* the tree needs to be to support all the salient features needed for the decision analysis of a given problem. To answer this we need to refer back to the essential components that are required before a decision problem can be fully specified. These are:

(1) the probability of receiving a particular reward;
(2) the rewards associated with each possible pair of decision rule and outcome.

Thus the tree has to be sufficiently refined for the full consequences of any possible unfolding of history will be known by the DM. In a problem represented by a tree it is essential that two different unfoldings of history giving rise to different consequences – and so in particular their associated rewards – are distinguished by different root to leaf paths. On the other hand, although it is sometimes computationally convenient or more transparent to express two outfoldings with the same sequence of decisions and associated distributions on consequences by different root to leaf paths it is not *technically* necessary to keep these separate. So the explanation has to be suited to the purpose of the analysis.

In the crime example above we have actually surreptitiously performed this simplifying combination. For example, if the tree is read strictly episodically, we have appeared to suggest that we decide to perform two tests simultaneously and not to use the result of one test to determine whether or not we do another. For completeness more subtrees could have been included that depicted different decisions labelling the choice of which of the two tests to perform first. However this is unnecessary because neither the rewards associated with any decision rule considered nor the probability of any subsequent events to these rules depends on the order these two investigations were done. If either of these rewards or probabilities differed, for example if there is time to take a decision about testing one sample *contingent* on the other it would be *essential* to distinguish these two unfoldings: see the example in Smith and Thwaites (2008a). Atoms of the event space associated with the larger tree can be combined into coarser atoms associated with the simpler tree because the expectations of all functions needed for the decision analysis require as inputs only probabilities associated with events in the simpler event space.

These subtleties arise naturally from a verbal description of a problem: a DM will often – quite appropriately – not mention any features of her problem which, in her own judgement, are obviously unnecessary or irrelevant details. However it is important for the analyst to be aware that when the DM implicitly censors her description like this it can sometimes restrict her world view. So in the second example above the DM may not have even considered the possibility that she might perform tests sequentially. If the analyst can make her aware of this and if at some time in the future the DM might want to compare the contingent decision rules of choosing to analyse a second piece of evidence depending on whether or not the first was successful it might well be a good idea for the analyst to encourage her to work with a larger tree and keep these two distinct histories separate from the beginning.

Such elements of the elicitation process make the decision analyst's task a challenging one. However it is exactly the knowledge that implicit constraints are hidden in any description given by the DM that enables a decision analyst to contribute to the DM's understanding of the limitations of her world view. The subsequent expansion of this world view can have a liberating effect on the client's creative reasoning. One well-tried way such necessary embellishments can be elicited is to make conditional independence queries about the story: a process described in detail in a later chapter.

So when choosing a particular tree as a framework for an explanation of how situations might unfold there is obviously a trade-off. On the one hand it is important to try to keep the tree as simple as possible – so that it is easier to read, explore and modify. On the other hand the tree needs to be sufficiently refined so that it can provide a rich enough framework for performing the initial decision analysis and to explore possible new options on it. Fortunately the sequential nature of the description encoded in a tree often allows us to split it up into subtrees, each giving part of the story. This enables us to focus on different elements of the description independently and so build up a picture of the whole by integrating smaller component elements. The separation of different components of a description of a problem is called a *credence decomposition*. There are many different types of credence decomposition each appropriate to different types of explanation. But a decomposition associated with the unfoldings of an episodic tree is a particularly useful one. It is then often the case that many of its situations are parallel to one another. This usually indicates that there are considerable amounts of conditional independences lying hidden in the DM's explanation. These independences often enable the problem to be re-expressed in alternative and topologically simpler graphical frameworks: see later in this book. But my own experience has been that the episodic tree, whilst being rather cumbersome, is also one of the most expressive of graphs to embody descriptions of how situations unfold. I therefore tend to fall back on this representation especially when other simpler but less expressive graphical frameworks appear to break down.

2.4.2 Bayesian game theory and rationality

A second way of assessing probabilities associated with other people's behaviour is more technical but sometimes necessary for certain types of one-off scenarios. Consider probabilities on the guilt subtree of the crime example. Here the prosecution could make the following bold premises and argue as follows:

- The jury itself will act as if it is a rational Bayesian DM assigning probabilities and choose a decision maximising their expected efficacy of the resulting consequences.
- The jury adopts as their own the story depicted by the prosecution's episodic tree.
- The jury will believe the probabilities presented by the forensic scientists concerning the sampling probability of the different sorts of matches being found, whether from the suspect or otherwise.
- The jury will add the edge probabilities associated with the initial tree. Within our example two such probabilities are the probability the jury assigns to the event that the brother entered the house, and before any evidence has been presented, the probability that the culprit is guilty.

To use this structure the DM will need to produce her own subjective probability distributions for the probabilities (random variables) the jury assign to the conditional events in the initial tree. This interesting inferential structure is widely analysed especially by economists in the discipline of Bayesian game theory. In application areas like the one above the second and third assumptions are often fairly secure. Usually the qualitative structure of the description – here the tree – is the easiest part of a model with which to find agreement between different sides (Smith, 1996; Smith and Allard, 1996).

Several probabilities, supported by established scientific argument or extensive sampling of relevant populations, will also be stable across players. However the first bullet is more fragile. Whilst in combat scenarios and some economic models the assumption of Bayes rationality is fairly well supported, in others – and especially domains which are fundamentally social and unscientific – this appears not to be the case: see Chapter 4. Furthermore it is tricky for a DM to produce good estimates of other people's probabilities because these probabilities can be distorted in practice by a myriad of biases. On the other hand a benchmarking exercise about how the DM believes a rational body would behave is often illuminating and there are some surprising examples of when its predictions come very close to reality. See for example Dodd *et al.* (2006); Gigerenzer (2002); Oaksford and Chater (2006, 1998); Smith (1980b).

We will now leave this example and return to some more simple trees which can be used to demonstrate certain useful techniques.

2.4.3 Feasibility and consequences

Trees have been used for many years to teach chess, especially within the Russian school, and this motivates a final example where probabilities concern the behaviour of others and provides a simple example where issues of the feasibility of a tree representation and the link with the definition of consequence arise.

Example 2.4. The edges of a chess tree emanating from the decision vertices correspond to the set of legal moves available to the player – the DM. The edges from the chance vertices correspond to legal moves available to her opponent. The situations are taken in their obvious historic order consistent with order of the moves of the game. The rules of chess are such that all games will be completed by a fixed time so all root to leaf paths are of finite length and the terminal consequences in the tree are a loss, draw or win.

The chess tree is a good example with which to illustrate certain points. The first is that although this is an entirely deterministic game with simple well-defined rules its computational complexity forces even computers to approximate the class of moves the opponent considers and make assessments of intermediate decisions. Because the number of edges from each situation is large – about 50 – and the average length of a game – the length of a typical game is about 60 moves – a game tree is gigantic and impossible for even the most powerful computer to analyse fully. Chess trees therefore have to be simplified. First their breadth must not be too large – considering only sensible α type edges for both the supported player and her opponent, for example disregarding moves leading to immediate loss of a major piece. This essentially restricts the DM's decision space to a small subspace of those open to her and assigns zero probability to many of the moves an opponent makes.

Second the depth needs to be limited and not projected forward until its completion. Therefore its leaves cannot always be labelled with a sure loss, draw or win but with a position a certain number of moves ahead. This leaf is then given a numerical scoring reflecting its promise. Thus even though the game is intrinsically deterministic the solutions used both by computers and humans use an approximating decision tree of the problem and something like an algorithm choosing the decision maximising the expected score of the promise of the position a certain number of moves ahead.

Computers and humans differ here. Computers are able to search forward and compare orders of magnitude more paths than a human. However current programs tend to score leaf positions in a rather naive bean-counting way. Humans search a much smaller space but try to compensate by evaluating promising combinations and likely responses and are much more refined in the way they assess the promise of the leaf positions. This book addresses the decision support of human not computer DMs. So the necessarily approximate and subjective framework must allow the identification of promising classes of decision rules, good assessments of the uncertainties in the system and good evaluations of the resources derived from the consequent possible positions that her chosen course of action might lead her to.

The Bayesian chess DM needs to assign probabilities to the opponent's possible moves. Chess-playing computer programs almost inevitably assume that their opponent – with probability one – will play the move that the computer calculates as optimal at that point. This uses a rather naive Bayesian game theory approach where the DM assumes with certainty that her opponent will act exactly as she would and reduces the problem to a deterministic one. On the other hand, human players will work on the assumption that the opponent will choose from a small selection of "good" moves – where the term good is defined by the supported player as ones she believes her opponent will consider playing. Different players will use different methods to assign these probabilities. Some, for example Kasparov, advocate an approach close to a Bayesian game theory one but this time enacted with some introduced uncertainty and restricted to these plausible moves. On the other hand others take a more behavioural approach accommodating information about the preferences of their perceived knowledge of the likes and limitations of their opponent. Thus note this comment by the celebrated player Korchnoi (2001). Knowing that were his opponent to play $d \times e6$ he would end up in a very poor position playing Black he writes.

"the move $d \times e6$ by no means suggests itself . . . the move appears to be a concession to Black."

He therefore decided to chance that his opponent would not find the move – which he did not – and subsequently Korchnoi went on to win the game.

This illustrates that even in simple mental games that are entirely deterministic, using past behavioural information to assess other's probabilities is not necessarily a poor one: see Kadane and Larkey (1982) for a passionate defence of this position. Certainly in less defined environments where assumptions of rational decision making are not credible, the behavioural assignment of probabilities is often the only practical course.

2.5 Rollback decision trees

In all but simple decision problems the decision maker needs to choose a sequence of good decisions, d_i, $i = 1, 2, \ldots, k$, based on the information they have collected by the time the decision d_i needs to be enacted. So let $\boldsymbol{X}_0$ denote the information available to the DM when she takes her first decision $d_1 \in D_1$ – the space of the first commiting decision – and $\boldsymbol{X}_1(d_1)$ denote the new information arriving after the decision d_1 has been committed to but before d_2. Note that if, for example d_1 was the decision of the extent DM sampled then $\boldsymbol{X}_1(d_1)$ – together with the outcome space $\mathbb{X}(d_1)$ in which it lies – could well depend on the decision d_1. The next decision d_2 can be chosen as a function of the previous decision committed to – d_1- and the information $(\boldsymbol{x}_0, \boldsymbol{x}_1)$ already gathered – that decision chosen from a space D_2 determined by what has happened so far. Again the decision space D_2 can depend both on d_1 and $(\boldsymbol{x}_0, \boldsymbol{x}_1)$. For example if d_1 were to perform exploratory surgery, then the possible courses of action considered when it were found that no tumour existed could be very different from those considered when a tumour was found.

Let $d^{(r)} \triangleq (d_1, d_2, \ldots, d_r)$ and $\boldsymbol{x}^{(r)} \triangleq (\boldsymbol{x}_0, \boldsymbol{x}_1, \boldsymbol{x}_2, \ldots, \boldsymbol{x}_r)$. Then continuing in this way we see that at the rth stage of the decision process $1 \leq r \leq k$ the DM needs to choose a decision $d_r \in D_r$ where both d_r and D_r are a function of $(d^{(r-1)}, \boldsymbol{x}^{(r-1)})$, leading to an outcome $\boldsymbol{x}_r \in \mathbb{X}_r$ where both $\boldsymbol{x}_r$ and $\mathbb{X}_r$ are a function of $(d^{(r)}, \boldsymbol{x}^{(r-1)})$ for $r \geq 2$. Such a sequence of decisions $\boldsymbol{d} = (d_1, d_2, \ldots, d_k)$ made as a function of the information gathered at each stage and the commitments made already is called a *decision rule*. The analyst needs to be able to facilitate the DM's wise choice of a decision rule in the light of the consequences such a sequence of committing decisions might have.

We have already encountered some simple decision rules in the examples above. Thus for example in the chess example a player's decision rule is the rule that specifies how she will plan to play given in response to all the possible moves open to her opponent. Thus her rth move d_r is chosen as a function of $(d^{(r-1)}, \boldsymbol{x}^{(r-1)})$; the moves she and her opponent have made so far. The blood-pooling example of Chapter 1 gives a much simpler setting. Here $\boldsymbol{X}_0$ is unknown, the DM then decides her pooling d_1. This leads to her observing $\boldsymbol{X}_1$ which tells her if the pool was positive. In the original statement of the problem, all pools discovered to be positive would have all their members tested individually. But we mentioned that the DM could also consider a second pooling d_2 of the positive group. In the criminal example above $\boldsymbol{X}_0$ corresponds to having obtained a partial match of the fingerprint before the police need to prosecute d_1. If they decide not to prosecute then they learn and do nothing more – so that $\mathbb{X}_1$ and D_2 are both null. On the other hand, if they choose to prosecute, although they can collect no further information before their next decision – so that $\mathbb{X}_1$ is again empty, they can decide which sampling to perform D_2 after which they will learn, on the basis of the evidence, whether or not the jury will find the suspect guilty.

Now although episodic – and especially historic – trees are very useful for describing decision problems and providing a framework for eliciting, depicting and exploiting as many parallel situations as possible, they are not so good as a framework for depicting the set of decision rules available to the DM. However if instead of introducing situations in the description in the order they happen we introduce them consistently with when the DM will *discover* what has happened then the decision tree can not only be used as a representation of the DM's problem but also double as a framework for the efficient calculation of an optimal policy. Such a tree is called an extensive form tree or a *rollback tree* and is one of the most popular tree-based depictions of a decision problem. Sometimes, like in the chess example above, the historic and rollback trees are the same. But this tends to be the exception rather than the rule.

We may suffer some loss by substituting a backwards induction decision tree for a causal tree. The edges expressed in the new graph may be associated with conditional probabilities which are expressed anti-causally. For example in a medical diagnosis we may well need to introduce symptoms before their causes – diseases – because the doctor usually sees the symptoms of a disease before the disease itself is confirmed. So this means that events depicted in a backwards induction tree by edges will have associated probabilities that will usually need to be calculated off-line using Bayes rule and the law of total probability – using as inputs their elicited causal counterparts. Nevertheless this is often a small price to pay for a graphical framework that supports the calculation of an optimal decision rule.

The extensive form decision tree bases the calculation of an optimal policy on the following definition of an optimal decision rule.

Definition 2.5. A *current judgement optimal* (cjo) decision rule d^* is a decision rule which assumes that in the future after a DM learns more she will continue to act optimally where this future optimality is defined using the DM's current beliefs.

Although cjo decision rules can be defined outside a Bayesian context, here the DM searches for a decision rule that maximises expected reward. To demand that the EMV DM chooses a cjo rule then simply requires her to plan her first decision assuming that she will choose any future decision so that she maximises her expected payoff. She uses her current probability model to work out what this expectation might be. Thus she calculates her revised expectations by formally conditioning on the events she believes might happen in the future. Note that under this assumption her future probability distribution of each possible future unfolding of what she might believe after discovering each possible out-turn of events can be calculated at the current time using Bayes rule and her *current* joint probability distribution.

If the DM assumes that her joint probability distribution over all unfoldings is not changed except by formally conditioning on what she will see then it can be proved that a Bayes decision rule must be a cjo decision rule (see e.g. Raiffa and Schlaifer (1961)). Nearly all Bayesian decision theory explicitly or implicitly assumes that it is appropriate to choose a cjo decision rule in this way. And such reasoning is certainly easy to justify to an auditor. After all what is more natural than for DM to justify her current plans on the assumption that – within the (probabilistic) framework of her current beliefs and understanding (what else could she use?) – she plans to act optimally in the future.

Thus standard Bayesian decision theory prescribes that an optimal decision rule is cjo. To illustrate how we can use a rollback tree together with this property to identify an optimal policy, consider the following example.

Example 2.6. A valued customer has told the DM that he is prepared to buy some speculative new machinery provided that it works immediately it is installed. It will work if part of the machinery is sufficiently flat. If she decides not to scan the machine – d_0 – and deliver immediately – a_1 – and the machinery does not work immediately it will be returned and she will obtain nothing. On the other hand if it works she will receive £10,000. She could decide not to scan but undertake an immediate overhaul of the item – decision a_2 – at a cost of £2,000. If she finds the part is not flat enough it will then cost a further £1,000 to fix it, but after this she knows the customer will be satisfied. There is also the possibility of performing either one scan – d_1 – or two scans – d_2 – at a total cost of £900 using a scanning device to check for a fault. Each scanning device will give independent readings conditional on whether or not the machine will fail. Prior to any scan the DM currently believes that the probability that the machinery will not be flat enough is 0.2. Any scanning device will indicate that the machinery is faulty given it is with probability 0.9 but indicates that it is faulty when it is not with probability 0.4.

Because this problem is a simple one it is possible to list all its decision rules. Note in the listing below that $-$ indicates a negative result of a scan and $+$ a positive one. Thus, for example, decision rule $d[4]$ denotes the decision to scan once and on obtaining a negative result to send off the product but on seeing a positive indication of a fault to overhaul the machinery before dispatching it.

$$
\begin{aligned}
d[1] &= (d_0, a_1) \\
d[2] &= (d_0, a_2) \\
d[3] &= ((d_1, -, a_1), (d_1, +, a_1)) \\
d[4] &= ((d_1, -, a_1), (d_1, +, a_2)) \\
d[5] &= ((d_1, -, a_2), (d_1, +, a_1)) \\
d[6] &= ((d_1, -, a_2), (d_1, +, a_2)) \\
d[7] &= ((d_2, --, a_1), (d_2, -+, a_1), (d_2, ++, a_1)) \\
d[8] &= ((d_2, --, a_1), (d_2, -+, a_1), (d_2, ++, a_2)) \\
d[9] &= ((d_2, --, a_1), (d_2, -+, a_2), (d_2, ++, a_1)) \\
d[10] &= ((d_2, --, a_1), (d_2, -+, a_2), (d_2, ++, a_2)) \\
d[11] &= ((d_2, --, a_2), (d_2, -+, a_1), (d_2, ++, a_1)) \\
d[12] &= ((d_2, --, a_2), (d_2, -+, a_1), (d_2, ++, a_2)) \\
d[13] &= ((d_2, --, a_2), (d_2, -+, a_2), (d_2, ++, a_1)) \\
d[14] &= ((d_2, --, a_2), (d_2, -+, a_2), (d_2, ++, a_2))
\end{aligned}
$$

To draw a rollback tree we introduce situations in the order they are enacted or observed. Thus the first decision faced by the DM is whether to perform 0, 1 or 2 scans, respectively denoted here by d_0, d_1, d_2. She will then observe how many if any of these scanning devices indicate that the machinery is faulty. On the basis of this evidence she must then decide whether or not to overhaul the machine: i.e. choose between a_1 and a_2. The last thing she will discover is whether or not the machinery will work. By this time she will know what her payoff will be for any course of action she has taken.

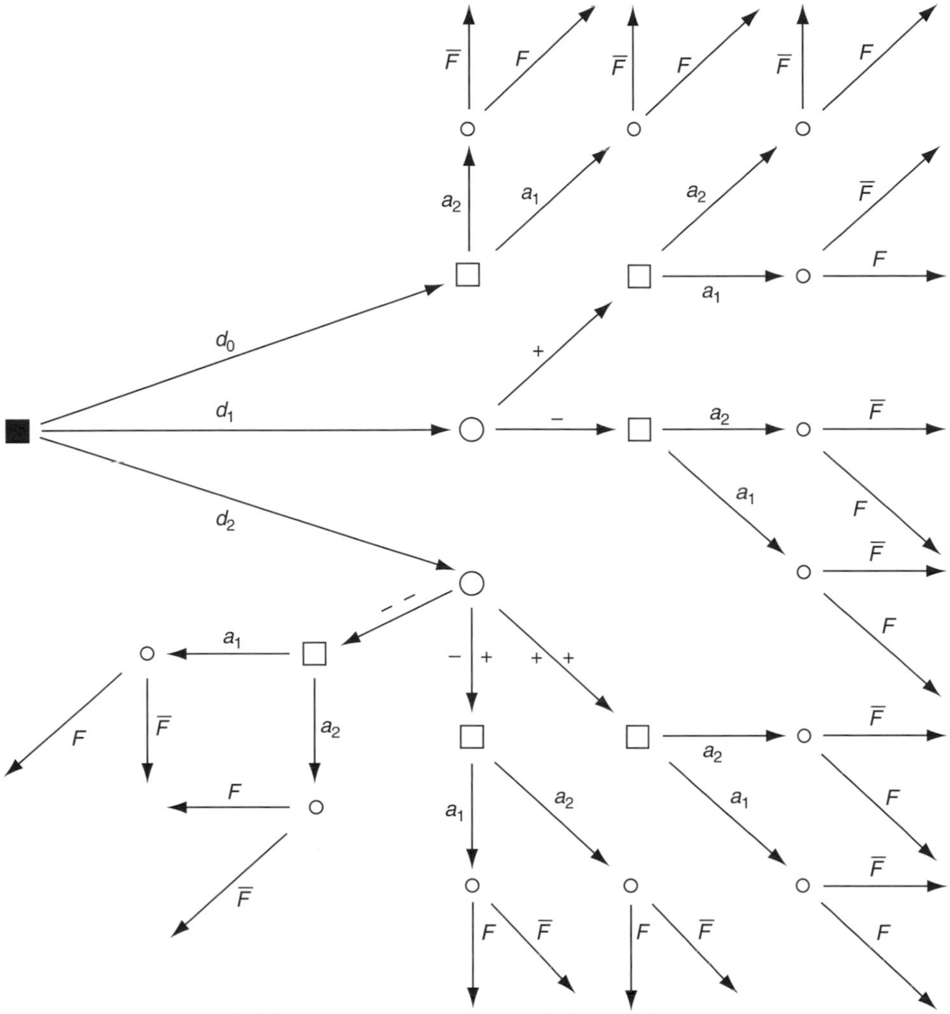

It can be seen that any decision rule can be associated with a subtree of a rollback tree.

Definition 2.7. The *decision subtree* $\mathcal{T}(d)$ of an extensive form tree $\mathcal{T}$ associated with a decision rule $d \in D$ is such that:

(1) The root of $\mathcal{T}(d)$ is the root of $\mathcal{T}$.
(2) All the root to leaf paths of $\mathcal{T}(d)$ are also root to leaf paths of $\mathcal{T}$.

(3) If a chance situation $v \in V(\mathcal{T}(d))$ – its vertex set – then the subtree will contain all edges emanating from v in its edge set $E(\mathcal{T}(d))$.
(4) If a decision situation $v \in V(\mathcal{T}(d))$ then the subtree contains exactly one edge emanating from it in its edge set $E(\mathcal{T}(d))$.

In the tree above, for example, the subtree below depicts the decision rule $d[4]$ that scans the item once and if it gives a positive result decides to send the machine off to the customer whilst if the scan gives a negative result then the DM chooses to overhaul the machine.

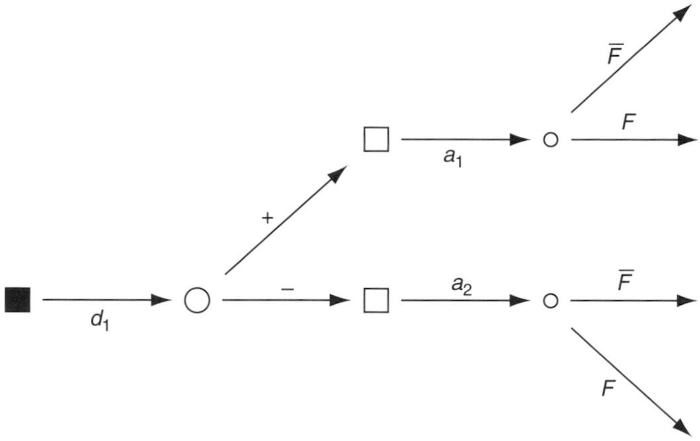

Because of the chronology of a rollback tree, provided there are no additional constraining conditions in a problem there is a one-to-one correspondence between its decision subtrees and its associated decision rules. This forms the basis of another useful property of a rollback tree: it can be used as a framework for calculating an optimal decision rule. Furthermore because it uses the property discussed above – that an optimal decision simply assumes that future decisions will be chosen optimally – it helps to explain in a transparent way to a DM and her auditor *why* that decision rule is optimal. This process of calculation: called an *extensive form analysis* or a *backward induction* algorithm is illustrated below using the example above.

We first need to embellish the tree with its leaf consequences and edge probabilities as discussed above. The leaf payoffs have been written in units of £1,000 at the tip of each associated leaf. Thus for example 7.5 is written at the tip of the root to leaf path $(d_1, +, a_2, \overline{F})$. This is the amount (in £1000s) of scanning the item – cost £500, obtaining a positive + indication of a fault, overhauling the machine at a cost of £2,000 but finding it had no fault and obtaining £10,000 from the customer: giving the DM a total payoff of

$$£7,500 = £10,000 - £2,000 - £500.$$

Other payoffs are calculated similarly. Notice in this example that the consequences are simply monetary so it makes sense to try to identify an EMV decision rule for this problem.

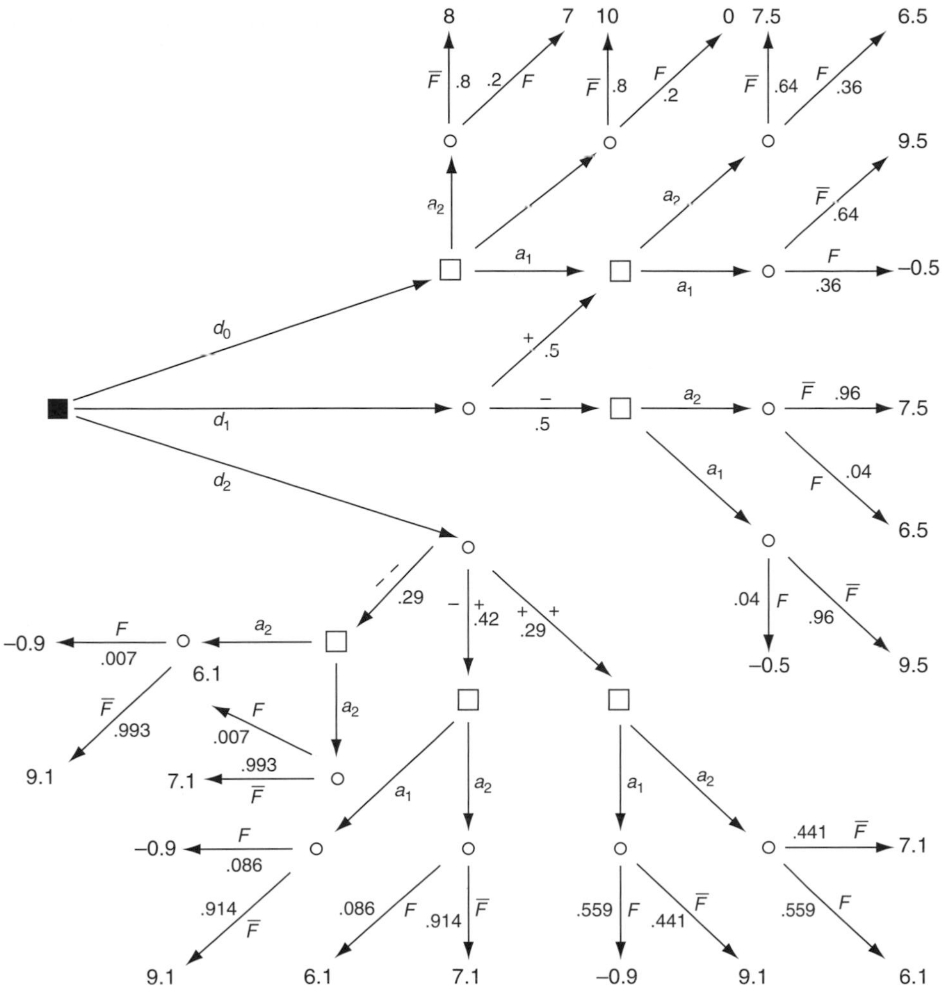

The next task is to label the edges from the chance nodes of the tree. For most rollback trees this will require some calculations because the tree is not episodic and so all edges are not necessarily labelling events in their natural causal order. Thus the edges $(d_0, a_2, \overline{F})$ and $(d_0, a_1, \overline{F})$ are associated with the event $\overline{F}$ that the machine is not faulty, given as $1 - 0.2 = 0.8$, whilst the probability associated with the events (d_0, a_2, F) and (d_0, a_1, F) is 0.2. So these probabilities can be put directly on their associated edge on the tree above.

On the other hand the DM's probabilities of a positive indication of a fault are only given *conditional* on whether or not a fault exists. But we see that the edge labelled $+$ corresponds to the probability she would assign to a fault being positively indicated before she learned whether or not a fault existed. So this is her marginal probability of detecting a fault. Fortunately this probability is straightforward to calculate from the law of total probability. Thus the probability the scan will indicate a fault is

$$\begin{aligned}P(+) &= P(+|\overline{F})P(\overline{F}) + P(+|F)P(F)\\ &= 0.4 \times 0.8 + 0.9 \times 0.2 = 0.5.\end{aligned}$$

It follows that the probability $P(-)$ the scan does not indicate a fault is $1 - 0.5 = 0.5$ also. Similarly, because the results of scan are independent we can calculate that $P(++)$, the probability that two independent scans both give positive results, and $P(--)$, the probability that two independent scans both give a negative result, are given respectively by

$$\begin{aligned}P(++) &= P(+|\overline{F})P(+|\overline{F})P(\overline{F}) + P(+|F)P(+|F)P(F)\\ &= (0.4)^2 \times 0.8 + (0.9)^2 \times 0.2 = 0.29,\\ P(--) &= P(-|\overline{F})P(-|\overline{F})P(\overline{F}) + P(-|F)P(-|F)P(F)\\ &= (0.6)^2 \times 0.8 + (0.1)^2 \times 0.2 = 0.29.\end{aligned}$$

By subtraction from one the DM can now calculate the probability of one scanner indicating a fault is $P(-+) = 0.42$.

The edges into the leaves of the tree not yet calculated denote the DM's probability of a fault given certain observations. So the appropriate probabilities associated to these edges are the conditional probabilities of a fault or not given the observation leading to that edge. These again need to be calculated, this time by Bayes rule. So for example

$$P(\overline{F}|+) = (P(+))^{-1}\, P(+|\overline{F})P(\overline{F}) = 0.64.$$

Placing the probabilities on the associated edges using this formula gives the fully embellished decision tree above.

Backwards induction can now be used on this rollback tree to find an EMV decision rule. We have already noted that any optimal rule of this type will be a cjo. So in particular this means that after observing the result of anything she might learn – in this case the result of any scans she might perform – the DM should choose a decision maximising her expected payoff. Assuming this note that if she chooses d_0 then she should choose a_1 because this has greater expected payoff (in £s), $10,000 \times .8 + 0 \times .2 = 8,000$ than the alternative a_2 which has expected payoff $8,000 \times .8 + 7,000 \times .2 = 7,500$. She therefore knows that she can ignore the possibility of deciding a_2 after choosing d_0. Therefore delete the subtree starting with the edge (d_0, a_2).

Similarly suppose the DM were to choose d_1 and observed a positive indication that a fault might be present. Then the expected payoff associated with subsequent choice a_1 can be calculated as 5.9 in units of £1,000s and for a_2 is 7.14. So clearly the DM will ignore a_1 – delete the subtree starting with this edge – and do a_2. So write the payoff 7.14 on this edge. Performing this operation for all the last decisions the DM might make produces the simplified tree given below.

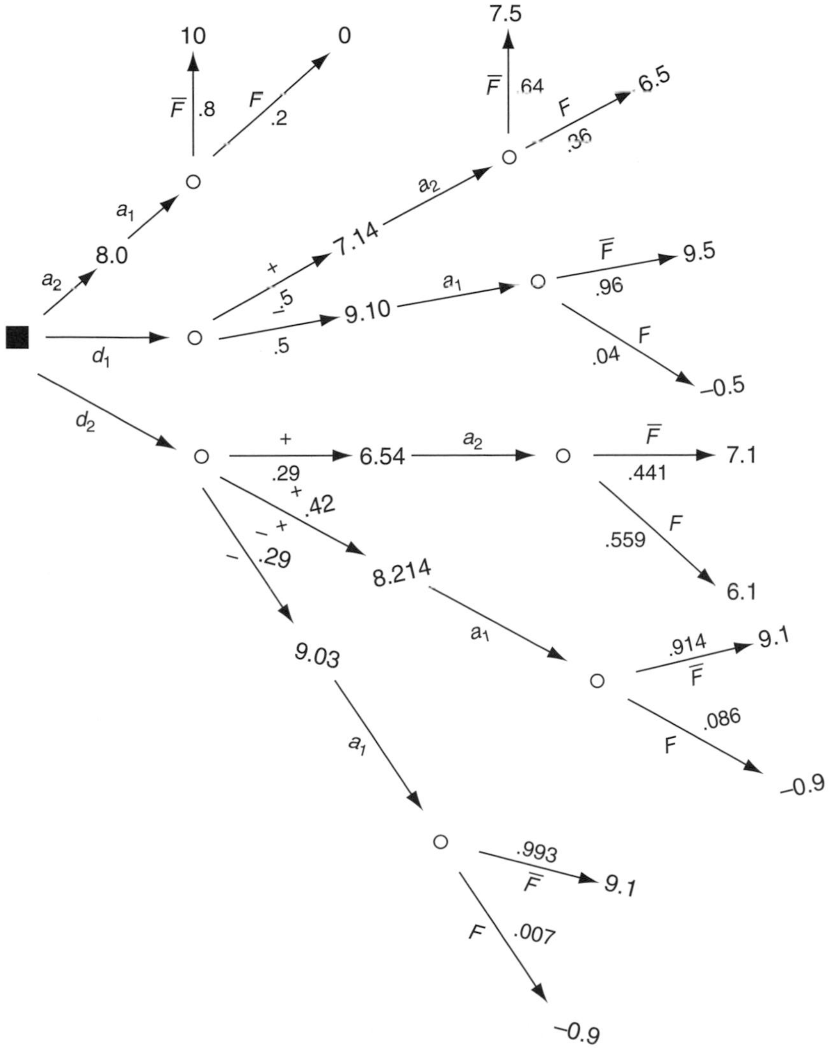

But now note that the expected payoff the DM receives if she chooses d_1 and subsequently acts optimally can also be calculated. This is just

$$.5 \times 7.14 + .5 \times 9.10 = 8.12$$

and if she chooses d_2 and subsequently acts optimally she receives

$$.29 \times 6.54 + .42 \times 8.24 + .29 \times 9.03 = 7.98.$$

We see that d_1 is best and subtrees beginning with edge d_0 or d_2 can be deleted from the tree without loss. The expected payoff associated with d_1 is then transferred to the root vertex.

The final tree is given below.

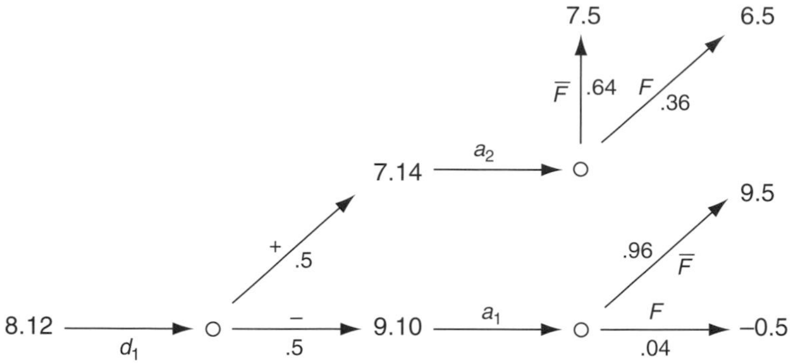

Note that the final tree depicts the decision rule $d[4]$ listed above with the associated expected payoff given at its root. Thus our EMV decision is to scan once and if a positive indication of a fault is indicated then overhaul the machine but otherwise send it off immediately.

This method of working backward from the leaves of the tree discovering the best action to take contingent on the past and then averaging over the associated expected payoffs to obtain the expected payoffs associated with the previous committing decisions can clearly be performed however many stages there are in the decision making process. Moreover the final tree obtained using the construction above will be the tree depicting an optimal decision. The rollback tree and this algorithm is currently coded in many pieces of decision support software. So once edge probabilities have been calculated the optimal decision rule of even very complex trees can be calculated almost instantaneously.

2.6 Normal form trees

Sometimes it is valuable to understand how the Bayes decision rule depends on some of the input probabilities. So for example, in the example above whilst being fairly confident about the probabilities of scanners given the machine is or is not faulty – after all these may well have been assessed by extensive previous experimentation studying the performance of the scanning device – the DM may well feel that the probability p she assigns to the current machine being faulty is much less secure. A normal form analysis is designed to determine which decision rules might be optimal under *some* value of the probability p of a particular causal event E and also to determine when to choose each of these candidate decisions as a function of p.

A normal form analysis avoids the use of Bayes rule, because it take the chance vertices in their historic order so that no reversing of conditioning is required. Instead it calculates the pair $(V_1(d[i]), V_2(d[i]))$ for each possible decision rule $d[i] \in D$. Typically there are a large number of such decision rules but in the simple example above there are just 14. The first component is the expected payoff associated with $d[i]$ if the analysed event E occurs and the second the payoff if it does not. Theoretically there is a tree underlying a normal form analysis which is closer to an episodic tree. The root of this tree has D edges emanating from it each labelled by one of the many decision rules DM could use. From each situation attached to a decision rule edge is a chance vertex with two edges labelling whether or not the identified event E has happened. We then follow this by a sequence of chance nodes introduced in the order in which these "symptoms" are observed under the decision rule defining that part of the decision tree. The subtree emanating from decision $d[4]$ in the example above is given below.

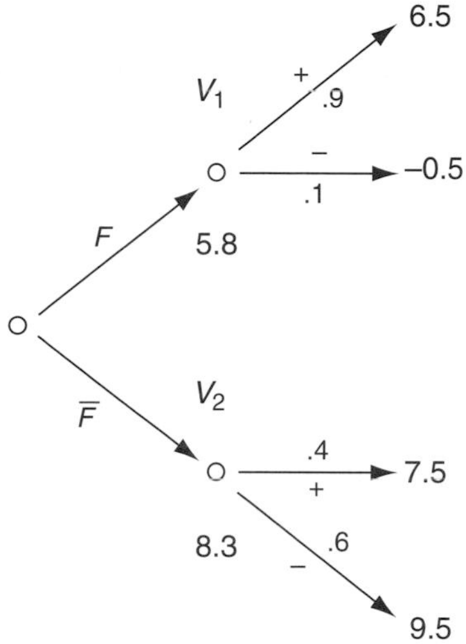

The terminal payoffs are particular to the decision rule. Here under $d[4]$ if a + is observed then the DM will overhaul and find the fault if it exists. So her payoff associated with the root to leaf path $(d[4], F, +)$ will be $10 - 0.5 - 2 - 1 = 6.5$ (in £1,000) – the payoff for delivering a faultless machine less the cost of scanning, overhauling, finding the fault and fixing it. This amount appears on the corresponding leaf of this event in the tree above.

Once all the leaf payoffs have been calculated these are placed on the tips of their associated leaves. These are summarised in the tables below.

Dec. rule	$d[1]$	$d[2]$	$d[3]$	$d[4]$	$d[5]$	$d[6]$	$d[7]$
V_1	0.0	7.0	−0.5	5.8	−0.2	6.5	−0.9
V_2	10.0	8.0	9.5	8.7	8.3	7.5	9.0

Dec. rule	$d[8]$	$d[9]$	$d[10]$	$d[11]$	$d[12]$	$d[13]$	$d[14]$
V_1	4.77	0.36	6.03	−0.84	4.84	0.44	6.1
V_2	8.78	8.14	7.82	8.38	8.06	7.42	7.1

Note that the expected payoff $\overline{G}(d[i])$ associated with decision rule $d[i]$ is given by

$$\overline{G}(d[i]) = pV_1(d[i]) + (1-p)\,V_2(d[i]) \tag{2.1}$$

where $0 \leq p \leq 1$ is the probability of the event that the underlying cause – here the fault – exists. It can immediately be seen from this table that many decision rules $d[i]$ have a lower expected payoff than another $d[j]$ whether or not the fault exists. Whenever both $V_1(d[j]) \leq V_1(d[i])$ and $V_2(d[j]) \leq V_2(d[i])$ decision rule $d[i]$ is said to *dominate* $d[j]$ and *strictly dominate* $d[j]$ when one of these inequalities is strict. When $d[i]$ (strictly) dominates $d[j]$ then

$$\overline{G}(d[i])(<) \leq \overline{G}(d[j])$$

for all possible values of p. If follows that $d[i]$ is never uniquely the best decision rule and $d[j]$ is at least preferred to it. In fact the only decision rules in our problem not strictly dominated by another are given in the table below.

Dec. rule	$d[1]$	$d[8]$	$d[4]$	$d[2]$
V_1	0.0	4.77	5.8	7.0
V_2	10.0	8.78	8.7	8.0

The two decisions $d[1]$ and $d[2]$ respectively are associated with immediate dispatch or immediate overhauling, decision $d[4]$ we have discussed above and $d[8]$ is the decision to use two scanners and to send off the product immediately unless the fault is indicated twice when the DM will overhaul the machine.

To check whether all four of these rules are optimal for at least some value of p we can plot V_2 against V_1 and note that for a fixed value of p

$$\overline{G}(d[j]) \geq \overline{G}(d[i])$$

if and only if

$$pV_1(d[j]) + (1-p)\,V_2(d[j]) \geq pV_1(d[i]) + (1-p)\,V_2(d[i])$$

$$\frac{p}{1-p} \geq -\frac{(V_2(d[j]) - V_2(d[i]))}{(V_1(d[j]) - V_1(d[i]))}.$$

For this inequality to hold the pair $(V_1(d[i]), V_2(d[i]))$ must lie on a line of slope $-p(1-p)^{-1}$ with smaller intersection with the V_1 axis than the point $(V_1(d[j]), V_2(d[j]))$. in Figure 2.1. This means in particular that only decision rules on the NE boundary of this graph can be optimal for some p. This boundary is called the Pareto boundary. This precludes not only the dominated decisions we have already identified as suboptimal but also decision rule $d[8]$. So $d[8]$ is not a Bayes decision for any value of p.

The set of values of p where each of these three decisions are optimal is easily discovered. Thus $d[1]$ – the decision to send off the machine immediately – is at least as good as $d[4]$ (and $d[2]$) whenever

$$\frac{p}{1-p}(0.0 - 5.8) \geq 8.7 - 10.0 \Leftrightarrow p \leq 0.183.$$

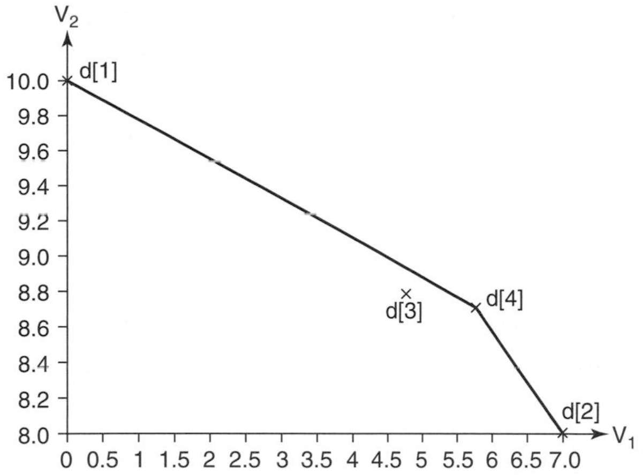

Figure 2.1. The Pareto boundary for the machine sales problem.

Decision rule $d[4]$ is optimal if $p \geq 0.183$ and

$$\frac{p}{1-p}(5.8 - 7.0) \geq 8.0 - 8.7 \Leftrightarrow p \leq 0.333.$$

Finally $d[2]$ is optimal if $p \geq 0.333$.

The analysis of this little example demonstrates a common phenomenon that only a relatively small subset of the decision rules could be optimal whatever the probability of a "causal event" is. The normal form analysis also gives ranges of the values of the uncertain probability p in which one of the candidate decisions might be optimal. Typically for moderate-sized discrete problems the neighbourhoods of p where a given decision rule is optimal are often quite wide. Sometimes a mis-specified probability will lead the DM to make the wrong decision. However even then the adverse consequences are usually not too bad. Thus in extensive form analysis we noted that $d[4]$ was optimal when $p = 0.2$. So suppose the DM's probability if it were elicited with more care is actually $p = 0.15$ so that the decision $d[1]$ is in fact the optimal one. In this case it is easy to calculate that the difference between the expected payoff using decision $d[1]$ and not $d[2]$ is

$$\overline{G}(d[1]) - \overline{G}(d[4]) = £1,000(8.500 - 8.265) = £235$$

which in the context of the amounts involved is not too dramatic a loss. This type of sensitivity analysis can therefore reassure the DM that the consequences of minor mis-specification of probabilities will not too great. Usually, provided that she is in the right ball park, she will choose a good if not totally optimal decision using the methods described above: see for example Raiffa (1968); Raiffa and Schlaifer (1961); Smith (1988a) for further discussion.

2.7 Temporal coherence and episodic trees*

Bayesian decision analysis proceeds assuming the cjo assumption. Although it is almost invariably used it actually represents quite a strong assumption. If the ramifications of a decision lead far into the future, it can really distort judgements. There are two problems with its adoption as a completely general principle. The first occurs when although the tree remains an appropriate description of the development of the scenario, because of unforeseen events the DM's edge probabilities may change not just because she has accommodated new information using Bayes rule but also because her appreciation of the problem deepens so transforming her underlying judgements. Several authors have appreciated this difficulty. For example Goldsein (1985); Goldstein and Wooff (2007) in a Bayesian context substitute the temporal sure preference condition which gives conditions that essentially treats an assessed probability at the current time as the DM's *expectation* of the value she would assign to that event in the future. The EMV decision rule actually remains unchanged under this hypothesis as do the utility maximising strategies discussed in the next chapter – if they are carefully augmented to take account of this phenomenon. So at least from a theoretical perspective violations of the cjo assumption of this type are not that critical.

However a second and more profound problem is that it is quite likely the DM's whole framework of thought about far distant events will change including her appreciation of the scope of possible events that might happen. This has already been illustrated several times. If this Damascus experience occurs – and over a long period of time we should hope it would for the supported DM – then the topology of her decision tree and her appreciation of which situations in it are parallel will almost certainly change in unexpected and possibly dramatic ways as she thinks creatively and outside the box. The inadequacy of the Bayesian paradigm can no longer be overcome by substituting some probabilities or adding more subtrees to existing leaves in a consistent way.

The consequences of such a potential change of constructs are extremely unpredictable but also in the medium to long term quite likely. For example in the first chapter we discussed a very simple case where a doctor, when confronted with symptoms which were extremely unlikely under any of the explanatory diseases she had considered as possibilities up to that point in time, could quite legitimately search for a radically new explanation of what she saw and propose this as an explanation.

More generally we have argued above that an applied Bayesian analysis should be accompanied by various diagnostic tests to check the continued validity of the model. These essentially treat the currently adopted Bayesian model as a null hypothesis and check whether the observed data is very unexpected under this model. If such surprising events happen, as in the example above, the DM is encouraged to rethink her model. The observed data may give insights which suggest she should discard what she thought to be true in favour of a different explanation. Such diagnostics are an essential part of a Bayesian analyst's toolkit. However their routine adoption means that the DM believes she might violate the cjo principle in the longer term. To my knowledge there is no formal way of adjusting the Bayesian paradigm to address this problem. But to deny its possibility is to deny the possibility to the DM of the sort of creative insights such an analysis is designed to provoke.

For further discussion on this and related issues see Goldstein and Rougier (2009); Shafer *et al.* (2000) and references therein.

However within the context of a decision analysis using the methodology advocated here this problem is not a big obstacle. Here the DM is seen as being facilitated by the analyst in presenting her best coherent argument, based on her current beliefs and facts currently accepted by all parties and using plausible principles such as cjo to explain them to an auditor. In practice all parties must accept that in the future the analyses will be further refined and sometimes completely replaced. But at the current time the arguments present a plausible and defensible position for her to take within the context of what is currently seen as justifiable both scientifically and within the norms and standards of current thought.

This subjective rationale for decision analysis which is both provisional and fashioned by the norms and scientific dogma of the society in which it is made is in my judgement a critical one to adopt if the outputs of a decision analysis are to carry any credibility. All analyses we consider here fall into this category. The decision analysis is for a particular time and for a particular time-limited purpose. The logical coherence we demand from the DM only concerns this limited domain. Of course we may hope that the analysis the DM performs for the instance she faces now will retain many of its features in future analyses of analogous problems. Indeed this is often found to be the case. But we do not demand this as a logical necessity within the methodology described below.

2.8 Summary

Event trees are an extremely useful framework for representing discrete decision problems. They provide a powerful descriptive representation about hypotheses concerning how situations unfold. Furthermore this representation can be used as a transparent framework for calculating optimal policies. Despite having been used as a decision analytic framework for a very long time, see e.g. Raiffa and Schlaifer (1961) and Raiffa (1968), in recent years their use has been rather neglected. One problem with event trees of all kinds is that nowhere in their topology is an explicit representation of hypotheses about dependence relationships between the state and measurement variables in the system. As we noted in the criminal example above such qualitative information can often be elicited. On the other hand these conditional independence relationships can be very easily represented by a Bayes net, influence diagram (see below) or other topological structures (see e.g. Covaliu and Oliver (1995); Jaegar (2004); Jensen *et al.* (2004); Smith and Anderson (2008) and references therein). We will discuss some of these representations in later chapters.

However we proceed there are a number of challenges and limitations facing a decision analysis:

(1) We saw in the criminal example above that the event space of even a relatively self-contained problem can quickly become very large. So in complicated problems efficient ways need to be developed to limit the space of decision rules so that this space is manageable. I briefly discussed this issue in the chess-playing example above. Appropriate simplifying representations are often

best determined by the context of the problem considered although a number of universal methods are helpful. But it is important to be aware that the current decision analysis and its associated model representation is likely to be a framework supporting the DM in analysing her problem creatively. At its best it is a useful but imperfect summary of the DM's current beliefs about the real problem at hand, a tool in need of regular reappraisal and modification as the DM develops her awareness of the opportunities and threats presented by her problem.

(2) It is often not possible to measure the efficacy of the consequences arising from a given course of action and the development it might give rise to by a single financial reward. Indeed the simple illustrations given above demonstrate that this scenario is the exception rather than the rule. Furthermore even when rewards are purely financial their consequences can rarely be well summarised by their expectation.

This last difficulty is the most pressing one to explain and address. Fortunately it is possible to generalise the EMV strategy in a very simple way to provide a framework for a methodology of decision analysis which is widely applicable and addresses all the inadequacies mentioned in the second point above. We address this issue in the next chapter.

2.9 Exercises

2.1 Consider the event tree having three situations $\{v_0, v_1, v_2\}$, leaves $\{v_3, v_4, v_5, v_6\}$ and edges $v_0 \rightarrow v_1$, $v_0 \rightarrow v_2$, $v_1 \rightarrow v_3, v_1 \rightarrow v_4, v_2 \rightarrow v_5, v_2 \rightarrow v_6$ Suppose that $\{v_1, v_2\}$are parallel situations with edge $v_1 \rightarrow v_3$ associated with $v_2 \rightarrow v_5$ and edge $v_1 \rightarrow v_4$ associated with $v_2 \rightarrow v_6$. Let $X_1 = 0$ if event $\{v_3, v_4\}$ occurs and $X_1 = 1$ if $\{v_5, v_6\}$ and let $X_2 = 0$ if event $\{v_3, v_5\}$ occurs and $X_2 = 1$ if $\{v_4, v_6\}$ occurs. Prove that $X_1 \amalg X_2$.

2.2 A patient is admitted to hospital on his thirtieth birthday suspected of having either disease A or disease B where the DM believes $P(A) = 0.4$ and $P(B) = 0.6$. If untreated the probability the patient dies is 0.8 but will otherwise recover and have a normal life expectancy of 80 years. The doctor has three actions open to her: d_0 = not to treat, d_1= give the patient a course of drugs, or d_2 = operate. Independently of the illness operating d_1 will kill the patient with probability 0.5 and d_2 with probability 0.2. If treatment d_1 is given, if it does not kill the patient, if disease A is present the treatment will have no effect or cure him with probabilities 0.5 but will have no effect when B is the disease. On the other hand if treatment d_2 is given and the patient survives the operation, it will cure him with probability 0.8 if he suffers type A condition and with probability 0.4 if he has illness B and otherwise will have no effect.

Draw a rollback tree of this problem and identify the DM's EMV decision rule when her reward is the expected number of lives saved.

2.3 An oil company has been given the option to drill in one of two oil fields F_1 and F_2 but not both. The company believes that the existence of oil in one field is independent of the other with the probability of oil in F_1 being 0.4 and in F_2 being 0.2. A net profit of \$770m is expected if oil is struck in A and \$1950m if oil is struck in B. The company can pay \$60m to investigate either but not both of the fields if it chooses. Whether or not it takes this option it can then choose not to drill (d_0) or drill F_i (d_i) $i = 1, 2$. The investigation is not entirely foolproof. The DM believe that when oil is present the investigators advise drilling with probability 0.8 and when oil is not present will advise drilling with probability 0.4. The cost of accepting the option on either field is \$310.

Draw the rollback tree of this problem and find the decision maximising the company's expected payoff.

2.4 Two years ahead a company called Mango will need to decide between one of three decisions: continue to market its current machine M_0 (d_0), market a new version M_1 of its old product (d_1) or market a replacement machine M_2 (d_2). The machines M_1 and M_2 can only be marketed if they have been successfully developed: events which are judged by Mango to have respective probabilities 0.9 and 0.6. The cost of developing M_1 is \$3m and M_2 is \$5m and the company can choose to develop either or both of M_1 and M_2. The expected net profits from M_0, M_1 or M_2 if successfully developed are \$2m, \$10m and \$18m.

Use a rollback tree to calculate the decision rule maximising the company's expected payoff.

3
Utilities and rewards

3.1 Introduction

Hundreds of years ago gamblers realised – especially in the contexts where large stakes were involved – that blindly using the EMV strategy could be disastrous. Consider the following example.

Example 3.1. A game has k stages, $k = 1, 2, 3, \ldots$, and costs £M to enter. At stage 1 the gambler has a stake $S_1 > 0$. At stage $k \geq 1$ if the game has not terminated the gambler can decide to either terminate it and to take away her stake or continue the game. If the game terminates she takes the stake S_k. On the other hand if she chooses to continue the game a fair coin is tossed. If a head results then her stake will be quadrupled (i.e. $S_{k+1} = 4S_k$) but if a tail appears the game will terminate and she leaves with nothing.

To find an EMV decision rule for this game the DM must calculate her expected reward $\overline{G}^*$ measured in £s under the best possible play d^* of the game and so assess whether she should actually buy in as a function of the cost M: choosing to play the game if $\overline{G} > M$. Suppose that under d^* she reaches stage k, $k \geq 1$. Her stake will have increased to $S_k = 4^{k-1}S_1$. She must now discover under d^* whether or not to take her current winnings or to continue where this decision is a function of k and M. Note that if she reaches stage k then her expected payoff if she terminates is S_k whilst her expected payoff if she continues is at least the expected payoff she would obtain were she to continue and stop at the next stage. To continue to play one more time has expected payoff $\frac{1}{2} \times 2S_k + \frac{1}{2} \times 0 = 2S_k > S_k$. So whatever the value of k and M the EMV decision rule d^* will have the property that she will continue to gamble on the next stage if she is still in the game. It is clear that this is simply the decision to gamble indefinitely. It is easy to check that the expected payoff for this strategy is infinite whenever M is finite. Sadly by playing this strategy the gambler will lose M unless the infinite sequence of tosses of the coin in the game all result in heads – an event with probability zero. So by following this strategy the gambler loses M with probability 1.

What are the reasons for this so-called St. Petersburg paradox, which demonstrates that even in situations where someone is only interested in a monetary reward a DM can be seriously misled by using an EMV strategy?

Well there are at least two. The first is that the mathematical term "expectation" should not be identified with the common usage of the word "expectation". The two meanings can be dramatically different. Thus in the example above to say that a rational gambler should "expect" to win an infinite amount – in any real sense of the term "expect" – is clearly preposterous. The identification of the mathematical term with its common usage can therefore be seriously misleading: especially when the rewards associated with different possible consequences differ widely. It follows that whilst it *sounds* good to choose a decision that maximises expected payoff – as defined mathematically – this is not necessarily a good thing to do!

The second problem is a more subtle one and is to do with our understanding of wealth. When X is some measurement variable and an increasing function f is nonlinear then it is not in general true that $\mathbb{E}(f(X)) = f(\mathbb{E}(X))$, i.e. the value we "expect" a function of the measurement to take is not necessarily the function evaluated at the value we "expect" the measurement to take. So by following the EMV strategy the DM implicitly uses the *scale* with which she measures her wealth. To appreciate the dramatic effect a different choice of scale can make consider the example above where the gambler tries to choose the decision ensuring the best *proportionate* increase in wealth from the initial stake S_1. In the game above if her return is $S = S_1^{\lambda}$ and she aims to choose the decision to maximise the expected value of λ then whatever the cost of the game it is easily checked that the EMV decision on payoff λ should take the stake and not gamble at all. For by gambling she risks a possibility of an infinite negative loss with nonzero probability; so her expected payoff associated with any gamble with this as a possibility is also $-\infty$.

So whether a DM chooses actual winnings, proportionate winnings or some other function of her gain will have a critical impact on how she will choose to act under an expectation-based strategy. We cannot proceed to define an optimal decision using probabilistic ideas without having first elicited the *scale* by which the gambler will choose to measure her success. However once this scale has been elicited it is demonstrated below that for most scenarios a rational DM should follow a transformed version of the EMV strategy. This is almost identical to the EMV strategy but on an elicited scale of reward called "utility". She then chooses a decision that maximises the expectation of her utility function of rewards. Within this more general framework it is also straightforward to address more complicated scenarios like those in the criminal case of the last chapter where consequences of any policy and its subsequent outcome cannot be measured simply in terms of a single attribute like money.

Now the generalisation described in this chapter is not universally applicable. It is therefore critical for an analyst to be able to appreciate when a decision problem should not be approached in the way we recommend for the rest of this book. To enable the reader to come to this judgement, in the ensuing sections we will discuss the sorts of preferences the DM must hold before this methodology is justified. A careful examination of these boundaries of applicability should convince you how widely applicable the Bayesian decision methodology is, as well as providing an awareness of when the methodology may mislead.

3.2 Utility and the value of a consequence

Begin by considering the machine dispatch example of the last chapter where the only consequence of interest to the DM can be effectively measured by the attribute of the amount of financial gain. The worst financial consequence considered in the example was one where the DM paid for two scans and still dispatched a faulty piece of machinery. This has an associated payoff of $-$£900. The highest financial reward the DM could attain would be when she spent nothing and dispatched a working machine: this would have a payoff of £10, 000. Note that the EMV strategy would force the DM to find an option that gave £4, 500 with certainty preferable to an option where the DM received $-$£900 with probability 0.5 and £10, 000 with probability 0.5 – the second option having an associated expected payoff of £4, 550. But can we really argue that all rational DMs should always want to commit themselves to this preference? For example when a company will become insolvent unless it obtains at least £4, 000 from this transaction then surely it cannot be asserted that they should gamble when by taking the less risky first option they ensure their survival. On the other hand were a company to need at least £8, 000 to survive a "rational" DM is likely to prefer the gambling second option.

Of course it is often the case that a DM's decisions are not as stark as this. Nevertheless it is quite rare for a rational DM to find all bets $b(x, y, \alpha)$ giving £x with probability α and £y with probability $1 - \alpha$, for all values of (x, y, α) where $x < y$ and $0 \leq \alpha \leq 1$ equivalent to one giving £r where $r = (1 - \alpha)x + \alpha y$ for sure – and this is what is demanded of an EMV DM.

A Bayesian decision analyst surmounts this difficulty by first eliciting an appropriate *scale* on which to measure her attributes. This is flexible enough to allow for the differing needs of the DM such as those illustrated above. This scaling of preferences is elicited by asking the DM to specify her preferences between two – possibly hypothetical – gambles which are relatively simple for the DM to assess and defend. The analyst then uses these statements as a secure basis from which to deduce what the DM's preferences should be over more complex distributions of rewards she might actually be faced with and which are much more difficult for her to evaluate. To be able to deduce how a "rational" DM's preferences on complicated gambles should relate to the simple ones will require a rational DM to acquiesce to follow certain rules. Of course the applicability of such a rule base – or axiomatic system – will in general depend on the needs and aspirations of the DM, the context of the problem and the demands of the decision making process. But there is one set of axioms which seems to be very compelling in a wide range of different scenarios and which forms the basis for most Bayesian decision analyses. Furthermore, because this axiomatic system – and its variants – has existed for a long time and is a provenly reliable framework in a wide range of domains an auditor will be inclined to accept them. We will outline one such version of this axiomatic system below.

It is first helpful to make more precise here what it means for a DM to prefer one decision rule $d[1]$ to another $d[2]$. We ask the DM to imagine that she is to give instructions to an agent to enact her preferences in a hypothetical market place. Whenever she states that she strictly prefers $d[2]$ to $d[1]$ she is stating that she is instructing this agent to exchange the option of enacting $d[1]$ for the option of enacting $d[2]$ if the trade becomes available. If two decisions are equally preferable then she is equally content for her agent to substitute one for the other or to retain the current one. The trade will occur only with other traders who have no more information on any of the events involved in any of the gambles than the DM.

In the last chapter we saw that each possible decision rule open to a DM may lead to one of a number of consequences. We first need to assume that it is possible to measure the success achieved by attaining a particular consequence by the value $\boldsymbol{r}$ taken by a vector of *attributes* $\boldsymbol{R}$ – where $\boldsymbol{R}$ takes values in $\mathcal{R}$. Thus in the machine delivery example the attribute r is one-dimensional: simply monetary return. In the court case the attribute vector must have at least two components: one measuring the successful conviction of the suspect and the other the financial cost of taking the suspect to court. The first axiom we use is the following.

Axiom 3.2 (Probabilistic determinism). If two decision rules $d[1]$ and $d[2]$ give rise to the same reward distribution over its attributes then the DM will find $d[1]$ and $d[2]$ equally preferable.

This is often a compelling rule for the DM to use. However – although this is the starting point of many axiomatisations – it is nevertheless by no means straightforward for an unaided DM to follow in practice. Nor is it universally applicable. Four practical difficulties are listed below.

(1) The DM needs to think hard to *identify* what the important consequences of her possible acts really are. The implications of certain things happening can be extensive and varied and the analyst may need to tease them out of the DM. In the analysis given in the last chapter of the example involving a potentially allergenic ingredient the company may need to reflect on not only the financial but also the legal conseqences and consequences on the reputation of the company of marketing a product that turns out to be allergenic. In the criminal example of Chapter 2 where the police have to decide whether or not take the suspect to court they may need to consider not only the financial implications of the associated forensic investigation and the resource cost of preparing this case but also indirect consequences of distracting staff from potentially more fruitful prosecutions, deterrence effects on future crime and so on. The choice of just how to define the scope of the analysis so that on the one hand it is full enough to appropriately address the main issues but on the other is sufficiently focused so that the DM is not overwhelmed, needs careful handling. In particular any analysis is open ended: it could always be more refined. Even in a very closed environment like the chess example of the last chapter when the problem is well defined at the time of writing, the development of an appropriate score of an unfinished game is seen as critical. But these challenges increase when the success of a policy is more open to interpretation, such as in the criminal example. And even when this elicitation is carefully performed there may be hidden consequences that are impossible for the DM to envisage now but will become apparent only later. For example the ethical dimension of coffee production – as perceived by the customer – has become an intrinsic consequence recently when previously producers took little notice of whether they could publicise this. It would be forgivable for a company not to have predicted in the 1980s that the ability to publicise the effective ethical nature of their product would be intrinsic to their future commercial survival.

(2) For a probability distribution to be associated to a vector of attributes, those attributes need to be a random vector. This means in particular that it must be possible for the vector of attributes to be defined in a *specific, measurable* and *time-limited* way. Consider a medical scenario. Then being "healthy" or "not healthy" is too ambiguously stated to be treated as an attribute measuring the impact of the consequences of any act because it is not a well-defined event. It therefore cannot admit a well-defined probability distribution over which the ensuing outcomes can be scored. The idea of being healthy needs to be measured by a more specific proxy such as "at a given time t of being able to clock up at least 2 miles on a walking machine in under 15 minutes" or "not this". In the example of the potentially allergenic product, how might the company define an attribute to reflect the extent of the "damage of its reputation"? In the court case, how do the police measure the deterrence effect of a successful prosecution of their case? The challenge here to the analyst is to help the DM to find a definition of an attribute vector that on the one hand captures the essential meaning of the implications of the ensuing consequences but on the other is measurable in the way illustrated above.

(3) Once a good vector of attributes for measuring the success of a natural consequence is found, the analyst must be aware that this proxy vector may no longer be appropriate if the space of decision rules to which it is applied is *extended*. For example suppose that an academic's natural consequence is the usefulness of her research to the wider academic community. One attribute which might be a good (albeit slightly imperfect) proxy for this consequence – applicable over the set of acts she would normally consider taking – would be to measure success by the count of the number of citations she receives from other academics. But if she substitutes this attribute for her natural consequence beyond the domain of decisive actions she would usually consider then both she and anyone auditing her performance in this way will be seriously misled. For example she would be encouraged to enter into explicit or implicit private agreements with other academics to always cite each other's works whenever there is the slightest link, to write papers with errors in them to induce many academics to cite the error, to write articles in over-researched areas which she just manages to publish before the hundreds of other academics who would have discovered the same results only a few months later or even to accept papers she referees only if the authors had cited her own papers. To decide to act in any of the ways above would score very highly under this attribute with high probability. However none of these policies would achieve a good natural consequence. Used to compare possible extensions of decision rules, natural consequences need to be appraised and the proxy measurable consequence checked for its continued applicability.

(4) One final issue is not associated with the meaning of the attributes but on the existence of a probability distribution that fully describes her beliefs: an issue we will defer to Chapter 4 where conditions leading to the existence of a subjective probability are discussed

Because of the third point, a DM is especially vulnerable when consequences of her actions are appraised by a third party in terms of the proxy attribute rather than the real consequences it is designed to measure. Technically this problem occurs when attributes measuring success of a policy are transformed into *targets*. However fortunately, with this caveat and when the comparability axiom discussed in Chapter 4 legitimately holds, in most scenarios a DM supported by an analyst can successfully characterise her reward distribution so that probabilistic determinism is appropriate: see illustration in this chapter and Chapter 6.

The next assumption needed is that the DM has a *total order* on the set of consequences, as measured by the values of the vector of attributes $\boldsymbol{R}$ might take after the possible decisions she could make. Thus if $\boldsymbol{r}_1$ and $\boldsymbol{r}_2$ are two possible vectors of values of her attributes, we assume that there are only three possibilities: $\boldsymbol{r}_1 \prec \boldsymbol{r}_2$ i.e. she strictly prefers $\boldsymbol{r}_2$ to $\boldsymbol{r}_1$, $\boldsymbol{r}_1 \sim \boldsymbol{r}_2$ i.e. she finds $\boldsymbol{r}_1$ equally preferable to $\boldsymbol{r}_2$ or $\boldsymbol{r}_1 \succ \boldsymbol{r}_2$ i.e. she strictly prefers $\boldsymbol{r}_1$ to $\boldsymbol{r}_2$. We shall write $\boldsymbol{r}_1 \preceq \boldsymbol{r}_2$ when $\boldsymbol{r}_2$ is at least as preferable as $\boldsymbol{r}_1$ and $\boldsymbol{r}_1 \succeq \boldsymbol{r}_2$ when $\boldsymbol{r}_1$ is at least as preferable as $\boldsymbol{r}_2$. For all vectors of attributes, the relation $\preceq$ must be *transitive* i.e. if $\boldsymbol{r}_1 \preceq \boldsymbol{r}_2$ and $\boldsymbol{r}_2 \preceq \boldsymbol{r}_3$ then $\boldsymbol{r}_1 \preceq \boldsymbol{r}_3$, *reflexive* i.e. $\boldsymbol{r} \preceq \boldsymbol{r}$ and have the property that if $\boldsymbol{r}_1 \preceq \boldsymbol{r}_2$ and $\boldsymbol{r}_2 \preceq \boldsymbol{r}_1$ then $\boldsymbol{r}_1 \sim \boldsymbol{r}_2$. In particular we assume that it is impossible for both $\boldsymbol{r}_2 \preceq \boldsymbol{r}_1$ and $\boldsymbol{r}_1 \prec \boldsymbol{r}_2$ to hold simultaneously.

With a single attribute of monetary payoff the assumption of preferential total order is usually immediate. Thus $r_1 \preceq r_2$ whenever $r_1 \leq r_2$: i.e. more money is at least as preferable to less – and clearly the real numbers and hence monetary reward measured in any units is totally ordered. But more generally we would also hope that the DM should at least be able to order – in terms of their desirability – the possible vector of values of her attributes that might happen as a result of her decisions. If she defines her reward space comprehensively enough then this should be possible although in some domains only a partial order over rewards is initially given.

For the DM to have a total order on attributes is not quite enough. It is also necessary to assume that she has a total order over decision rules she might make, both hypothetical ones and real ones. By Axiom 3.2 any such decision rule d can be identified with the probability distribution over its attributes. In particular preferring one decision rule $d[1]$ over another $d[2]$ can be identified with preferring the probability distribution $P[1]$ of attributes associated with $d[1]$ to the probability distribution $P[2]$ of attributes associated with $d[2]$. We use the same notation for preferences between decision rules and reward distributions as used on reward, for example writing $d[1] \preceq d[2]$ to read "the DM finds decision rule $d[2]$ at least as preferable to $d[1]$" and $P[1] \preceq P[2]$ to read "the DM finds the distribution of rewards $P[2]$ at least as preferable as the distribution $P[1]$".

Axiom 3.3 (Total order). The DM has a total order of preferences she considers over all probability distributions, both real and hypothetical over the space $\mathcal{R}$ of all values that the random vector $\boldsymbol{R}$ of attributes might take.

Note that we are *not* assuming the DM is currently aware of what these preferences are: after all being able to compare two complicated betting schemes without appropriate tools to help could be overwhelming. What we *are* assuming is that this total order over distributions on attributes *exists* and given the appropriate tools she can discern them. The analyst's task is to help her discover these.

There are at least two good reasons for requiring the total order axiom. First is simply a pragmatic one. To be able to compare all pairs of decisions $(d[1], d[2])$ with each other *requires* us to be able to assert that either $d[1] \preceq d[2]$ or $d[2] \preceq d[1]$: otherwise we are saying these are incomparable and so cannot be unambiguously compared. Furthermore to guarantee that d^* can be the best of an arbitrary set of possible decisions requires that there are no cycles. For example suppose three decision rules $(d[1], d[2], d[3])$ gave us that $d[1] \prec d[2]$ and $d[2] \prec d[3]$ but $d[3] \prec d[1]$. How can we state which of these is best? No one decision is better than the other two. We are therefore forced into only partial answers to the questions we might like to ask.

Second, once we equate the idea of a strict preference of a decision $d[2]$ to a decision $d[1]$ with an instruction to an agent to exchange $d[2]$ with $d[1]$ this will force the agent to be exposed to engaging in an infinite cycle of exchanges whenever there are three decisions satisfying $d[1] \prec d[2]$, $d[2] \prec d[3]$ and $d[3] \prec d[1]$. Furthermore if these preferences were significant ones then because $d[1] \prec d[2]$ she should be prepared to forfeit a small amount r of reward to switch from $d[1]$ to $d[2]$, forfeit r to switch from $d[2]$ to $d[3]$ and forfeit r to switch from $d[3]$ back to $d[1]$ again! Furthermore the agent is exploited by traders no more informed than she is. So the agent can be used as a reward pump giving away $3r$ in each cycle of transaction whilst cycling round holding the same reward. So within the free market of agent transactions we have used as a basis for defining preferences it a *necessary* condition for this to lead to sensible outcomes in general is that no such cycles can occur and that preferences are transitive.

In my opinion transitivity is a compelling *requirement* for a workable definition for the rationality of a DM of the type considered in this book. However there are many decision problems where transitivity cannot in general be guaranteed. This happens most readily when the DM is a group with different objectives or beliefs. A common example of the loss of transitivity can occur when decisions are made by majority voting. Thus suppose three members M_1, M_2 and M_3 of a collective choose their decisions between options $d[1], d[2], d[3]$ by the majority of votes they obtain and their preferences are as follows.

$$M_1 : d[1] \prec d[2] \prec d[3],$$
$$M_2 : d[2] \prec d[3] \prec d[1],$$
$$M_3 : d[3] \prec d[1] \prec d[2].$$

Then M_1 and M_3 – and hence the collective – have preference $d[1] \prec d[2]$, M_1 and M_2 – and hence the collective – have preference $d[2] \prec d[3]$, and finally M_2 and M_3 – and hence the collective – have preference $d[3] \prec d[1]$. This is called Condorcet's paradox. Another example of a similar but stronger inconsistency is Arrow's paradox. It can be proved that there is only one way of guaranteeing that pairwise preferences between members of a collective can be expressed as a function of member's pairwise preferences always leads to transitive group preferences. This is when the group's preferences completely coincide with those of one of the members: i.e. that there is a dictatorship. So in this sense there can be no nontrivial rational combination of simple pairwise preferences. For these and other reasons in this book we have restricted ourselves to problems where there is one responsible agent: that is a person or group who share a single utility function and a single probability distribution. Interesting discussions of group decision making beyond the scope of this book are reviewed in French and Rios Insua (2000).

Given that the DM is prepared to cede that she has a total order on her preferences, there are several possible answers to how she can be helped to discern what these preferences are over complicated betting schemes. The simplest one is given below where a few extra gambles are added to those in the agent's market of decision rules.

Notation 3.4. Let $B(\boldsymbol{r}_1, \boldsymbol{r}_2, \alpha)$ denote a betting scheme that gives the value of attributes $\boldsymbol{r}_2$ with probability α and $\boldsymbol{r}_1$ with probability $(1-\alpha)$ where $\boldsymbol{r}_1 \prec \boldsymbol{r}_2$.

Note that from distributional equivalence $\boldsymbol{r}_1 \sim B(\boldsymbol{r}_1, \boldsymbol{r}_2, 0)$ and $\boldsymbol{r}_2 \sim B(\boldsymbol{r}_1, \boldsymbol{r}_2, 1)$. It is also clear that whenever $\alpha_1 < \alpha_2$, any rational DM will have preferences which satisfy

$$B(\boldsymbol{r}_1, \boldsymbol{r}_2, \alpha_1) \prec B(\boldsymbol{r}_1, \boldsymbol{r}_2, \alpha_2) \tag{3.1}$$

because $B(\boldsymbol{r}_1, \boldsymbol{r}_2, \alpha_2)$ has a higher probability of delivering the more preferable consequence $\boldsymbol{r}_2$ than $B(\boldsymbol{r}_1, \boldsymbol{r}_2, \alpha_1)$.

Suppose that the DM can identify the worst possible value $\boldsymbol{r}^0$ of her attribute vector and the best possible consequence $\boldsymbol{r}^*$. Clearly when consequences are monetary this is trivial: for example in the machine dispatch problem above – in units of £1,000 – $r^0 = 0.9$ and $r^* = 10$. In many other scenarios this is also a simple task. We now face the DM with some simple hypothetical bets and assume the following.

Axiom 3.5 (Weak Archimedean). For all consequences $\boldsymbol{r} \in \mathcal{R}$ there is a (unique) value $\alpha(\boldsymbol{r})$ where $0 \leq \alpha(\boldsymbol{r}) \leq 1$, for which

$$\boldsymbol{r} \sim B(\boldsymbol{r}^0, \boldsymbol{r}^*, \alpha(\boldsymbol{r})).$$

Note that when a DM follows an EMV strategy on the one-dimensional attribute r

$$\alpha(r) = (r^* - r^0)^{-1} r.$$

So in particular the EMV strategy satisfies this axiom. But also note that by allowing $\alpha(r)$ to be *any* increasing function of r on to the closed interval [0, 1] there is much more flexibility than one demanding that this is linear as in the EMV strategy.

Notice that if $\alpha(\boldsymbol{r})$ exists then it must be unique. For if $\boldsymbol{r} \sim B(\boldsymbol{r}^0, \boldsymbol{r}^*, \alpha_1(\boldsymbol{r}))$ and $\boldsymbol{r} \sim B(\boldsymbol{r}^0, \boldsymbol{r}^*, \alpha_2(\boldsymbol{r}))$ where $\alpha_1(\boldsymbol{r}) < \alpha_2(\boldsymbol{r})$ then under the total order axiom,

$$B(\boldsymbol{r}^0, \boldsymbol{r}^*, \alpha_2(\boldsymbol{r})) \preceq B(\boldsymbol{r}^0, \boldsymbol{r}^*, \alpha_1(\boldsymbol{r}))$$

which would contradict (3.1).

So this axiom will only be broken if for some value of attributes $\boldsymbol{r}$, $\boldsymbol{r}^0 \prec \boldsymbol{r} \prec \boldsymbol{r}^*$ there exists an $\alpha(\boldsymbol{r})$, $0 \leq \alpha(\boldsymbol{r}) \leq 1$, such that for all α such that $0 \leq \alpha < \alpha(\boldsymbol{r})$

$$\boldsymbol{r} \succ B(\boldsymbol{r}^0, \boldsymbol{r}^*, \alpha(\boldsymbol{r}))$$

for all α such that $\alpha < \alpha(\boldsymbol{r}) \leq 1$

$$\boldsymbol{r} \prec B(\boldsymbol{r}^0, \boldsymbol{r}^*, \alpha(\boldsymbol{r}))$$

but for which

$$\boldsymbol{r} \not\sim B(\boldsymbol{r}^0, \boldsymbol{r}^*, \alpha(\boldsymbol{r})).$$

Thinking in terms of monetary payoff or other continuous one-dimensional measures of consequence it is difficult to imagine situations when the DM would want to express such refined judgements. However the axiom is not universally applicable. When consequences have several components and the nature of these is that the DM judges one component infinitely more important than another then she may not want to obey this rule.

Example 3.6. The consequences of a medical treatment are judged to be its cost and its success. The worst consequence $\boldsymbol{r}^0$ is one that is expensive and the patient certainly dies. The best $\boldsymbol{r}^*$ treatment is one that is cheap and the patient certainly survives. A third treatment $\boldsymbol{r}^1$ is equally expensive but ensures survival. Now suppose a doctor is faced with a treatment but which is certainly expensive but the patient will certainly survive. Clearly $\boldsymbol{r}^0 \prec \boldsymbol{r}^1 \prec \boldsymbol{r}^*$. The gamble $B(\boldsymbol{r}^0, \boldsymbol{r}^*, \alpha(\boldsymbol{r}))$ can be related to a treatment which is cheap and the patient survives with probability α and is expensive and the treatment kills the patient with probability $1 - \alpha$. The doctor may well legitimately argue that she prefers $\boldsymbol{r}^1$ to any treatment regimes $B(\boldsymbol{r}^0, \boldsymbol{r}^*, \alpha(\boldsymbol{r}))$ for all $\alpha, 0 \leq \alpha < 1$ – however close to 1 the value α is because she views the survival of a patient as infinitely more important than the cost of treatment. In this case $\alpha(\boldsymbol{r})$ does not exist.

Thus when used with problems with many different types of consequence this axiom needs to be checked. Loosely stated it is necessary for all component attributes to be comparable in terms of their relative benefit. Incidentally in examples like the one above under the demands of budgetary discipline the DM will often be forced to have Archimedean preferences so that she is prepared to choose a cheaper treatment if the probability it kills the patient is sufficiently minute.

Great clarity comes when the preferences of a DM satisfy the Archimedean axiom, for then every value $\boldsymbol{r} \in \mathcal{R}$ of a vector of attributes can be identified with a unique real number $\alpha(\boldsymbol{r})$ where $0 \leq \alpha(\boldsymbol{r}) \leq 1$. The larger the value of $\alpha(\boldsymbol{r})$ the more desirable is the attribute $\boldsymbol{r}$. The rescaling of the possible high-dimensional attribute vector $\boldsymbol{r}$ on to a single real number $\alpha(\boldsymbol{r})$ allow a generalisation of payoff that makes the EMV strategy applicable to a much wider class of problems.

Definition 3.7. For each possible consequence $\boldsymbol{r}$ call $\alpha(\boldsymbol{r})$ the DM's *utility function referenced to* $(\boldsymbol{r}^0, \boldsymbol{r}^*)$.

When the attribute is one-dimensional then the utility function references to (r^0, r^*) can be elicited directly from the DM expressed as an increasing real-valued function whose

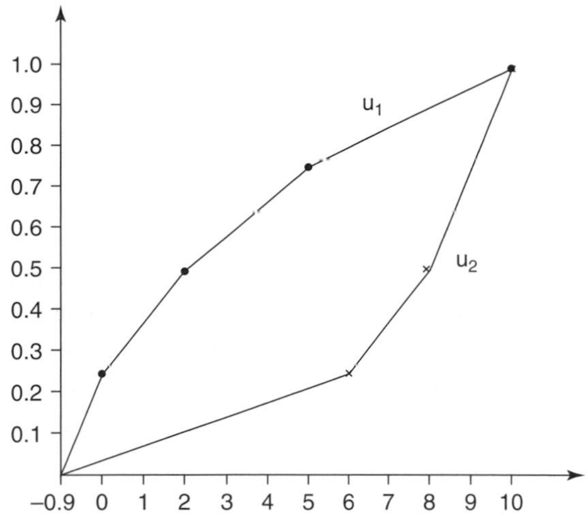

Figure 3.1. Two utility functions $U_1 = \alpha_1$ and $U_2 = \alpha_2$.

domain is the closed interval $\left[r^0, r^*\right]$. For example in the machine dispatch example of the last chapter consequences r are simply measured by payoff, in units of £1,000, with the worst possible outcome being $r^0 = -0.9$ and the best $r^* = 10$. Clearly we can expect any rational DM to set $\alpha(-0.9) = 0$ and $\alpha(10) = 1$. She will also chose $\alpha(r)$ to be increasing in r, $-0.9 \leq r \leq 10$ – so that the larger the probability of the better outcome the more she is prepared to forfeit with certainty. Figure 3.1 gives two possible elicited utility functions $u_1(r) = \alpha_1(r)$ and $u_2(r) = \alpha_2(r)$. The first $\alpha_1(r)$ has a non-increasing slope and so is called a *risk averse* utility function. A DM often has a risk averse utility and reflects a disinclination to gamble. Thus under $\alpha_1(r)$ note that the maximum the DM is prepared to forfeit in a 50–50 gamble between -0.9 and 10 is 2 whilst for an EMV DM this would be 4.55. These are quite common and represent a DM's preference. The form of the utility $\alpha_2(r)$ is much rarer, has a non-decreasing derivative and is called *risk seeking*. Here the DM would trade a certain return of 8 (in £1,000) for a 50–50 gamble between -0.9 and 10 perhaps because 10 is much more useful to her than 8 for example to pay off a creditor.

To prove the main result of this section the DM needs to be prepared to follow one more rule.

Axiom 3.8 (Weak substitution). If P_1, P_2 and P are any three reward distributions, then for all probabilities β, $0 < \beta < 1$

$$P_1 \sim P_2 \Leftrightarrow \beta P_1 + (1 - \beta)P \sim \beta P_2 + (1 - \beta)P.$$

In general this is a very benign demand of the DM. For example suppose the DM states she is indifferent to having an orange or an apple. Then one consequence of this axiom is that the DM can conclude that if she will be indifferent between the option of receiving an apple with probability β and otherwise receiving a banana and the option of of receiving an orange with probability β and otherwise a banana. My own view is that this is the most compelling of rules of rational choice. Sadly it is the rule most commonly violated by unaided DMs, even ones who are otherwise reasonable and well trained, especially when combined with probabilistic determinism. This unfortunate phenomenon will be discussed below. It is therefore certainly unwise to assume this axiom holds when *describing* the behaviour of many unaided DMs: an issue which is a big headache for Bayesian game theorists. This also impinges on other problems within the scope of this book. For example issues associated with the rationality of the chess opponent or the jury member which form the basis of choosing appropriate probability of other's acts can be seriously undermined by this phenomenon.

When appropriate, the weak substitution axiom can be used to extend the definition of the DM's utility to much more complicated gambles than those receiving a particular consequence with certainty. Thus suppose that all possible values of attributes $\boldsymbol{r}_1, \boldsymbol{r}_2, \ldots, \boldsymbol{r}_n \in \mathcal{R}$ that might arise after using a certain decision rule, lie between the worst value $\boldsymbol{r}^0$ and the best value $\boldsymbol{r}^*$. Suppose also that we have elicited the utility $\alpha(\boldsymbol{r}_i)$ referenced to $(\boldsymbol{r}^0, \boldsymbol{r}^*)$ and also elicited the DM's probability mass function Q over possible consequences where a reward $\boldsymbol{r}_i$ happens with probability $q_i \geq 0$, $1 \leq i \leq n$ where $\sum_{i=1}^n q_i = 1$. The weak substitution axiom now allows us to substitute for the betting scheme obtaining $\boldsymbol{r}_1$ with certainty, the equivalent betting scheme that gives the worst consequence $\boldsymbol{r}^0$ with probability $1 - \alpha(\boldsymbol{r}_1)$ and the best $\boldsymbol{r}^*$ with probability $\alpha(\boldsymbol{r}_1)$. Call this new betting scheme Q_1. But now note that $\boldsymbol{r}_2$ in Q_1 can be replaced by the the gamble that gives $\boldsymbol{r}^0$ with probability $(1 - \alpha(\boldsymbol{r}_2))$ and $\boldsymbol{r}^*$ with probability $\alpha(\boldsymbol{r}_2)$. Continue in this way until we find a distribution Q_n preferentially equivalent to Q where all possible consequences have been replaced by their equivalent bet between the extreme consequence $\boldsymbol{r}^0$ and $\boldsymbol{r}^*$. The gamble Q_n has the useful property that it can only result in two possible outcomes: $\boldsymbol{r}^0$ or $\boldsymbol{r}^*$. Summing the probabilities over events that might giving rise to the best outcome gives us that Q is therefore equally preferable under probabilistic determinism to a gamble with distribution

$$Q_n = \begin{cases} \boldsymbol{r}^* \text{ with probability } \alpha(Q) \\ \boldsymbol{r}^0 \text{ with probability } (1 - \alpha(Q)) \end{cases}$$

where

$$\alpha(Q) = \sum_{i=1}^{n} \alpha(\boldsymbol{r}_i) q_i$$

But this is simply the mathematical formula for the expectation of the utility referenced to $(\boldsymbol{r}^0, \boldsymbol{r}^*)$ under the distribution Q!

It follows that under the rationality axioms above we can deduce from the real-valued function $\alpha(\boldsymbol{r})$ the DM's preferences over *any* possible distribution Q over a finite set of consequences whose attributes are all at least as good as $\boldsymbol{r}^0$ and no better than $\boldsymbol{r}^*$. For each possible decision rule d the DM simply calculates her expected utility $\alpha(Q_d)$ associated

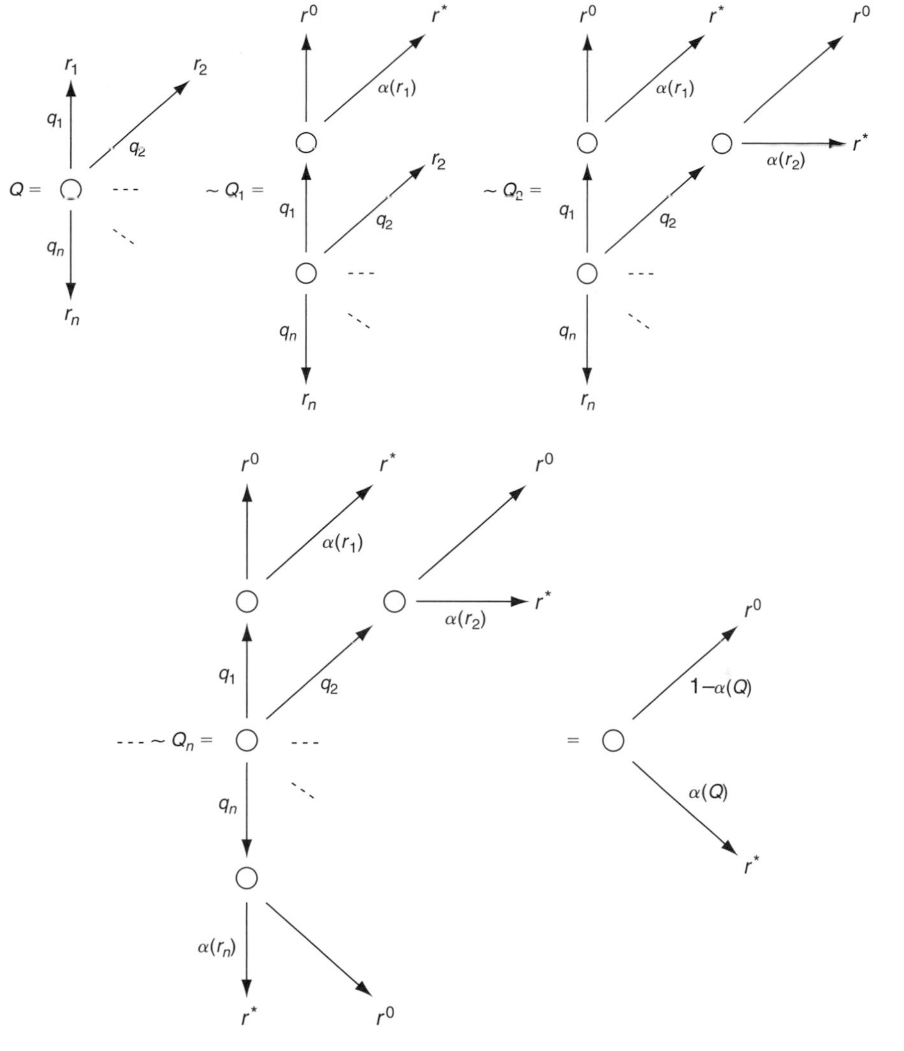

with its distribution over attributes Q_d. The total order property and (3.1) now allow us to conclude that for any two decision rules $d[1]$ and $d[2]$ giving rise to respective finite distributions on attributes $Q[1]$ and $Q[2]$, $Q[1] \preceq Q[2]$ if and only if $\alpha(Q[1]) \leq \alpha(Q[2])$.

3.2.1 Revisiting the dispatch normal form tree with a nonlinear utility function

It is a simple matter to perform a new analysis of a tree with its monetary payoffs – or more generally in problems whose consequences can be measured by a one-dimensional attribute – given at the tips of the tree. So consider the dispatch example of the last chapter.

Simply take the client's utility graph and transform each consequence r by $\alpha(r)$. Suppose $\alpha(r)$ corresponds to the first risk averse utility function U_1 given in Figure 3.1. From this graph the analyst can evaluate the utilities of all possible terminal payoffs in the table below. By substituting these utilities for their associated payoffs and then using the backwards induction algorithm described in Chapter 2 the DM will discover that the optimal decision rule is still decision $d[4]$. You are asked to do this in Exercise 3.6. On the other hand under the risk-seeking utility U_2 the optimal expected utility is obtained by $d[1]$ – the decision to dispatch the machine immediately giving the largest gain with probability 0.8 but risking obtaining nothing.

Terminal payoff r	−0.9	−0.5	0.0	6.1	6.5	7.0
Utility $U_1(r)$	0.0	0.111	0.25	0.805	0.825	0.85
Utility $U_2(r)$	10.0	8.0	9.5	8.7	8.3	7.5
Terminal payoff r	7.1	7.5	8.0	9.1	9.5	10
Utility $U_1(r)$	0.855	0.875	0.9	0.955	0.9725	1
Utility $U_1(r)$	8.78	8.14	7.82	8.38	8.06	7.42

(3.2)

3.2.2 Strong substitution and Alias's paradox*

Some useful properties can be deduced about the preferences of a DM who satisfies the axioms above. Let P_1, P_2 and P be any three reward distributions and $P_1 \precsim P_2 \Leftrightarrow \alpha(P_1) \leq \alpha(P_2)$. Let $Q_1 = \beta P_1 + (1-\beta)P$ and $Q_2 = \beta P_2 + (1-\beta)P$, for $\beta, 0 < \beta < 1$. Then since expectation is linear under mixing the expected utilities $\alpha(Q_i), i = 1, 2$ are given by

$$\alpha(Q_i) = \beta\alpha(P_i) + (1-\beta)\alpha(P)$$

we have that then for all $\beta, 0 < \gamma < 1$

$$P_1 \precsim P_2 \Leftrightarrow \alpha(Q_1) \leq \alpha(Q_2) \Longleftrightarrow Q_1 \precsim Q_2$$

This property is called the strong substitution or independence property and is sometimes assumed as an axiom. One advantage of this stronger assumption is that the assumption used in weak substitution that there is a best and worst outcome of the attributes is unnecessary: the two reference utilities can be *any* attributes $(\boldsymbol{r}^0, \boldsymbol{r}^*)$ with the property that $\boldsymbol{r}^0 \prec \boldsymbol{r}^*$ – see Exercise 3.1.

There are many celebrated examples of a DM having preferences that break this property and are clearly irrational. For example after the 9–11 attack in the US an insurance company produced a policy which essentially took one of its products $Q[1]$ – which gave insurance cover for damage to property caused by a variety of circumstances including acts of terrorism – and rewrote it into a second product. This second product $Q[2]$ cost the same as the first but would *only* pay out when damage was caused by terrorism. Suppose the consequence space is a successful claim r^* or receiving no support r^0. Suppose the policy is only active until the first claim, let the event of that valid claim being because of a terrorist attack be t or not $\bar{t}$ and another valid claim be o or not $\bar{o}$. The DM's preferences between these two policies can then be represented as below.

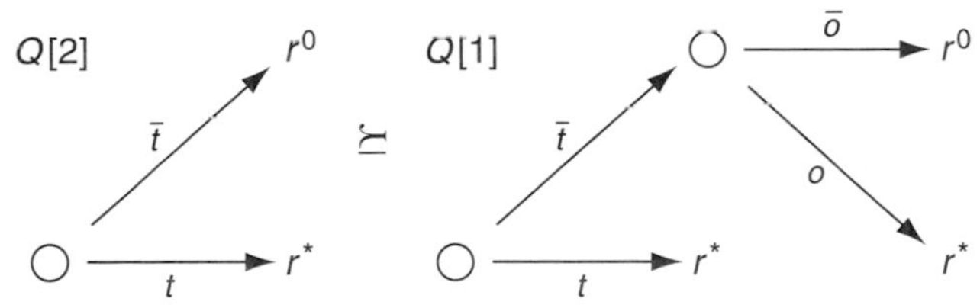

Note here that by the weak substitution axiom the DM implicitly finds r^0 equivalent to obtaining r^* if another claim happens when terrorism has not caused a claim: an equivalence that could not occur unless the probability of non-terrorism but substance for another claim were zero. So this choice is not supported by a DM following our axioms who thinks claiming for other reasons is a real possibility. Furthermore ordinary logic tells us it would be stupid to buy the second product rather than the first. This argument did not prevent many customers buying the second product!

However there are some scenarios where the violation of the strong substitution property is less obviously irrational. The following example is adapted from one given by Kahenman and Tversky (1979) of a preference paradox first identified by Allias, where payoffs are given in £s.

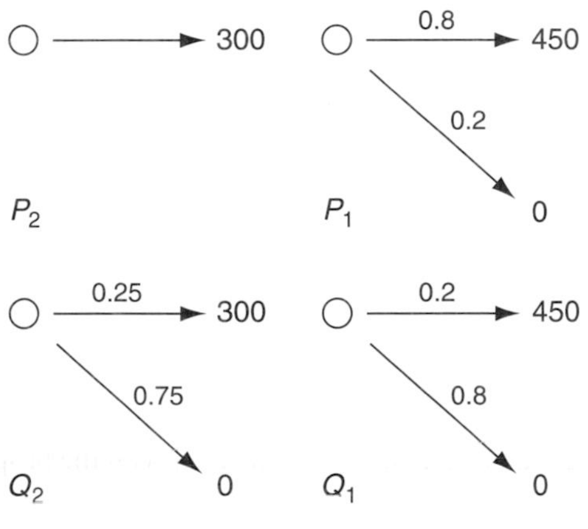

In practice many DMs state the preferences $P_2 \succ P_1$ but $Q_2 \prec Q_1$. One argument sometimes presented to support these choices is that the DM is unwilling to gamble on P_1 and lose when she could get the substantial guaranteed amount by choosing P_2. However in the second two bets she argues that she is likely to win nothing anyway so she may as well take bet Q_1 which has a comparable probability of winning and gives a significantly higher reward in preference to Q_2. However the assumption of probabilistic determinism forces preferences between Q_1 and Q_2 to be the same as

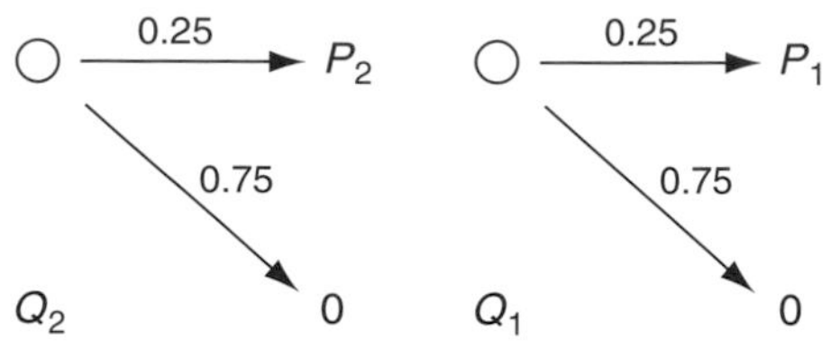

i.e. being given a gamble with only a probability 0.25 of success and then if success is achieved of being given a choice between P_1 and P_2. The strong substitution property therefore demands that the DM prefer Q_2 to Q_1 if P_2 is preferred to P_1 and Q_1 to Q_2 if P_1 is preferred to P_2.

There are some interesting features of this example. First if directly faced with the two betting schemes above then the justifying argument for the second preference fails because it would commit the DM to risk the embarrassment of losing a certain reward of £300 if she succeeded in the first part of the gamble. So presumably her argument would incline her to the Bayesian choice of Q_2 over Q_1. Indeed if in this sequential betting scheme her current preferences would not only violate the strong substitution property but also force her to choose a decision rule which was not cjo: i.e. was not consistent with the belief that she would choose optimally in the future.

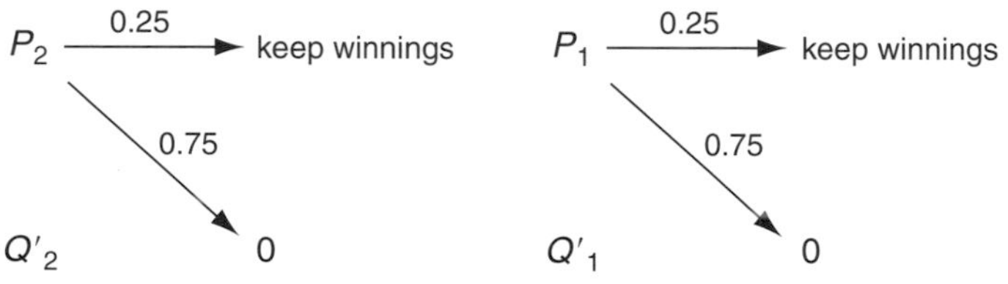

On the other hand, when presented with the two distributionally equivalent bets ($Q_2 \sim Q'_2$, $Q_1 \sim Q'_1$) but where losses with high probability occur *after* the result of the bets, after choosing P_2 the DM is no longer certain of the reward she takes away. Therefore the rationale given above would incline her to prefer Q'_1 over Q'_2. So her preferences, governed by the sort of logic given above, would violate the probabilistic determinism axiom, i.e. two different gambles with the same associated reward distributions are not necessarily considered by this DM as equivalent.

A second point here is to note that the axiomatic system described above works best when there is a sense of the effect of consequences rolling forward. Thus when eliciting the $\alpha(r)$ such that $r \sim B(r^0, r^*, \alpha(r))$ it is often helpful to elicit this not in terms of a "reward" r which is a final point but in terms of a useful resource to use in future acts and to guard against certain eventualities. In the betting example above if the DM thinks of r^* in terms of a future resource their agent may later choose to invest an amount that may be needed to support some mischance rather than simply bank, then the DM would probably prefer P_1 to P_2 as well as Q_1 to Q_2. Note that our definition of preference in terms of a market for exchanging decision options is consistent with this type of elicitation. If this elicitation is not appropriate to the context however then the basic assumption of probabilistic determinism may not be appropriate. Happily in my experience the times when an aided DM is unhappy to follow the rule of probabilistic determinism have been rare. Incidentally note that the different historic trees give a useful framework for addressing this paradox.

3.3 Properties and illustrations of rational choice

3.3.1 Independence of reference points

Suppose the DM is in a context where she is happy with the axioms above. One uneasy question might be to ask whether the precise choice of possible best $\boldsymbol{r}^*$of worse $\boldsymbol{r}^0$ case scenarios had an effect on the rescaling of consequences into utility. Suppose the DM decides she is only interested in decisions giving rise to reward distributions P whose consequences $\boldsymbol{r}$ such that $\boldsymbol{r}^* \prec \boldsymbol{r}^{**} \preceq \boldsymbol{r} \preceq \boldsymbol{r}^{00} \prec \boldsymbol{r}^0$ lie between a different pair of reference points. What would happen if she used the pair $(\boldsymbol{r}^{00}, \boldsymbol{r}^{**})$ to reference the utility and eliciting $\boldsymbol{r} \sim b(\boldsymbol{r}^{00}, \boldsymbol{r}^{**}, \beta(\boldsymbol{r}))$ instead of $(\boldsymbol{r}^0, \boldsymbol{r}^*)$ and eliciting $\boldsymbol{r} \sim b(\boldsymbol{r}^0, \boldsymbol{r}^*, \alpha(\boldsymbol{r}))$? By definition

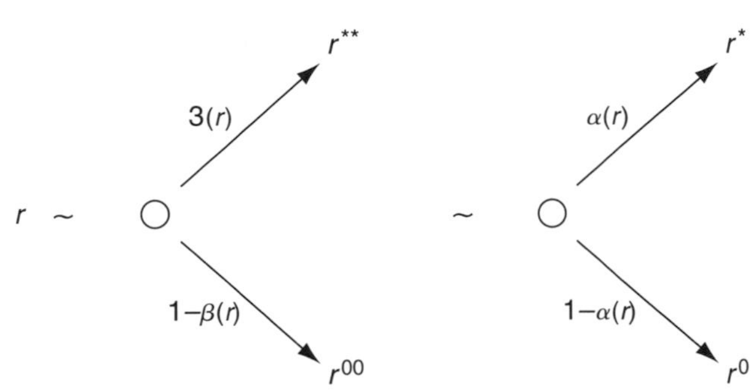

where

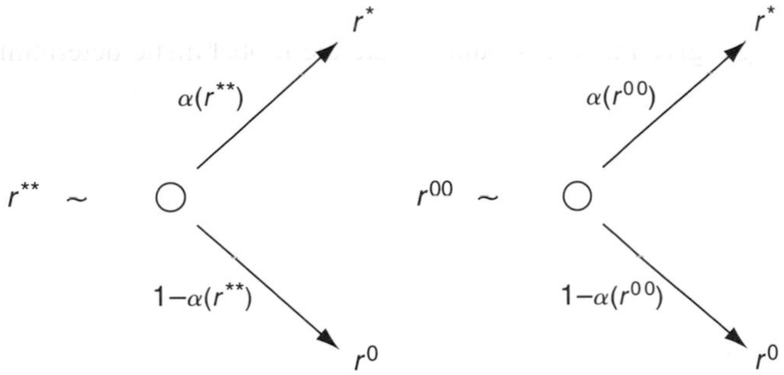

So using the weak substitution axiom it follows that on substituting for $\boldsymbol{r}^{**}$ and $\boldsymbol{r}^{00}$ into the first gamble

$$\boldsymbol{r} \sim b(\boldsymbol{r}^0, \boldsymbol{r}^*, \{\beta(\boldsymbol{r})\alpha(\boldsymbol{r}^{**}) + (1 - \beta(\boldsymbol{r}))\alpha(\boldsymbol{r}^{00})\}).$$

Therefore

$$\begin{aligned}\alpha(\boldsymbol{r}) &= \beta(\boldsymbol{r})\alpha(\boldsymbol{r}^{**}) + (1 - \beta(\boldsymbol{r}))\alpha(\boldsymbol{r}^{00}) \\ &= \alpha(\boldsymbol{r}^{00}) + \beta(\boldsymbol{r})\{\alpha(\boldsymbol{r}^{**}) - \alpha(\boldsymbol{r}^{00})\}.\end{aligned}$$

Since by definition $\boldsymbol{r}^{**} \succ \boldsymbol{r}^{00}$, $\alpha(\boldsymbol{r}^{**}) > \alpha(\boldsymbol{r}^{00})$. So $\alpha(\boldsymbol{r})$ and $\beta(\boldsymbol{r})$ are related by a *strictly increasing linear transformation* for all $\boldsymbol{r}^{**} \preceq \boldsymbol{r} \preceq \boldsymbol{r}^{00}$. Furthermore this strictly increasing linear function is *uniquely* determined by $\left(\boldsymbol{r}^{00}, \boldsymbol{r}^{**}\right)$ because it must be chosen so that the two points $\beta(\boldsymbol{r}^{00}) = 0$ and $\beta(\boldsymbol{r}^{**}) = 1$.

It follows using the linear equation above that the two expected utilities $\beta(P_1)$ and $\beta(P_2)$ referenced to $\left(\boldsymbol{r}^{00}, \boldsymbol{r}^{**}\right)$ satisfy

$$\begin{aligned}\beta(P_1) \leq \beta(P_2) &\Leftrightarrow \\ \beta(P_2) - \beta(P_1) \geq 0 &\Leftrightarrow \\ \sum_{i=1}^{n} \beta(\boldsymbol{r}_i)\{p_{2i} - p_{1i}\} \geq 0 &\Leftrightarrow \\ \sum_{i=1}^{n} \alpha(\boldsymbol{r}_i)\{p_{2i} - p_{1i}\} \geq 0 &\Leftrightarrow \\ \alpha(P_1) \leq \alpha(P_2).\end{aligned}$$

So when $P_1 \preceq P_2$ whether we reference the utility to $\left(\boldsymbol{r}^{00}, \boldsymbol{r}^{**}\right)$ or to $\left(\boldsymbol{r}^0, \boldsymbol{r}^*\right)$ the utility functions reflect the same order of preference. In particular they will both take their highest value at the same distribution. So in this sense and from a technical point of view it should not matter how we choose the reference points; the utility function of a rational DM will reflect the same preferences. And if a utility function $U(\boldsymbol{r}) : \boldsymbol{r} \in \mathcal{R}$ is defined to satisfy $U(\boldsymbol{r}) = a + b\alpha(\boldsymbol{r})$, and $\boldsymbol{r} \sim b(\boldsymbol{r}_1, \boldsymbol{r}_2, \alpha(\boldsymbol{r}))$ then

$$U(\boldsymbol{r}) = \alpha(\boldsymbol{r})U(\boldsymbol{r}_2) + (1 - \alpha(\boldsymbol{r}))U(\boldsymbol{r}_1). \tag{3.3}$$

So in particular, for any $\{\boldsymbol{r} \in \mathcal{R} : \boldsymbol{r}_1 \prec \boldsymbol{r} \prec \boldsymbol{r}_2\}$,

$$\alpha(\boldsymbol{r}) = \frac{U(\boldsymbol{r}) - U(\boldsymbol{r}_1)}{U(\boldsymbol{r}_2) - U(\boldsymbol{r}_1)}. \tag{3.4}$$

A decision which produces a distribution over attributes maximising this expectation is called the *Bayes decision under utility U*. So the range chosen to measure utility U is unimportant. If there is an obvious lowest and highest reward then most mathematicians would set $U(\boldsymbol{r}) = \alpha(\boldsymbol{r})$. However for less mathematical people often find a utility score lying between 0 and 100 is more natural than between 0 and 1: they are often more used to obtaining scores which lie in this range. In this context it is often good to communicate results using $U(\boldsymbol{r}) = 100\alpha(\boldsymbol{r})$ instead.

The implications above are profound and will form the basis on which the rest of the book is developed. However to illustrate this theorem we begin by discussing three simple examples: the first simply illustrating how decisions can be linked to distributions and hence appraised for their rationality.

Example 3.9. Sections of the circumference of a betting wheel are coloured red, blue, white or green. The proportion of the circumference which is coloured red is a quarter, white a quarter, blue a third and green a sixth. The gambler needs to guess where a free-spinning pointer attached to the centre of the wheel will come to rest. The gambler's payoff matrix for this game is given below.

BET \ OUTCOME	red	white	blue	green
bet red [$d(1)$]	4	0	−1	−1
bet white [$d(2)$]	0	4	−1	−1
bet blue [$d(3)$]	−1	−1	2	1
bet green [$d(4)$]	−1	−1	1	2

Each of three gamblers $G(1), G(2)$ and $G(3)$ is asked to give their preferences over the bets above. $G(1)$ states that she prefers $d(2)$ to $d(1)$ and $d(3)$ to $d(4)$. $G(2)$ states that she prefers $d(1)$ to $d(4)$ and $d(3)$ to $d(4)$. $G(3)$ states that she prefers $d(1)$ to $d(4)$ and $d(4)$ to $d(3)$. If all gamblers have an attribute which is money and the higher the payoff the better, which of these gamblers is rational in the sense above?

In this example we can calculate the distribution of consequences associated with each decision.

BET \ OUTCOME	−1	0	1	2	3	4
$d(1)$	0.5	0.25	0	0	0	0.25
$d(2)$	0.5	0.25	0	0	0	0.25
$d(3)$	0.5	0	$0.166'$	$0.33'$	0	0
$d(4)$	0.5	0	$0.33'$	$0.166'$	0	0

Clearly $G(1)$ breaks the probabilistic determinism axiom because the payoff distributions of the attributes of $d(1)$ and $d(2)$ are the same but she prefers $d(1)$ to $d(2)$. More subtly defining

DIST.\OUTCOME	−1	1	2
P	0.6	0.2	0.2
P_1	0	1	0
P_2	0	0	1

we note that

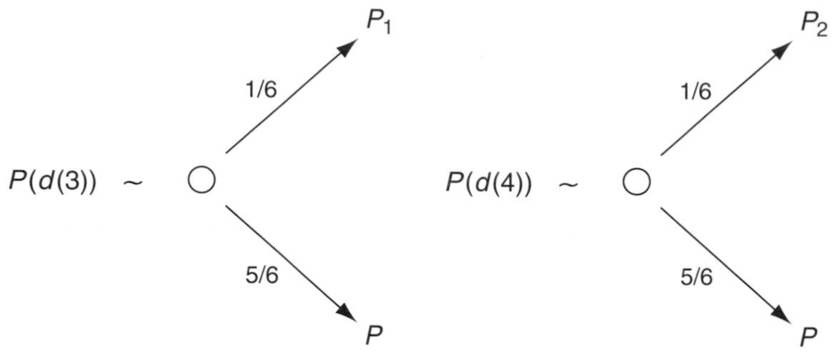

so, by the second axiom, since $P_1 \prec P_2$ we should be able to conclude $P(d(4)) \prec P(d(3))$. So $G(3)$ is not a Bayesian rational DM. To demonstrate that $G(2)$ is expected utility maximising, all we need to do is find a utility function consistent with her preferences. It can be checked, for example, that a linear utility function gives rise to $G(2)$'s preference order. So there is no evidence for her not being rational.

Example 3.10. In the SIDS court case example of Chapter 1 the evaluated outcomes are associated with four combinations of outcomes and decision:

$(0, 0)$: finding the suspect innocent when she is innocent,

$(0, 1)$: finding the suspect innocent when she is guilty,

$(1, 0)$: finding the suspect guilty when she is innocent,

$(1, 1)$: finding the suspect guilty when she is guilty.

How should a rational jury decide when their shared posterior probability of the suspect's guilt is p^*?

Let $U(i,j)$, $i,j = 0, 1$ denote the juror's utility function. Then a rational jury should choose the expected utility maximising decision: i.e. find the suspect guilty if

$$U(1,1)p^* + U(1,0)(1-p^*) > U(0,1)p^* + U(0,0)(1-p^*).$$

Noting that we can expect that the jury will prefer getting a decision right rather than wrong we can safely assume that $U(1,1) > U(0,1)$ and that $U(0,0) > U(1,0)$. It follows that the above inequality rearranges into finding the suspect guilty when

$$p^*(1-p^*)^{-1} > A$$

where

$$A = \frac{U(0,0) - U(1,0)}{U(1,1) - U(0,1)}.$$

So a rational jury should decide the suspect is guilty if their posterior odds of guilt is above a threshold A. Note that if they find a correct conviction and a correct acquittal equally preferable then $U(0,0) = U(1,1)$ and if they find wrongful conviction of an innocent suspect a worse outcome than the acquittal of a guilty suspect – which might be expected in most European courts – also $U(0,1) > U(1,0)$. So under these two cultural conditions we can expect $A > 1$ when the posterior probability of guilt would need to be greater than $1/2$ before the jury would contemplate convicting. Other than this there are no logical constraints on the jury. Their decision will depend on their particular interpretation of "reasonable doubt" which could depend on both the seriousness of the crime as well as the constitution of the jury chosen at random from a population.

A further point is illustrated by this simple example. First recall that the posterior odds of guilt are the prior odds of guilt multiplied by the likelihood ratio as given by equation (1.6). So the jury will convict if

$$\frac{P(\text{Evidence}|\text{ Guilty})}{P(\text{Evidence}|\text{ Innocent})} \geq \frac{(1-p)}{p}A.$$

This means that even if it were possible to observe a particular jury's decisions over a wide number of similar but independent cases, it would not be possible to deduce the probability or the utility of the group given their behaviour. A DM could only learn about the ratio of A and the prior odds: these two components are otherwise hopelessly confounded.

The next example characterises an optimal decision rule under a nonlinear utility in a selection context.

Example 3.11. The DM's sole customer has told you that he will buy a large number of one of the types of computer $(M(1), M(2), \ldots, M(n))$ that DM manufactures but has yet to decide which one. The DM believes that the probability he will choose computer $M(i)$ is $p(i)$, $1 \leq i \leq n$. The DM must decide how the amount $d(i)$ of the total T (in £s) of earmarked development money she spends on computer $M(i)$, so that $\sum_{i=1}^{n} d(i) = T$ when her utility function for choosing the allocation of expenditure as $\boldsymbol{d} = (d(1), d(2), \ldots, d(n))$ is $U[\boldsymbol{d}] = A \log d(i^{*}) + B$, $A > 0$, where $M(i^{*})$ is the computer chosen by the customer at some future date. How should she allocate her resources?

The DM's expected utility $\overline{U}[\boldsymbol{d}]$ for choosing the allocation $\boldsymbol{d}$ is $\overline{U}[\boldsymbol{d}] = A\overline{V}[\boldsymbol{d}] + B$ where

$$\overline{V}[\boldsymbol{d}] = \sum_{i=1}^{n} \log d(i).p(i)$$

so that to find a DM's Bayes decision $\boldsymbol{d}^{*}$ it is necessary to find a vector of allocations $\boldsymbol{d}^{*}$ to maximise $\overline{V}[\boldsymbol{d}]$. We will use the well-known result that if a function f has the property that $\frac{d^2f}{dx^2} < 0$ for all x then Jensen's inequality (Grimmet and Stirzaker, 1982) implies

$$\sum_{i=1}^{n} p(i)f[x(i)] \leq f\left(\sum_{i=1}^{n} x(i)p(i)\right)$$

where $\sum_{i=1}^{n} p(i) = 1$ and $p(i) > 0,\ 1 \leq i \leq n$.

Now let $f(x) = \log(x)$ and $x(i) = d(i)\,(Tp(i))^{-1}$ in the inequality above. This gives

$$\sum_{i=1}^{n} p(i) \log\left[\frac{d(i)}{Tp(i)}\right] \leq \log\left(\sum_{i=1}^{n} \frac{d(i)}{T}\right) = \log(1) = 0.$$

Since $\sum_{i=1}^{n} d(i) = T$, by defining $d^*(i) \triangleq p(i).T$ this inequality can be rewritten as

$$\overline{V}[\boldsymbol{d}] = \sum_{i=1}^{n} p(i) \log[d(i)] \leq \sum_{i=1}^{n} p(i) \log[d^*(i)] = \overline{V}[\boldsymbol{d}^*]$$

where $\boldsymbol{d}^* = (d^*(1), d^*(2), \ldots, d^*(n))$ and $d^*(i)$ is defined above, is a possible resource allocation. Hence it is optimal for the DM to divide the research allocation proportionately to the probability she gives to the customer choosing each given computer. Note here that if instead the DM's utility function is $U[\boldsymbol{d}] = Ad(i^*) + B, A > 0$ then she should choose an entirely different strategy of allocating *all* her resources to develop the computer she believes is most probable for the customer to choose.

The derivation of the results above are the simplest but not the most general. There are various extensions possible to this underlying development. If we assume the strong rather than the weak substitution property as an axiom it is possible to demonstrate that even when there is no least preferable vector of attributes or most preferable vector of attributes a utility function can still be defined and that a rational DM should still choose an expected utility maximising decision. The proof of this is well documented and appears in Exercise 3.1.

3.4 Eliciting a utility function with a dimensional attribute

3.4.1 The midpoint method

The midpoint method has been found to be a useful way of quickly eliciting a utility function U with a one-dimensional attribute. Let $r(0)$ denote the lowest possible reward obtainable from the class of decisions considered and $r(1)$ the highest. The first step of the process of utility elicitation simply begins by eliciting $r(1/2)$ which is the value for which

$$r(1/2) \sim 0.5r(0) + 0.5r(1).$$

Having discovered this 50–50 gamble point this can be used in the second step to determine the next points

$$r(1/4) \sim 0.5r(0) + 0.5r(1/2)$$

and

$$r(3/4) \sim 0.5r(1/2) + 0.5r(1).$$

In the third step the reward space is again divided, first calculating

$$r(1/8) \sim 0.5r(0) + 0.5r(1/4)$$

and then the other three such midpoints. Continuing in this way, successively dividing up the reward space and finding a "midpoint" between two previously elicited points next to one another, note that by definition $r(k/2^n)$ has utility $k/2^n$ for any positive integer n. After n steps of this process we will have therefore found the utility of all $r(k/2^n)$ and integer $k, 0 \leq k \leq 2^n$. Since the DM's function U is increasing, for any reward $r \in [r(k/2^n), r(\{k+1\}/2^n)]$

$$k/2^n \leq U(r) \leq \{k+1\}/2^n.$$

It follows that for all rewards r

$$|U(r) - U_n(r)| \leq 2^{-n}$$

where U_n is the linear interpolation on the points $r[k/2^n]$. So U is uniformly close to U_n for a sufficiently large value of n. Note that this in turn implies that for all decisions d, the expected utility functions associated with all decisions satisfy $\left|\overline{U}(d) - \overline{U}_n(d)\right| \leq 2^{-n}$ where $\overline{U}_n(d)$ denotes the expected utility associated with U_n. Therefore with accurate elicitation, the evaluation of the efficacy of any decision can be made arbitrarily good using this method.

In practice, because U is often differentiable and slowly varying, the bounds above turn out to be extremely good. I have only rarely needed to set $n > 3$ and often $n = 2$ suffices: an elicitation of just 3 points. Note that this method only uses 50–50 gambles so that the DM is not asked to consider events with small probability.

3.4.2 Elicitation using characterised properties

Of course there is a measurement error associated with any elicitation. However the analyst can try to keep these errors as small as possible. First, all the rewards in elicitation gambles should be chosen so that they are as close as possible to each other. Note that in the midpoint method the early stage elicitations most violate this practice and so tend to be most subject to elicitation error. Second it is good practice only to compare a reward with two other attainable rewards. So it is usually good practice to make the bound $r(0)$ as large as possible and $r(1)$ as small as possible. Sometimes removing clearly suboptimal decision rules from the decision space D enables the reference points $r(0)$ and $r(1)$ to be much closer together. Third it is especially helpful for the simple betting schemes considered to have a real practical analogue within the context or to be calibrated against ones that have. Thus for example an insurance policy or general investment protocols are often extremely valuable in accessing a company's current attitude to risk which in turn informs the form of their utility function.

Finally a company's policy can sometimes *characterise* a utility function and this makes it much easier to elicit reliably. For example the DM might want her utility function on symmetric gambles not to be a function of her current wealth. This would demand that, for all values of reward r and $h > 0$,

$$x \sim \alpha(h).(r+h) + (1-\alpha(h)).(r-h)$$

where the elicitation probability α must not depend on r. From our definition of U this in turn implies that

$$U(r) = \alpha(h).U(r+h) + (1-\alpha(h)).U(r-h).$$

After some algebra this can be shown to imply – see Exercise 3.6 – that U – normalised so that $U(r(0)) = 0$ and $U(r(1)) = 1$ where $r(0)$ is the lowest and $r(1)$ the highest possible reward – must take one of the three forms

$$U(r) = \begin{cases} (\exp(\lambda r(1)) - \exp\lambda r(0))^{-1}(\exp(\lambda r) - \exp\lambda r(0)) & \text{when } \alpha < 1/2 \\ r\{r(1) - r(0)\}^{-1} & \text{when } \alpha = 1/2 \\ (\exp(\lambda r(1)) - \exp\lambda r(0))^{-1}(\exp\lambda r(0) - \exp(-\lambda r)) & \text{when } \alpha > 1/2 \end{cases} \tag{3.5}$$

where $\lambda > 0$. Because risk aversion is common usually $\alpha > 1/2$ and the last form is appropriate. Note that the advantage of being able to make this characterisation is that there is now only one parameter λ to elicit and this can be obtained from one indifference statement. Furthermore unlike in the initial stages of a midpoint method elicitation, λ – and hence the whole utility function – can be elicited from a single comparison of betting preferences between gambles with reward of only moderately different size.

The utility function given above is not the only one to have a characterisation, there are several other important classes, see for example Keeney and Raiffa (1976) and von Winterfeldt and Edwards (1986).

3.5 The expected value of perfect information

Many decision rules involve taking preparatory samples before a final committing decision. Usually such initial information gathering has an associated cost – sometimes financial – sometimes because the delay caused by sampling reduces the possible scope of future acts. For example in the dispatch problem the DM had the opportunity of gathering information about whether a fault existed by scanning the product a number n of times. Here it would be useful for the DM to learn the number of scans it is worth even considering to perform as a function of their cost and which options can be immediately discarded as being suboptimal.

One useful bound that can help to limit the extent to which a DM should rationally consider gathering information is the expected value of perfect information. This is calculated by appending to the set of gambles considered a further set of hypothetical gambles. These are ones that for the same cost of each potential sampling decision perform an experiment that tells the DM the precise situation. So for example in the dispatch example for the decision to take two independent scans and then act appropriately we also consider the hypothetical option of paying the same amount and performing an experiment telling the DM precisely whether or not a fault existed.

The reasons this hypothetical alternative is worth considering is firstly that it is usually simple to calculate its associated expected utility. This is because this utility will depend only on the probability of the event of interest – here whether or not a fault exists – a value directly elicited – and second the certain cost of that experiment. Secondly the hypothetical option is clearly no worse than its actual counterpart. So if the easy-to-evaluate hypothetical option is worse than some other real option then it automatically follows that any equivalent real option is suboptimal and not worth considering as a possible viable option.

Return to the dispatch example. For simplicity assume that the DM's utility function is really linear in payoff so that the EMV strategy is the right one to use. The expected payoff associated with immediate dispatch given that the probability of a fault is p is easy to calculate from the initially given information as $10(1-p)$. On the other hand if sampling at a cost of c gave perfect information then clearly the DM should overhaul if a fault exists – an event of probability p – is found giving a pay-off of 7 or dispatch if no fault exists – an event with probability $1-p$ – with an associated payoff 10. It follows that the perfect information sampling is only better than immediate despatch if

$$(7-c)p + (10-c)(1-p) < 10(1-p) + 0p \Leftrightarrow c > 7p.$$

Thus no sampling scheme – no matter how informative – is worth considering if its cost is greater than $7p$.

There are two further points to notice here. First the same idea can be used when the DM has any utility function, not just a linear one. So for example in the illustration above all she need do is to substitute $U(7-c)$ for $(7-c)$, $U(10-c)$ for $(10-c)$, $U(10)$ for 10 and $U(0)$ for 0 in the equation above to discover her new inequality for c. Thus substituting and rearranging tells the DM that any option for costing c where c is such that

$$\frac{p}{1-p} < \frac{U(10) - U(10-c)}{U(7-c)}$$

should be disregarded. Second, if the DM has already identified the decision that is best of the ones considered so far then this can be the reference gamble that the hypothetical schemes must better. It follows that in the dispatch example when $p = 0.2$ we could instead use the expected pay-off associated with the policy to scan once rather than the expected pay-off associated with immediate dispatch, when comparing the efficacy of different exploratory experiments or different numbers of scans to those already considered. I have found that routinely calculating such bounds early in a decision analysis is often surprisingly helpful in paring away many decision rules that initially and superficially appear promising.

3.6 Bayes decisions when reward distributions are continuous

To find an EMV decision rule when variables in a problem are continuous uses exactly analogous procedures as in the discrete case. Thus the DM simply needs to choose a decision to minimise expected loss or equivalently maximise expected pay-off. Explicitly if the consequences $\boldsymbol{\theta} \in \Theta$ given observations $\boldsymbol{y} \in \mathcal{Y}$ are believed by the decision maker to have a joint density $p(\boldsymbol{\theta}|\boldsymbol{y})$ then following an EMV strategy simply entails, for each possible observation $\boldsymbol{y} \in \mathcal{Y}$ choosing a decision $d(\boldsymbol{y}) \in D$ where D is the decision space so as to minimise the expected loss

$$\overline{L}(d) = \int_{\boldsymbol{\theta} \in \Theta} L(d, \boldsymbol{\theta}) p(\boldsymbol{\theta}|\boldsymbol{y}) d\boldsymbol{\theta}$$

where $\overline{L}(d)$ denotes her expected loss or, equivalently, maximise

$$\overline{R}(d) = \int_{\boldsymbol{\theta} \in \Theta} R(d, \boldsymbol{\theta}) p(\boldsymbol{\theta}|\boldsymbol{y}) d\boldsymbol{\theta}$$

where $\overline{R}(d)$ denotes her expected payoff. Of course the arguments for specifying distributions consistently with their causal order apply in this case as strongly as they did in the discrete setting of the last chapter. So it is usually better to elicit $p(\boldsymbol{\theta})$ and $p(\boldsymbol{y}|\boldsymbol{\theta})$ and use the absolutely continuous version of Bayes rule to infer $p(\boldsymbol{\theta}|\boldsymbol{y})$. Illustrations of how Bayes rule is applied to continuous problems will be postponed to Chapter 5. So at this point let us assume that $p(\boldsymbol{\theta}|\boldsymbol{y})$ has already been calculated and the DM needs to identify her Bayes rule.

Just as in discrete problems, for most scenarios it is necessary for the DM to deviate from EMV decision making and to incorporate her preference using her elicited utility function. You will observe that the elicitation of a utility function U described above did not use that the distribution of the rewards was discrete. All the methods described there would work equally well if the rewards were continuous. Integrals simply need to be substituted for the sums in these expression It can also be shown (see for example De Groot (1970); Robert (2001)) that, with the addition of a technical measurability condition, in order to satisfy the axioms given in that section the DM would need to be expected utility maximising. In the context of $\boldsymbol{\theta}$ being absolutely continuous this would correspond to finding a decision rule $d^*(\boldsymbol{y})$ which maximises $\overline{U}(d)$ where

$$\overline{U}(d) = \int_{\boldsymbol{\theta} \in \Theta} U\left(R(d, \boldsymbol{\theta})\right) p(\boldsymbol{\theta}|\boldsymbol{y}) d\boldsymbol{\theta}.$$

It is argued below that for most practical purposes a utility function will be bounded. When this is the case U can be assumed to take values between 0 and 1 with 0 corresponding to the worst possible reward and 1 to the best. Note that in the case when rewards are one-dimensional U will be nondecreasing in reward r and, by definition, the EMV DM will have a utility function which is linear in r.

It is not unusual for the DM to need to make decisions not about the parameters $\boldsymbol{\theta}$ themselves but about a vector $\mathbf{Z}$ of future variables taking values in $\mathcal{Z}$ whose distribution is dependent on them. In the next chapter we will see that most models will have the property that $\mathbf{Z} \amalg \boldsymbol{Y}|\boldsymbol{\theta}$: i.e. all relevant information about the vector $\mathbf{Z}$ contained in what is observed is transmitted through $\boldsymbol{\theta}$. In this case when $\mathbf{Z}$ is absolutely continuous

$$\overline{U}_z(d) = \int_{z \in \mathcal{Z}} U_z\left(R_z(d,z)\right) p(z|\mathbf{y}) dz$$

and when the future random vector of interest $\mathbf{Z}$ is discrete

$$\overline{U}_z(d) = \sum_{z \in \mathcal{Z}} U_z\left(R_z(d,z)\right) p(z|\mathbf{y})$$

where the density or mass function $p(z|\mathbf{y})$ in each case is given by

$$p(z|\mathbf{y}) = \int_{\boldsymbol{\theta} \in \Theta} p(z|\boldsymbol{\theta}) p(\boldsymbol{\theta}|\mathbf{y}) d\boldsymbol{\theta}.$$

In the continuous context optimal decisions of an expected utility maximising DM often lead her to make decisions about a parameter vector $\boldsymbol{\theta}$ that link closely to classical point estimates. However the Bayesian methodology allows such estimates to be adjusted in response to the beliefs, needs and priorities of the DM, as reflected through her utility function U and her prior distribution.

3.7 Calculating expected losses

We find that when both d and θ are one-dimensional, taking values on the real line and losses are symmetric and increasing in $|d-\theta|$, so that d is a simple estimate of θ then Bayes decisions of an EMV DM often turn out to be familiar summary statistics associated with the posterior distribution.

Example 3.12 (Quadratic loss). Here $L(d,\theta) \triangleq (d-\theta)^2$. Then if $\mu(\mathbf{y})$ denotes the mean of the posterior distribution of θ and $\mathbb{E}$ represents expectation under density $p(\boldsymbol{\theta}|\mathbf{y})$ then

$$\begin{aligned}\overline{L}(d) &= \mathbb{E}(d-\theta)^2 = \mathbb{E}(\{\theta-\mu(\mathbf{y})\} - \{d-\mu(\mathbf{y})\})^2 \\ &= \mathbb{E}(\{\theta-\mu(\mathbf{y})\})^2 - 2\{d-\mu(\mathbf{y})\}\,\mathbb{E}(\{\theta-\mu(\mathbf{y})\} + \{d-\mu(\mathbf{y})\}^2) \qquad (3.6)\end{aligned}$$

by the linearity of expectation. Since

$$\mathbb{E}(\{\theta-\mu(\mathbf{y})\} = \mathbb{E}(\theta|\mathbf{y}) - \mu(\mathbf{y}) = 0$$

by definition whilst the first term in equation (3.6) is the posterior variance of θ – which we shall denote by $\sigma^2(\mathbf{y})$ – we have that, provided $\sigma^2(\mathbf{y})$ exists,

$$\overline{L}(d) = \sigma^2(\mathbf{y}) + \{d - \mu(\mathbf{y})\}^2 .$$

Thus since $\{d - \mu(\mathbf{y})\}^2 \geq 0$ and is zero only when $d = \mu(\mathbf{y})$, the Bayes decision d^* – i.e. the choice of decision minimising $\overline{L}(d)$ – is the posterior mean $\mu(\mathbf{y})$. Furthermore from the equation above we see that the expected loss on taking this decision is the posterior variance $\sigma^2(\mathbf{y})$.

Example 3.13 (Absolute loss). When $L(d, \theta) = |d - \theta|$ then, provided $\boldsymbol{E}(|\theta|) < \infty$ it can be shown that the Bayes decision is the median of the posterior distribution of θ (see e.g. De Groot (1970) and Exercise 3.11). Note that in this case the associated expected loss does not have a closed form.

Example 3.14 (Step payoff). Here we assume for $b > 0$, the payoff $R_b(d, \theta)$ is given by the *step payoff*

$$R_b(d, \theta) = \begin{cases} 1 \text{ when } |d - \theta| \leq b \\ 0 \text{ when } |d - \theta| > b \end{cases} \tag{3.7}$$

so that when $|d - \theta| \leq b$ the estimate d is considered satisfactory whilst otherwise it is not. In this case

$$\overline{R}_b(d) = \int_{\theta=d-b}^{d+b} p(\theta|\mathbf{y}) d\theta = P(d + b|\mathbf{y}) - P(d - b|\mathbf{y}) \tag{3.8}$$

where $P(\theta|\mathbf{y})$ is the posterior distribution function of θ. Differentiating and setting to zero we therefore see that any Bayes decision d^* must satisfy

$$p(d^* + b|\mathbf{y}) = p(d^* - b|\mathbf{y}).$$

In particular for unimodal densities when such a solution is unique because it can be shown that the Bayes decision must exist solving this equation gives the Bayes decision. Finding this solution is usually a simple task when p can be written in closed form (see Smith (1980a) and Exercise 3.7). It is easily checked that when $p(\theta|\mathbf{y})$ as $b \to 0$ the Bayes decision $d^*(b)$ associated with this payoff distribution tends to the mode $m(\mathbf{y})$ of the posterior density. Furthermore whenever $p(\theta|\mathbf{y})$ is unimodal and symmetric about a mode $m(\mathbf{y})$ then, whatever the value of $b > 0, d^* = m(\mathbf{y})$. From equation (3.8) note that $d^*(b)$ is simply the midpoint of the interval of length $2b$ of maximum posterior probability, i.e. the midpoint of a *minimum length credibility interval* of length $2b$ whose *significance* is $\overline{R}(d)$.

Of course, as in the discrete problems illustrated in the last chapter, loss functions need not have this simple symmetric form: see below.

3.7.1 Expected utility maximisation for continuous problems

We saw above that EMV decision rules under symmetric payoffs gave optimal decisions that corresponded to the well-known summaries of location such as means medians and modes. The associated expected payoff associated with these optimal decisions also corresponded to familiar measures of uncertainty like variance and certain credibility intervals. However if the DM has a nonlinear utility then her optimal decisions can be seen to provide interesting trade-off between various features of the posterior density.

Example 3.15. Suppose DM's density over rewards r given she takes decision d is normally distributed with mean $\overline{r}(d)$ and variance $\sigma_r^2(d)$. Suppose your utility function U is given by

$$U(r) = 1 - e^{-\lambda r}$$

where the size of the parameter $\lambda > 0$ reflects the size of DM's risk aversion, i.e. how reluctant she is to take a gamble. Then since $M_R(-\lambda) = \mathbb{E}(e^{-\lambda r})$ is the moment generating function of a normal $N(\overline{r}(d), \sigma_r^2(d))$ variable we have that

$$\mathbb{E}(e^{-\lambda r}) = \exp\left(-\lambda \overline{r}(d) + \frac{1}{2}\lambda^2 \sigma_r^2(d)\right).$$

Thus

$$\overline{U}(d) = 1 - \exp\left\{-\lambda\left(\overline{r}(d) - \frac{1}{2}\lambda \sigma_r^2(d)\right)\right\}$$

is maximised when $\overline{r}(d) - \frac{1}{2}\lambda\sigma_r^2(d)$ is maximised over $d \in D$. Notice if $\sigma_r^2(d)$ is not a function of d then DM will choose d to maximise her expected reward $\overline{r}(d)$ – i.e. agree with the EMV decision rule – whilst if the expected reward $\overline{r}(d)$ is the same for all decisions she will choose d to minimise the variance of her reward. When both $\overline{r}(d)$ and $\sigma_r^2(d)$ depend on d her optimal choice will trade off decisions with a high expected payoff with decisions with more certain return. The larger the risk aversion the larger the parameter λ and the more weight is put on choosing a decision ensuring low levels of uncertainty.

Example 3.16. Suppose the payoff associated with the pair $d \in D, \theta \in \Theta$ is given by (3.7) $b > 0$ so that we are rewarded. Then if U is *any* strictly increasing function of reward then

$$\overline{U}(d) = \{U(1) - U(0)\}\, \overline{R}_b(d) + U(0)$$

where $\overline{R}(d)$ is given in (3.7) and $U(1) - U(0) > 0$. So a decision maximising any such expected utility is one maximising $\overline{R}(d)$ as discussed in the example above. It is easily checked that if and only if a payoff function can take only one of two values will the utility maximising decision be invariant to the DM's utility function, as it is above. So in this sense, if we demand estimates not to depend on a client's gambling preferences as encoded through her utility function then we are forced into using a zero–one payoff structure such as the one above.

There are many examples of how formal Bayes decisions made under an appropriate utility function give automatic and defensible ways of explaining why a rational person should want to trade off future gain with future risks and is now intrinsic to the study of risky decision making in financial markets. (See Jorion (1991); Santos (2002)). In later chapters of the book we show how the DM's decisions under the choice of appropriate utility functions enable her to balance the achievements under many criteria in commonly encountered complex problems with multifaceted reward structures.

We conclude with an example that gives the expected payoff of a normal distribution under a conjugate loss function.

Example 3.17. Suppose θ has a Normal distribution with posterior mean μ and variance σ^2, the reward distribution is given by $R(d,\theta)$ where

$$R(d,\theta) = \exp -\{1/2(\theta - d)^2\} \tag{3.9}$$

and the DM's utility function $U(r) = r^{\lambda^{-1}}$ is the *power utility*. Note that if $\lambda = 1$ then the DM follows an EMV strategy, as $\lambda > 1$ becomes larger she becomes increasingly risk averse, i.e. more concerned to avoid small rewards, whilst as $\lambda < 1$ becomes smaller she becomes more and more risk seeking, placing increasing emphasis on trying to obtain a high reward. By definition

$$\begin{aligned}\overline{U}(d) &= \int_{-\infty}^{\infty} \exp -\{1/2\lambda^{-1}(\theta - d)^2\}(2\pi\sigma^2)^{-1/2} \exp -\{1/2\sigma^{-2}(\theta - \mu)^2\} d\theta \\ &= (2\pi\lambda)^{1/2} \int_{-\infty}^{\infty} (2\pi\lambda)^{-1/2} \exp -\{1/2\lambda^{-1}(\theta - d)^2\}(2\pi\sigma^2)^{-1/2} \\ &\quad \times \exp -\{1/2\sigma^{-2}(\theta - \mu)^2\} d\theta.\end{aligned}$$

Note that the integral above is just the formula for the density of the sum of two independent normal random variables θ and D having respective means and variances (μ, σ^2) and $(0, \lambda)$. Standard distributional results tell us that this convolution has a normal density with mean μ and variance $\sigma^2 + \lambda$. It follows that

$$\begin{aligned}\overline{U}(d) &= (2\pi\lambda)^{1/2}(2\pi(\sigma^2 + \lambda))^{-1/2} \exp -\{1/2(\sigma^2 + \lambda)^{-1}(\mu - d)^2\} \\ &= \lambda^{1/2}(\lambda + \sigma^2)^{-1/2} \exp -\{1/2(\sigma^2 + \lambda)^{-1}(\mu - d)^2\}.\end{aligned}$$

Clearly the Bayes decision is the one which chooses d so as to maximise the value of the exponential term, i.e. to choose $d^* = \mu$, whence

$$\overline{U}(d) = (1 + \sigma^2/\lambda)^{-1/2}.$$

Notice here that DM expects to be most content when she is highly risk averse, i.e. when λ is large, and least optimistic about the result when λ is small.

3.8 Bayes decisions under conflict*

So far in this book we have illustrated how various different components of a problem can be drawn together within a Bayesian analysis. Expected utility theory then prescribes that the DM should choose her decision in light of the potential average reward it might bring as measured by its expected utility. At first sight it therefore looks as if the Bayesian DM will always choose to compromise between different objectives. However this is far from the case. When the posterior information she receives genuinely represents several different possible explanations of the studied process – and in Chapter 6 we will encounter several situations where this will occur naturally – then the Bayesian paradigm not only provides a rich enough semantics to express this conflict but the theory developed above also describes how the DM should best address her problem. We begin by considering two very simple scenarios in a one-dimensional problem where the DM essentially needs to decide how to estimate the value of a parameter θ.

Example 3.18. Suppose the DM's density $p(\theta|\mathbf{y})$ after accommodating any information $\mathbf{y}$ she has available takes nonzero values only for θ in the interval $[-2, 10]$, is continuous and has the "two-tent" shape given by

$$p(\theta|\mathbf{y}) = \begin{cases} 0.2(2+\theta) & \text{when } -2 \leq \theta \leq -1 \\ -2\theta & \text{when } -1 < \theta \leq 0 \\ 0.032\theta & \text{when } 0 < \theta \leq 5 \\ 0.032(10-\theta) & \text{when } 5 < \theta \leq 10 \end{cases}$$

(Figure 3.2). This has modes at $\theta = -1, 5$ and an antimode at $\theta = 0$. The mode at -1 is higher than the one at 5 but the density dies more steeply at -1 than it does at 5. This density represents a belief that θ is four times more likely to be positive than negative, but if it is negative it is easier to predict accurately. Suppose the DM's reward distribution is given by the step payoff $R(d, \theta)$ defined above so that any Bayes decision $d^* \in [-2, 10]$ and from the analysis in that example must satisfy $p(d^* - b|\mathbf{y}) = p(d^* + b|\mathbf{y})$. When $0 < b \leq 1$ from

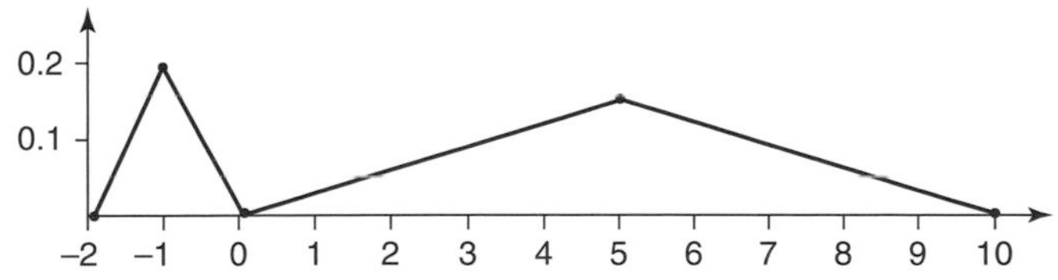

Figure 3.2. A two-tent density.

the graph of this density it is clear that this equation is satisfied if and only if $d^* = -1$, $d^* = 5$ and at a point near zero. It is easily checked that the solution near zero defines a local minimum of the expected payoff. It follows that the only two possible candidate decisions are $d^* = -1$ or $d^* = 5$. For each $0 < b \leq 1$ it is easy to calculate the expected payoffs $\overline{R}_b(-1)$ and $\overline{R}_b(5)$. It is easy to calculate the expected utility where

$$\overline{R}_b(-1) = 2b \times 0.2(1-b) + 0.2b^2 = 0.2b(2-b),$$
$$\overline{R}_b(5) = 2b \times 0.032(5-b) + 0.032b^2 = 0.032b(10-b).$$

It follows from some simple algebra that $\overline{R}_b(-1) > \overline{R}_b(5)$ if and only if $0 < b < \frac{10}{21}$. So if the tolerance for a decision to be acceptable is smaller than $\frac{10}{21}$ then the DM should guess near the highest posterior mode -1. However if the tolerance is greater than $\frac{10}{21}$ she should choose the decision 5 where more of the posterior weight is concentrated in its neighbourhood. So depending on the needs of the analysis she will choose quite different decisions.

The example above may look a little contrived. Here is a setting where a similar phenomenon is observed but which is commonly encountered in practice.

Example 3.19. The posterior density of θ is a discrete mixture

$$p(\theta|\mathbf{y}) = \sum_{i=1}^{m} \pi_i p_i(\theta)$$

where $p_i(\theta)$ is a normal density with mean μ_i and variance σ_i^2 and the probability weights $\sum_{i=1}^{m} \pi_i = 1$, $\pi_i \geq 0$, $i = 1, 2, \ldots, m$ are the posterior probabilities that each of the different normal densities are appropriate: see for example Chapter 5 and Fruthwirth-Schnatter (2006). Under the reward function (3.9) and the power utility function given in Example 3.17 above, the expected utility is

$$\overline{U}(d) = \sum_{i=1}^{m} \pi_i \overline{U}_i(d)$$

where, from the last example in the last section for $i = 1, 2, \ldots, m$

$$\overline{U}_i(d) = \lambda^{1/2}(\lambda + \sigma_i^2)^{-1/2} \exp -\{1/2(\sigma_i^2 + \lambda)^{-1}(\mu_i - d)^2\}.$$

This expected utility can have a very complex geometry, being proportional to a mixture of normal densities. In particular it can exhibit up to m local maxima if the means μ_i of the different mixture components are far enough apart from one another. Consider the simplest nontrivial example where $m = 2$ and $\sigma_1^2 = \sigma_2^2$ and $\pi_1 = \pi_2 = 1/2$ where without loss assume $\mu_1 < \mu_2$. Then it is easy to check Smith (1979b) that $\overline{U}(d)$ will have one local maximum at the posterior mean $\overline{\mu} = 1/2(\mu_1 + \mu_2)$ of θ when

$$\mu_2 - \mu_1 \leq 2\sqrt{\sigma^2 + \lambda}.$$

So if the means of the two component densities are close enough together then the Bayesian DM should choose to compromise. Note that as her risk aversion – and hence λ – increases she is increasingly inclined to compromise. Furthermore it is straightforward to check that under the condition above, whatever the value of π_1, $\overline{U}(d)$ has exactly one maximum so a compromise is necessary albeit weighted towards the mean of the component with the higher probability. However if the means of the different components of the mixture are sufficiently separated so that

$$\mu_2 - \mu_1 > 2\sqrt{\sigma^2 + \lambda}$$

then the posterior mean $\overline{\mu}$ is a local *minimum* of $\overline{U}(d)$, i.e. the worst decision in its neighbourhood. There are two equally good Bayes decisions d_1^* and d_2^* in this case where d_1^* lies in the open interval $(\mu_1, \overline{\mu})$ and d_2^* lies in the interval $(\overline{\mu}, \mu_2)$. As $\mu_2 - \mu_1 \to \infty$ (or $\sigma^2 + \lambda$ tends to 0) then $d_1^* \to \mu_1$ and $d_2^* \to \mu_2$. Thus as the means increasingly diverge from one another the DM should act as if she believed one or other of the models and choose a decision increasing close to either μ_1 or μ_2. Thus under these conditions she essentially chooses one of the component densities and acts (approximately) to optimise the expected utility associated with that component. Because each component has equal weight she is indifferent between each of these choices. When the probabilities are different and the distance between the two means is sufficiently large to cause this type of bifurcation then it is easily checked the Bayesian DM chooses a decision close to the mean of the component with the highest probability weight.

The last example demonstrates just how expressive decision making under the Bayesian paradigm can be, in particular how it can express when compromise between two competing explanations is optimal and alternatively when the DM should choose a decision that approximately commits to the possibility of one of the alternative explanations. Note also that the qualitative instructions of when to do this are compelling ones. It is also worth mentioning that the types of qualitative advice underlying this normal example and a particular shape of reward function do not depend on these assumptions and that – under certain regularity conditions – similar deduction can be made for all mixtures of symmetric distribution under symmetric reward, see Smith (1979b). The sorts of bifurcating decision making, although currently much neglected, appears widely when dealing with complex decision problems, either due to the existence of competing objectives or competing different explanations of evidence we will meet subsequently in this book.

We end this section with an example of a decision problem where the payoff function is asymmetric so that under-estimation is penalised in a different way from over-estimation: a very common scenario. We will see that in these scenarios the appropriate choice of decision can be very different from simply opting for a decision close to a best estimate of the quantity of interest.

Example 3.20. Suppose that the DM's rewards function $R_{a,b}(d,\theta)$ takes the form

$$R_{a,b}(d,\theta) = \begin{cases} a & d < \theta + b \\ 1 & |d - \theta| \leq b \\ 0 & d > \theta - b \end{cases} \tag{3.10}$$

where $0 < a < 1$ and $b > 0$. Here the DM obtains her best reward if she chooses a decision d within a distance b of θ. However if she over-estimates outside this region she obtains her worst reward whilst if she under-estimates she receives an intermediate reward a. So she needs to reconcile the competing objectives of estimating θ well but not over-estimating θ. First assume that she is an EMV DM. In Exercise 3.8 you are asked to prove that if the posterior density $p(\theta|\mathbf{y})$ of θ is continuous then all local maxima and minima d^* – and so in particular any interior Bayes decision of the expected reward $\overline{R}_{a,b}(d)$ – must satisfy the equation

$$(1 - a)\, p(d^* + b|\mathbf{y}) = p(d^* - b|\mathbf{y}).$$

When $\theta|\mathbf{y}$ has a normal distribution with mean μ and variance σ^2 then, taking logarithms of this equation and cancelling constants give us

$$\begin{aligned} \log(1 - a) - 1/2\sigma^{-2}(d^* + b - \mu)^2 &= -1/2\sigma^{-2}(d^* - b - \mu)^2 \\ 2\sigma^2 \log(1 - a) &= (d^* + b - \mu)^2 - (d^* - b - \mu)^2 \\ 2\sigma^2 \log(1 - a) &= 4b(d^* - \mu) \end{aligned}$$

so

$$d^* = \mu + \frac{\sigma^2 \log(1 - a)}{2b}.$$

This seems to make good sense. Noting that $\log(1 - a)$ is always negative but $\sigma^2, b > 0$ by definition the DM is instructed to choose a decision less than μ adjusting downwards in an attempt to avoid a maximum penalty because of over-estimation. The larger the DM's uncertainty σ^2 or the intermediate reward is the lower the decision should be. On the other hand the larger the maximum distance b of a decision needs to be from θ to get a reward, the closer to μ the decision d^* should be. You are asked in Exercise 3.8 to show that if the DM has a risk averse utility then the value of a is implicitly increasing in her risk aversion. It follows that the more risk averse the DM is the more conservative and lower her choice of d^*. In particular notice that the normality of her posterior always induces the DM to compromise between the two objectives in a way that varies smoothly with changing values of the hyperparameters of the reward function, her utility and her posterior mean and variance.

We will see in Chapter 5 that in simple conjugate Bayesian normal analyses where the variance is estimated the posterior density of the next observation is Student t. So Student t posterior distributions are commonly encountered. When the DM is rewarded as above then her Bayes decision rule can be quite different from the one described above, even though the Student t density looks very similar to the normal density. This is because a normal distribution gives a very low probability of getting an estimate very wrong: a property not shared by the Student t. For clarity we will illustrate this different bifurcating phenomenon using the simplest distribution form the Student t family – the Cauchy density. In Exercise 3.12 you are asked to confirm that the decision making of general Student t's closely follow the form arising from the Cauchy described below.

Example 3.21. Suppose a DM has a reward function (3.10) but that $p(\theta|\mathbf{y})$ is Cauchy with mode/median μ and spread parameter σ^2 and

$$p(\theta|\mathbf{y}) = (\pi\sigma)^{-1}(1 + \sigma^{-2}(\theta - \mu)^2)^{-1}. \tag{3.11}$$

Then writing $\delta = d^* - \mu$ all interior maxima of the associated expected payoff must therefore satisfy

$$(1 + \sigma^{-2}(\delta + b)^2) - (1 - a)(1 + \sigma^{-2}(\delta - b)^2) = 0$$

which on dividing by the parameter a rearranges to

$$\delta^2 + 2\lambda\delta + \tau^2 = 0$$

where $\lambda = b\left(2a^{-1} - 1\right)$ and $\tau^2 = \left(b^2 + \sigma^2\right) > 0$. This is a quadratic equation in δ and hence d^*. It is easily checked that the expected reward is always decreasing for large values of δ. It follows that if there are two roots, i.e. if $\lambda^2 > \tau^2$, then the larger root

$$\delta = (\lambda^2 - \tau^2)^{1/2} - \lambda \Leftrightarrow d^* = \mu + (\lambda^2 - \tau^2)^{1/2} - \lambda$$

is the maximum. As in the normal case, since $\tau^2 > 0$ this local maximum is less than μ. So by choosing this local maximum the DM will always tend to under-estimate. But this local maximum may well not be a global one and therefore may not be a Bayes decision. For example if $\lambda^2 < \tau^2$ – and this will be automatic if the spread parameter σ^2 is large enough or the tolerance b for a successful estimate is small enough – then the expected return is strictly decreasing on the whole real line. Then formally there is no Bayes decision unless the decision space is closed bounded below when the Bayes decision is this bound. Interestingly even when no such lower bound exists there is a decision not in the decision space which may well be a feasible practical decision, namely $d = -\infty$. This can be interpreted as taking the sure return of a by not gambling on θ. As μ and σ^2 change in the light of data this conservative decision will continue to be best until σ^2 becomes sufficiently small that $\overline{R}_{a,b}(d^*) > a$ when the DM suddenly decides that she is sure enough of getting near θ to gamble and choose the interior maximum.

I hope these simple examples have convinced you that Bayesian optimal decision making accommodates a very rich collection of ways to respond to information (see also Harrison and Smith (1979) and Smith *et al.* (1981)). The precious property this decision making has is that it is *justifiable* as being perfectly rational. As illustrated above the reasons for the decisions being made are completely embedded in the solution and furthermore these reasons usually sound compelling. For further discussions of the application of such qualitative insights used in more complex scenarios see Dodd *et al.* (2006); Smith (1980b, 1981); Smith *et al.* (2008b).

3.8.1 The choice of utility functions and stable decisions

Much of the early work on decision theory was highly theoretical. Within this framework it was common to address utility functions that were unbounded on the reward space and usually convex. It was noticed that in particular by assuming convexity the mathematics became much simpler. For example it could often be shown that Bayes decisions were unique. We have seen above that one price of this simplifying assumption was to develop a decision theory no longer expressive enough to embody ideas of compromise and set this against commitments to certain alternatives: a behaviour wise decision makers appeared to exhibit.

There are two fundamental theoretical issues with the use of convex loss functions on random variables with unbounded support which should disincline the DM from their use. The first is a simple one. These utility functions imply that the DM is much more concerned not to make terrible decisions than bad ones. This theoretical assumption is simply an implausible one to apply to most scenarios.

However it was later appreciated that convex utilities gave rise to insurmountable theoretical difficulties if used for problems with unbounded reward distributions. This is because the optimal decisions are then totally dependent on the DM's specification of the tails of her subjective distribution: something we argue below cannot be elicited with accuracy. The type of problem is illustrated below.

Example 3.22. Suppose that $L(d,\theta) = (d - \theta)^2$ and U is the identity function so that the DM is an EMV. We showed above that if her subjective density is $p(\theta)$ then her decision is then the posterior mean μ. Now suppose that the prior $q(\theta)$ has been elicited and approximates so that the DM's genuine prior density $p(\theta)$ is unknown but where

$$d_V(p,q) = 1/2 \int |p(\theta) - q(\theta)|\, d\theta = \sup_{A \subset R} |P(A) - Q(A)| \leq \varepsilon$$

where $\varepsilon > 0$ is small. In this sense we know the elicited prior probability of any event A has been elicited to an accuracy of ε. Experience has shown that this level of accuracy is the most we could hope for in any direct elicitation: see later in this chapter. It is easily checked that all the densities

$$p(\theta|\lambda) = (1 - \varepsilon)\, q(\theta) + \varepsilon h(\theta|\lambda)$$

where $h(\theta|\lambda)$ is any probability density with mean $\mu_\varepsilon(\lambda) = \mu + \varepsilon^{-1}(\lambda - \mu)$ is such that $d_V(p,q) \leq \varepsilon$ and the mean of $p(\theta|\lambda)$ is λ. It follows that however small ε is there is a density p in the neighbourhood of q whose mean is any value λ. It is also easy to check using this construction that the expected utility difference between what the DM should get were she to use her true density is arbitrarily large. So if the analyst admits that her elicitation cannot be perfect he cannot usefully advise the DM what she should do.

Similar problems exist for other domains of application with unbounded reward distributions and are particularly acute in problems like the one above when the loss function is convex. For an excellent and extensive discussion of this see Kadane and Chuang (1978). One might hope that if data is incorporated into the prior then these problems will disappear. But although this can sometimes help – see the discussion of robustness in Chapter 8 – if the tails of the sampling distribution are at all uncertain or are not exponential, even with appropriate choices of prior family these problems will persist. And we argue below that the specification of small probabilities such as those appearing in such sample distributions are exactly those that tend to be unreliable to elicit. So if a Bayesian insists on using a convex utility on variables the supports of whose densities are unbounded then it is usually essential for her to input extra information about tail events: a feature which is almost impossible to elicit accurately directly.

However the problems illustrated above are largely avoided when a DM has a bounded utility function, despite these being able to exhibit a much wider range of geometrical features. To see this suppose the DM's utility function $U(d,\boldsymbol{\theta})$ is bounded. Without loss rescale it to take values between 0 and 1. Now suppose we approximate a DM's genuine prior density $p(\boldsymbol{\theta})$ by $q(\boldsymbol{\theta})$ and let $\overline{U}_p(d)$ and $\overline{U}_q(d)$ denote the DM's expected utility with respect to her genuine and approximating density respectively. Then

$$\begin{aligned}
\sup_{d\in D} \left|\overline{U}_p(d) - \overline{U}_q(d)\right| &= \sup_{d\in D} \left| \int U(d,\theta)\,(p(\boldsymbol{\theta}) - q(\boldsymbol{\theta}))\,d\boldsymbol{\theta} \right| \\
&\leq \sup_{d\in D} \left| \int |U(d,\theta)|\,|p(\boldsymbol{\theta}) - q(\boldsymbol{\theta})|\,d\boldsymbol{\theta} \right| \\
&\leq \sup_{d\in D} \int |p(\boldsymbol{\theta}) - q(\boldsymbol{\theta})|\,d\boldsymbol{\theta} \triangleq d_V(p,q).
\end{aligned}$$

Let d_p^* and d_q^* denote respectively the DM's Bayes decision with respect to the elicited prior q and the genuine prior p. Then, provided the prior is elicited accurately so that $d_V(p,q) \leq \varepsilon$, where ε is small, it follows that

$$\begin{aligned}
\overline{U}_p(d_q^*) &\geq \overline{U}_q(d_q^*) + (\overline{U}_p(d_q^*) - \overline{U}_q(d_q^*)) \\
&\geq \overline{U}_q(d_p^*) - \varepsilon \\
&\geq \overline{U}_p(d_p^*) + (\overline{U}_q(d_p^*) - \overline{U}_p(d_p^*)) - \varepsilon \\
&\geq \overline{U}_p(d_p^*) - 2\varepsilon.
\end{aligned}$$

Therefore in this sense, when U is bounded, the expected utility score associated with the approximate elicited density is almost as good as would be obtained if elicitation had been exact. Of course it might still be the case that the decisions d_p^* and d_q^* are quite different from each other. This phenomenon is illustrated in the last example of the last section. But this will only be because, on the DM's genuine expected utility scale, these two decision scores are almost the same and she is therefore almost indifferent between them. From both a mathematical and a practical standpoint this is the best we can realistically hope for from an approximation. In the next chapter we show that this property's closeness in variation distance between two posterior distributions will nearly always be satisfied whatever prior is chosen provided that certain regularity conditions are met and assessments are based on very large samples with known sampling distributions. It follows from the above that decisions made under these conditions are also stable in the sense described above.

There is a large literature on weaker measures of closeness than variation distances which lead to closeness in decision in the bounded case above: see French and Rios Insua (2000) for an excellent survey of some of this work. It is also possible to define measures of closeness of one decision from another which do not demand that the expected utilities are close everywhere but only in the region of an optimal decision: see Kadane and Chuang (1978). But even under this weaker requirement it can be shown that decisions made using convex utility functions are often unstable to infinitessimal prior mis-specification.

3.9 Summary

At least from a formal point of view, we have demonstrated in this chapter that with a few important exceptions the DM should be encouraged to choose a decision that maximises her subjective expected utility. This not only helps her to choose wisely but also gives a framework through which she can explain why she plans to act in the chosen way. The paradigm is very rich and, provided non-convexity of utilities is assumed, allows the DM's optimal decision rules to respond to the subtle combinations of the sometimes conflicting features of the problem she faces.

However there are several practical problems with actually implementing this methodology.

(1) The actual structure of a problem is often complex and needs to be elicited. We have discussed one framework – the tree – that can be used for eliciting and analysing many-faceted problems. However this framework can become very cumbersome for large problems. Alternative methods will be explored later in this book, especially in Chapter 7.

(2) When the space of attributes is more than one-dimensional direct implementation of the types of method described here, whilst formally defensible, are difficult to implement in practice without introducing biases or overwhelming the DM with choices she may well have difficulty making. More structuring is therefore usually required before the idea presented in this chapter can be effectively applied in high-dimensional scenarios. Fortunately there is now a very large literature

on how to elicit a utility function for larger and more complex decision problems. This last topic has such practical importance I will dedicate one whole chapter to it.

(3) Although we have now discussed a DM's utilities we have assumed that her probabilities have been given. Of course this is not the case in practice. These values need to be elicited from her. For such an elicitation to be successful we first have to understand exactly what we plan to mean by a subjective probability. We then need to learn how to elicit these probabilities in a way that minimises potential biases. These two issues will be addressed in the next chapter. Then in Chapter 5 we will proceed to discuss how sampling information can be drawn into such probability specifications to make them more useful and more defensible.

3.10 Exercises

3.1 Prove the strong substitution principle: the two reference utilities can be *any* attributes $(\boldsymbol{r}^0, \boldsymbol{r}^*)$ with the property that $\boldsymbol{r}^0 \prec \boldsymbol{r}^*$.

3.2 Prove that an EMV DM has a linear utility function.

3.3 Your DM's utility function on a payoff $r \geq 0$ is given by $U(r) = 1 - \exp(-\lambda r)$. She must decide between two decisions d_1 and d_2. Decision d_1 gives a reward of £1 with probability $\frac{3}{4}$ and £0 with probability $\frac{1}{4}$. Decision d_2 gives a reward £r with probability $2^{-(r+1)}$ where $r = 0, 1, 2, 3, \ldots$. Prove that the DM will find d_1 at least as preferable to d_2 if and only if $\lambda \geq \log 3 - \log 2$. Explain this result in terms of risk aversion.

3.4 Your company Bigart insures paintings and has been invited to take over the insurance of a competing company Disart. Disart has recently gone into receivership, having previously quoted a yearly premium r_y, paid in full at the beginning of the insurance period, for a piece of artwork A. The insurance promises to pay an amount r^* the first time the artwork A is stolen in that year and your company has assessed the probability of this event happening to be p. Assume that insurance will need to be renegotiated if a claim is made in any year, and assume zero inflation and that your company's utility function takes the form as in Exercise 3.3 where $\lambda > 0$. Prove that your company should take over those insurance policies on the article A if

$$x_y > \lambda^{-1}[\log\{(1 - p) + p\exp(\lambda x^*)\}].$$

3.5 Three investors I_1, I_2, I_3 each wish to invest \$10,000 in share holdings. Four shares S_i, $i = 1, 2, 3, 4$ are available to each investor. S_1 will guarantee 8% interest over the next year. Share S_2 will give no return with probability 0.1, 8% interest with probability 0.5 and 16% with probability 0.4. S_3 pays 5% with probability 0.2 and 12% interest with probability 0.8. Share S_4 pays nothing with probability 0.2 and 16% with probability 0.8. An investor can also buy a portfolio P of shares which invests \$5,000 in both S_1 and S_4.

Investor I_1 has the preferences $P \prec S_1$ and $P \prec S_3$, investor I_2 states preferences $S_4 \prec S_3 \prec S_1$ and investor I_3 has the preferences $S_4 \prec S_1 \prec S_2$

Find the payoff distribution associated with the five investment opportunities above. Assuming each investor's utility is an increasing function of her payoff, identify which of these investors is expected utility maximising. For any who are not, demonstrate which axiom she breaks and for any who is utility maximising, write down a utility function consistent with these choices.

3.6 Perform the re-analysis of the dispatch tree with the new utility function described in (3.2).

3.7 A DM's reward distribution is known up to the specification of a binary random variable θ and she uses an EMV strategy. Her rewards $r(d, \theta)$ distribution for four possible decisions $\{d_1, d_2, d_3, d_4\}$ and the two possible values $\{0, 1\}$ of θ are given in the table below.

θ	d_1	d_2	d_3	d_4
0	0	3	4	10
1	10	2	4	0

Using a normal form analysis or otherwise, without having elicited her probability distribution on θ, identify which decisions could be optimal for the client if she uses an EMV strategy, and determine the values of $P(\theta = 1)$ when each decision is optimal. You have not yet elicited her utility function but know that this utility function is strictly increasing in monetary reward; which additional decision might she choose if she was using a CME strategy? Find a utility function and a $P(\theta = 1)$ when this additional decision is optimal.

3.8* Let $D_x U(x), D_x^2 U(x)$ represent respectively the first and second derivative of $U(x)$ with respect to x and assume that $U(x)$ is strictly increasing in x and for all x. Suppose the DM also tells you that if she is offered a choice to add to her current fortune an amount x for certain and she finds this equivalent to a gamble which gives a reward $x + h$ with probability $\alpha(h)$ and $x - h$ with probability $1 - \alpha$, $h \neq 0$, $\alpha(h)$ might depend on h but does not depend on x. Let $\lambda(h)$ be given by

$$\lambda(h) = \frac{1 - 2\alpha(h)}{h\alpha(h)}$$

Prove that

$$\lambda(h)\left[\frac{U(x) - U(x-h)}{h}\right] = \left[\frac{U(x+h) - 2U(x) + U(x-h)}{h^2}\right].$$

Hence or otherwise prove that her utility function $U(x)$ must either be linear or satisfy

$$\lambda D_x U(x) = D_x^2 U(x)$$

where $\lambda = \lim_{h \to 0} \lambda(h)$. Using that for any continuous differentiable function $f(x), D_x \log f(x) = \frac{D_x f(x)}{f(x)}$ or otherwise, prove that $U(x)$ must either be linear or take the form

$$U(x) = A + B\exp(\lambda x), \qquad \lambda \neq 0.$$

Hence prove the characterisation result given in Section 3.2.

3.9 A DM's payoff $R_b(d,y)$ where $b > 0$ is given by (3.7). She believes that Y has a Gamma distribution whose density $p(y)$, $y \geq 0$ is given by

$$p(y) = \frac{\beta^\alpha}{\Gamma(\alpha)} y^{\alpha-1} \exp\{-\beta y\}$$

where $\alpha > 1$ and $\beta > 0$, which has mode $m = \frac{\alpha-1}{\beta}$. Find her Bayes decision d^* under the payoff function $R_b(d,y)$ as an explicit function of m and b.

3.10 Prove that all interior local maxima of the expected payoff (3.10) given in Example 3.20 must satisfy

$$(1-a)\,p(d^* + b) = p(d^* - b)$$

when $p(\theta)$ is the DM's posterior density which is continuous on the real line. Show that if the DM is risk averse then the local maxima of the DM's expected utility satisfy the equation above but with the parameter a substituted by the parameter a'. Write down a' as a function of the DM's utility function U and a.

3.11 Show that if a payoff function $R(d,\theta)$ on a one-dimensional parameter θ is bounded between 0 and 1 and is decreasing in $|d-\theta|$ then its expected payoff function $\overline{R}(d)$ can be expressed as

$$\overline{R}(d) = \mathbb{E}\left(\overline{R}_B(d)\right)$$

where $\overline{R}_b(d)$ is the expected payoff associated with the step payoff $R_b(d,\theta)$ given in (3.7). Hence show that if the density of θ is symmetric and unimodal with mode at 0 then any EMV decision under such a loss function will be 0. Further show that this is still the case if the DM has any utility function strictly increasing in payoff.

3.12 An EMV DM has a reward function (3.10) and $p(\theta|\mathbf{y})$ is the Student t density given in (5.17). Prove that all the interior maxima of the associated expected payoff must satisfy a quadratic equation. Hence carefully explain how the DM's Bayes decision reflects the different values of the hyperparameters.

3.13 A DM has a linear utility and a *ramp payoff* $R(d,\theta)$ function on a one-dimensional parameter θ.

$$R(d,\theta) = \begin{cases} 1 & |\theta - d| \leq b \\ (c-b)^{-1}\left(c - |\theta - d|\right) & b < |\theta - d| \leq c \\ 0 & |\theta - d| > c. \end{cases}$$

Show that any EMV decision d^* under this payoff function with respect to the continuous posterior distribution function $F(\theta)$ must satisfy

$$F(d^* + c) - F(d^* + b) = F(d^* - b) - F(d^* - c).$$

Describe this condition in terms of areas under the density of θ. Prove that if the density of θ is strictly increasing to its unique mode m and then strictly decreasing then the equation above has a unique solution. Prove in this case that the EMV decision as $c \to \infty$ tends to the posterior median of θ.

4
Subjective probability and its elicitation

4.1 Defining subjective probabilities

4.1.1 Introduction

So far we have taken the concept of a subjective probability as a given. But what *exactly* should someone mean by a quoted probability? There are three criteria that such a definition needs satisfy if we are not to subverting the term "probability" for another use:

(1) In circumstances when a probability value can be taken as "understood" by a typical rational and educated person our definition must correspond to this value.
(2) The definition of subjective "probability" on collections of events should satisfy the familiar rules of probability, at least for finite collections of events.
(3) The magnitude of a person's subjective probability of an event in a decision problem must genuinely link to her strength of belief that the event might occur. For consistency with the development given so far in this book it would be convenient if this strength of belief were measured in terms of the betting preferences of the owner of the probability judgement.

To satisfy the first point above recall that there are various scenarios where the assignment of probabilities to events are uncontentious to most rational people in this society. It is therefore reasonable to assume that the DM's subjective probabilities agree with such commonly held probabilities. For example most people would be happy to assign a probability of one half to the toss of a fair coin resulting in a head. Two slightly more general standard probabilistic scenarios where common agreement exists are as follows.

Example 4.1 (Balls in a bag). The DM is told that a well-mixed bag contains exactly r white balls and $n - r$ black balls. The event $E_r(n)$ in question is that a random draw from this bag would result in a white ball being chosen. An educated rational DM can be expected to assign a probability $\frac{r}{n}$ to $E_r(n)$.

Example 4.2 (Betting wheel). A wheel of unit circumference – the circumference indexed by $a \in (0, 1)$ – has a half-open interval – or arc – $E_p(a) = (a, a + p_E]$, $0 < a \leq 1, 0 \leq p_E \leq 1$, around its circumference of length p_E which is marked white, whilst the rest of the circumference is marked black. The centre of the wheel is attached to a frame with a point x marked next to its circumference. The wheel is freely spun about its centre from a random starting position. The event $E_p(a)$ is said to have occurred if and only if a point in $E_p(a)$ lies next to the marker point x. Most educated DMs would be happy to assign a probability p_E to $E_p(a)$ – the proportion of the circumference marked white. Note that this assignment is the same regardless of the value of a: i.e. wherever the arc of length p_E is positioned around the wheel.

Note that an auditor is likely to be very sceptical of a rationale for action that explicitly or implicitly contradicts these assignments of probability to these events. So usually it will be necessary for any person involved in a decision making process to ensure her statements are consistent with these. This is a helpful starting point because it not only enables us to define certain consensus probabilities but also provides us with a benchmark with which to *measure* someone's subjective probabilities about other events where no such consensus exists. Analogously to the development of preference elicitation given in Chapter 3, the simplest way of making this comparison is to invent a hypothetical market place.

In this market an agent of the person providing the probabilities will trade lotteries with other agents no more informed than her concerning various events of interest together with lotteries on certain standardising events like those illustrated above. In his seminal book, Raiffa (1968) used balls in a bag to standardise probabilities. Here we will use the betting wheel as a reference scale with which to measure a person's probability.

Thus suppose that to proceed with a decision analysis an analyst needs to elicit the DM's subjective probability of the event A that a particular batch of chemicals will turn out to be contaminated. The DM is asked to compare her preferences between a gamble winning a prize if and only if A occurs and a set of gambles winning the same prize if $E_p(x)$ occurs for selected values of p and a fixed value of x, $0 \leq x \leq 1$, in the betting wheel gamble described above, which is of course independent of the elicited event.

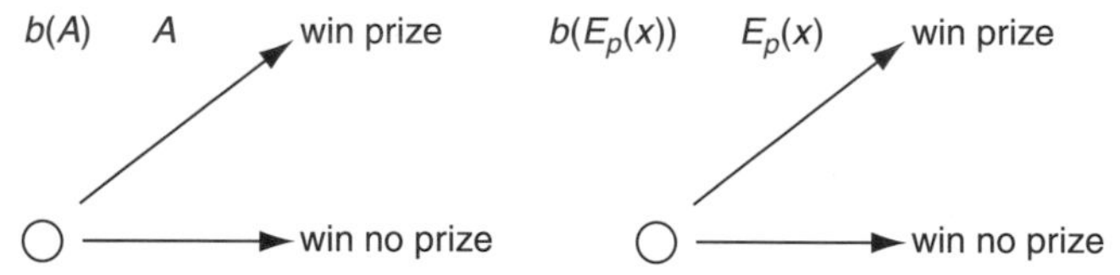

Here, to conduct this mind experiment, the DM needs to believe the lottery ticket is worth winning. For technical reasons discussed later for an accurate elicitation the prize from the lottery should ideally not impact on the attributes of the DM's utility function on the event in question: see below.

Starting with a fixed value of p such as $p = \frac{1}{2}$ the DM is then asked whether she prefers the first or second gamble. If she prefer the first then p is increased until $b(A) \sim b(E_p(x))$. If she prefers the second then p is reduced until $b(A) \sim b(E_p(x))$.

4.1.2 Coherence of subjective probability

Note that from the rules of rationality in Chapter 3 for all x, x', $0 < x, x' \leq 1$,

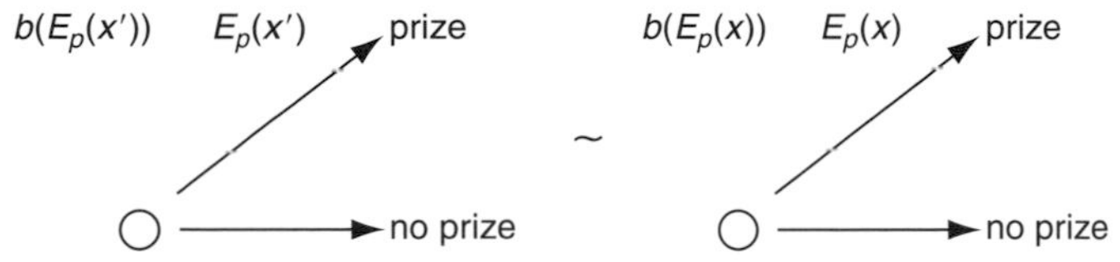

If the arc length around the circumference $p = p_1 + p_2$ for $0 \leq p \leq 1$ and $0 \leq p_1, p_2$ the half-open arc

$$(x_1, x_1 + p] = (x_1, x_1 + p_1] \cup (x_2, x_2 + p_2]$$

where $x_2 = x_1 + p_1$. It follows that when an agent simultaneously holds the two lotteries $b(E_{p_1}(x_1))$ and $b(E_{p_2}(x_2))$ then this is *logically* equivalent to her holding a lottery $b(E_p(x_1))$ where p, p_1, p_2, x_1, x_2 are defined above. So in particular for these calibrating subjective probabilities

$$P(E_p(x_1)) = P(E_{p_1}(x_1)) + P(E_{p_2}(x_2)) = P(E_{p_1}(x_1)) + P(E_{p_2}(x_1)). \tag{4.1}$$

Axiom 4.3 (Comparability). For every event A considered by the DM there is a *unique* p such that $b(A) \sim b(E_p)$.

In my opinion this axiom that a subjective probability can be defined and is unique is the most critical and disputable axioms of Bayesian inference. Suppose a DM is *actually* making gambles in a real market. In Domain 1 gambles concern events which no competing agent has more information about than she. In Domain 2 she is ill-informed compared to many competing agents. In Smith (1988a) I argued that she could reasonably argue that a gamble in Domain 2 to which she assigns a probability p was more risky to her, and so strictly less preferable to hold than a gamble in Domain 1 with the same probability p of that event. In such a scenario the Bayesian paradigm may then need to be generalised. This can be done but leads to a somewhat more complex and less developed inferential methodology than the Bayesian one: based on belief functions (Shafer, 1976; Shafer and Vovk, 2001) or upper and lower probability (Fine, 1973; Walley, 1991). These interesting and important generalisations of the methodology expounded here are unfortunately beyond the scope of this short book.

However many statisticians – see e.g. Bernardo and Smith (1996) and O'Hagan and Forster (2004) – have argued that a DM should follow the axiom above even when a person is seriously uninformed. There are strong if not compelling arguments for this view. Moreover in practice at least for the types of decision problems discussed in this book, it is often the case that a person is content to obey the comparability axiom. In the example given here the probability is elicited from an informed DM or expert or an expert in a non-competitive environment and these will be scenarios when the axiom of comparability is most compelling.

When this axiom holds then the event A will be assigned a subjective probability which is a single number between 0 and 1. It is then easy to demonstrate that subjective probabilities elicited in this way need to obey the familiar rules of probability for an agent to have well-defined preferences that will enable her to trade in lotteries. Thus suppose the person believes that the events A_1 and A_2 could not happen together – i.e. that the events are disjoint. Then if an agent holds the two lotteries $b(A_1)$ and $b(A_2)$ simultaneously this is logically equivalent to holding the single lottery $b(A_1 \cup A_2)$ giving the prize if and only if the event $A_1 \cup A_2$ that one of A_1 or A_2 happens. But by definition

$$b(A_1) \sim b(E_{p_1}(x_1)), \quad b(A_2) \sim b(E_{p_2}(x_1)), \quad b(A_1 \cup A_2) \sim b(E_{p_1+p_2}(x_1))$$

so again by definition

$$P(A_1) = P(E_{p_1}(x_1)), \quad P(A_2) = P(E_{p_2}(x_1)), \quad P(A_1 \cup A_2) = P(E_{p_1+p_2}(x_1)).$$

It follows that

$$P(A_1 \cup A_2) = P(A_1) + P(A_2). \tag{4.2}$$

If the DM assigns probabilities such that disjoint events A_1 and A_2 are such that $P(A_1 \cup A_2) \neq P(A_1) + P(A_2)$, then if the agent is allowed to hold more than one lottery at a given time her trading preferences will depend on how logically equivalent combinations of lotteries are communicated. This dependence is a property that the DM should try to avoid because her probabilities on certain events would then not be consistently defined. Furthermore, some additional regularity conditions would force the agent, when facing certain sequences of trades in the hypothetical market, to lose for sure (see e.g. Bernardo and Smith (1996) for a construction of such a betting scheme). So if the results of the hypothetical experiment are to make any sense then the elicited subjective probability must satisfy the property that subjective probabilities of disjoint event add in the sense of (4.2).

Note that the definition above ensures that $P(A) \geq 0$ for any elicited probability $P(A)$ of an event A. Furthermore if A is certain to occur the only rational assignment to this event is $P(A) = 1$ since the prize is won for certain on the betting wheel only if the whole of its circumference is coloured white.

Definition 4.4. Call a DM's probability assignments *coherent* if her probability assignments extend to provide a finitely additive probability measure P over the field $\mathcal{F}$ generated by all finite unions, intersections of complements of its elicited events: i.e. for all such events $A \in \mathcal{F}$, $P(A) \geq 0$, $P(\Omega) = 1$, where Ω is the exhaustive event = "something occurs" and for all disjoint $A_1, A_2 \in \mathcal{F}$,

$$P(A_1 \cup A_2) = P(A_1) + P(A_2).$$

Henceforth it is assumed that any DM has coherent probabilities over the events in the space salient to her problem. Of course this does not mean that *elicited* probabilities necessarily exactly satisfy this rule of probability even if the underlying beliefs of the DM do. There may well be measurement biases or small rounding errors which have distorted the actual quoted probabilities so that they are not coherent even when the DM wants to obey the comparability axiom. We will discuss some of these phenomena below. But when incoherences do occur, for the reasons given above, we henceforth assume that it is possible to persuade the DM to adjust the elicited probabilities so that they satisfy the extension above.

Conditional probability can also be defined within this lottery framework through a construction called a called off bet. Thus suppose we are interested in eliciting the conditional probability of A_2 given that A_1 occurs. The obvious interpretation of this conditional gamble is a lottery that is enacted if event A occurs and delivers the prize if B is subsequently discovered to have happened. Thus we compare lotteries of the form

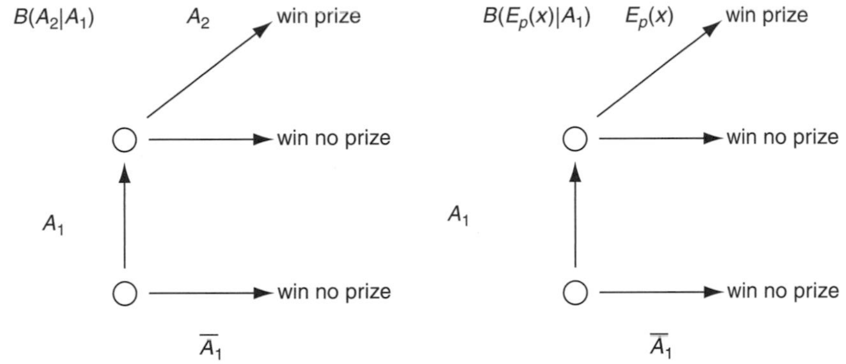

and find the value p_{A_2} such that $B(A_2|A_1) \sim B(E_{p_{A_2}}(x)|A_1)$ as before. It is set as an exercise below to check that by defining subjective conditional probabilities in this way ensures that the coherent DM will assign her conditional probabilities over finite sets of events so that the law of total probability and Bayes rule is satisfied by these assignments (see also Bernardo and Smith (1996); Lad (1996)).

It follows that by defining subjective probability in this way, with the caveats stated above, probabilities on a finite set of events will continue to satisfy all the familiar rules of probability. In particular the coherent DM will satisfy the second point given at the beginning of this chapter and in this sense the term "probability" is not perverted. This does not mean that the elicitation task is easy. Two difficulties need to be addressed. The first is a theoretical one of ensuring that the elicitation process is not confounded with the decision analysis itself. The second is the practical psychological one of finding ways of eliciting a probability which as far as possible avoid corrupting the transmission of a subjective probability. We begin with the theoretical issue.

4.2 On formal definitions of subjective probabilities

*4.2.1 The no-stake condition and elicitation bias**

There is an interesting technical point which demonstrates that probability elicitation must be performed with care if it is going to be faithful to a DM's beliefs. It also suggests that expert probabilities elicited remotely should be treated with caution. Thus suppose a person is expected utility maximising with a nonlinear utility U. However also suppose that the attributes of U have a different distribution depending on whether an event A occurs than if the event does not – event $\overline{A}$. Then it has been known at least since Ramsey (1931) that the nonlinearity of the utility and the dependence of the attribute distribution on whether or not A occurs can introduce a bias in the measurement of a subjective probability in a sense illustrated below. This issue is addressed in Kadane and Winkler (1988). I include a summary of their points below and a new discussion about how this can be addressed using a utility function with more than one attribute.

Thus suppose event A is the event that the DM wins a contract and she herself is the expert from whom the subjective probability p of A is elicited. Suppose she has a utility function with attribute vector $\boldsymbol{r} = (r_1, r_2, \ldots, r_n)$. Denote her density (or mass function) of these attributes given A occurs by $\pi(\boldsymbol{r}|A)$ and if A does not occur by $\pi(\boldsymbol{r}|\overline{A})$. Let the prize in the elicitation lottery give an additional reward s_n on the last attribute but leave the other attributes unchanged. Write $\boldsymbol{r}^+ = (r_1, r_2, \ldots, r_n + s_n)$ to represent this revised reward.

The expected utility $\overline{U}(A)$ associated with the lottery $b(A)$ and the expected utility $\overline{U}(q)$ associated with the lottery $b(E_q(x))$ for some $0 < x \leq 1$ are then respectively given by

$$\overline{U}(A) = p \int U(\boldsymbol{r}^+)\pi(\boldsymbol{r}|A)d\boldsymbol{r} + (1-p)\int U(\boldsymbol{r})\pi(\boldsymbol{r}|\overline{A})d\boldsymbol{r},$$

$$\overline{U}(q) = q\left\{p\int U(\boldsymbol{r}^+)\pi(\boldsymbol{r}|A)d\boldsymbol{r} + (1-p)\int U(\boldsymbol{r}^+)\pi(\boldsymbol{r}|\overline{A})d\boldsymbol{r}\right\}$$
$$+ (1-q)\left\{p\int U(\boldsymbol{r})\pi(\boldsymbol{r}|A)d\boldsymbol{r} + (1-p)\int U(\boldsymbol{r})\pi(\boldsymbol{r}|\overline{A})d\boldsymbol{r}\right\}.$$

The elicited probability quoted by the DM is the value q such that the lottery with the prize depending on A occurring is equally preferable to the one where the same prize is obtained if $E_q(x)$ occurs. From Chapter 3, since the DM is rational she will be indifferent between these two lotteries when $\overline{U}(A) = \overline{U}(q)$.

After a little rearrangement this gives us an elicited probability q satisfying the log odds identity

$$\frac{q}{(1-q)} = \lambda \frac{p}{(1-p)} \tag{4.3}$$

where

$$\lambda = \frac{\int \left\{U(\boldsymbol{r}^{+}) - U(\boldsymbol{r})\right\} \pi(\boldsymbol{r}|A)d\boldsymbol{r}}{\int \left\{U(\boldsymbol{r}^{+}) - U(\boldsymbol{r})\right\} \pi(\boldsymbol{r}|\overline{A})d\boldsymbol{r}}.$$

We note that the elicited probability $q = p$ – the DM actual probability – if and only if $\lambda = 1$. Now clearly if $\pi(\boldsymbol{r}|A) = \pi(\boldsymbol{r}|\overline{A})$ – a property called the *no-stake condition* – holds, i.e. the person believes that whether or not A occurs will not affect the value of her attributes, then automatically $\lambda = 1$. However in a decision analysis we might expect a DM to believe that A will have a bearing on her success as measured by her utility function. Why else would she be interested in A? And even when the probability is elicited from an expert, that expert might have a stake in whether or not the event happens. For example, if the DM's policy depends on climate change, an expert on this topic whose probability the DM might want to adopt as her own may well have funding dependent on whether or not a certain predicted change in climate takes place. Again the no-stake condition will be violated and the elicited probability will be prone to a bias if that person is rational. Furthermore examination of the formula above will demonstrate that such biases can be significant even when the stakes of the lottery used in the elicitation are small.

To ensure that $\lambda = 1$ in all such scenarios when the no-stake condition is violated requires that $U(\boldsymbol{r}^{+}) - U(\boldsymbol{r})$ is not a function of $\boldsymbol{r}$ but only s_n. This is clearly true if U is linear and has one attribute – so that the person is an EMV decision maker. However we have argued above that this is a very particular scenario, and will often not hold true even approximately. Fortunately when the form of the utility function is known to the analyst, for example when that person is also the DM, it is usually possible to construct a lottery so that this condition is met by her utility function U. Thus suppose that when elicitation rewards are ignored the DM's utility function has $n-1$ attributes $\boldsymbol{r}^{-} = (r_1, r_2, \ldots, r_{n-1})$. Concatenate to this vector an attribute r_n which is independent of the other attributes and whether or not A occurs so

$$\pi(r_n|\boldsymbol{r}^{-}, A) = \pi(r_n|\boldsymbol{r}^{-}, \overline{A}) \triangleq \pi(r_n) \tag{4.4}$$

and has the property that

$$U(\boldsymbol{r}^{+}) - U(\boldsymbol{r}) \triangleq u(s_n, r_n) \tag{4.5}$$

is a function of s_n and r_n only. Note in particular that if

$$U(\boldsymbol{r}) = (1 - k_n)U^{-}(\boldsymbol{r}^{-}) + k_n U_n(r_n)$$

where all functions are functions of their arguments only – for example when the attribute r_n is value independent of the other attributes with criterion weight k_n – see Chapter 6 –

condition (4.5) will hold since then

$$U(\boldsymbol{r}^+) - U(\boldsymbol{r}) = k_n \left(U_n(r_n + s_n) - U_n(r_n)\right).$$

When this is the case $U^-(\boldsymbol{r}^-)$ functions as the DM's utility function if elicitation is not enforced. This is because by definition and (4.4) the expectation $\overline{U}_n$ of U_n is constant over all distributions arising from all decision the DM might take. So choosing a decision from this class to maximise $\overline{U}(\boldsymbol{r})$ is the same as choosing one to maximise $\overline{U}^-$.

The challenge for the analyst is now to construct an elicitation lottery whose prize is a function of the concatenated attribute r_n *only*. This will then ensure that both (4.4) and (4.5) hold. Obviously the choice of an appropriate attribute with this property depends on the context of the decision analysis. But a typical attribute to concatenate here might be a prize of a lottery ticket which – if it is winning – gives the DM money to donate to a charity of her choice. Then

$$\lambda = \frac{\displaystyle\int u(s_n, r_n)\pi(r_n)dr_n \int \pi(\boldsymbol{r}^-|A)d\boldsymbol{r}^-}{\displaystyle\int u(s_n, r_n)\pi(r_n)dr_n \int \pi(\boldsymbol{r}^-|\overline{A})d\boldsymbol{r}^-} = 1.$$

The moral here is that it is usually formally possible to elicit a DM's subjective probability unambiguously using the method described above but the prize in the elicitation lottery needs to be chosen with care for the elicitation not to be liable to a systematic bias.

However the liability to bias is more critical when the analyst has no access to the subject's utility function or the design of the lottery. This is the usual scenario when probabilities from a remote expert are adopted by the DM as her own. Distortions can be especially acute if these ideas are used for modelling behaviour of markets or in game theory scenarios where the owners of subjective probabilities are typically inaccessible and where sometimes their probabilities can only be deduced from their behaviour. For a good discussion of these issues see Kadane and Winkler (1988).

Even in these scenarios the systematic form of the bias – as described by (4.3) can sometimes allow recalibration through the estimation of the bias term λ, see below.

So when probabilities are elicited in the way described above by definition these will agree on probabilities of events associated with betting wheel – a property required by the first point in the introduction of this chapter. Provided that the elicitation is performed with care in most scenarios we can expect the elicited probabilities to satisfy some familiar rules associated with probabilities assigned to disjoint events. The nature of the elicitation also ensures that in a genuine sense, these probabilities are increasing in the expert's certainty that the event will actually occur.

There are several other ways to define and measure subjective probabilities. One is to use a scoring rule: a technique described later in this chapter. Another is to use the construction of promissory notes (Bernardo and Smith, 1996; De Finetti, 1980; Goldstein and Wooff, 2007). Both these methods have advantages and disadvantages over the method described above. For more detailed comparisons of these different methods see Bernardo and Smith (1996).

4.2.2 Eliciting continuous densities

Obviously eliciting a prior over a continuous distribution is technically far harder than eliciting the probabilities of a finite set of single events. In particular without some sort of continuity assumptions it will be impossible to obtain accurately with respect to variation distance. However, provided the DM is prepared to state that her prior density is smooth in a sense described later in the book, close approximations can be obtained by eliciting a moderate number of probabilties of events in the space and then extending these to the whole space.

The simplest way of eliciting a prior is to choose one from within conjugate families, see the next chapter – or to choose one from a mixture of such densities as illustrated in the next chapter. The choice of certain conjugate families looks rather arbitrary. However these can often be characterised by qualitative invariance properties they exhibit and so are at least checkable: see Chapter 9. Also results are given later in the book to suggest that when the DM's prior is chosen from such a class provided the DM's genuine prior obeys certain properties, then inference after information from data is accommodated is robust to elicitation errors made using this method as demonstrated in Chapters 5 and 9. An extensive review of density elicitation is given in O'Hagan *et al.* (2006). An alternative is to use non-parametric priors like Dirichlet processes Ghosh and Ramamoorthi (2003) or Gaussian processes Rasmussen and Williams (2006). These are also characterised by certain properties they possess and allow somewhat more flexible learning. These are beyond the scope of this introductory text but are well documented in the references above and O'Hagan and Forster (2004).

4.2.3 Countable and finite additivity*

In the above it was noted that subjective probabilities on a field of event can be expected to satisfy the finite additivity axioms. When an event space is finite, this is sufficient for subjective probability to correspond to a probability space in the usual sense. However when the event space is infinite this is not so, for we need that if $\{E_n : n \geq 1\}$ is an *infinite* sequence of disjoint events then

$$P(\cup_{n\geq 1} E_n) = \sum_{n\geq 1} P(E_n). \tag{4.6}$$

There are many examples of finitely additive probability measures that do not satisfy (4.6): see Exercises 4.2, 4.3 and 4.4 below. For example a useful family of finitely additive but not countably additive probability measures defined on an exchangeable sequence of real random variables are ones satisfying the A_n property, see e.g. Hill (1988).

Finitely additive distributions can exhibit useful invariance properties especially on $\mathbb{R}^n$ or the space of positive definite matrices which can be argued are natural ones for the ignorant or uninformed DM to hold. And such priors with no apparent information in them have a superficial attraction. However they have strong downsides as well. The usual probability formulae like Bayes rule and the law of total probability need no longer hold: see Exercise 4.4. Inferences tend not to be robust in any normal sense – see for example

Wasserman (1996). Furthermore conditioning does not retain its natural interpretation. For example if we were to know the value of a random variable X taking values on the real line $\mathbb{R}$ then if we learn the value x of X then surely we will know for certain whether or not $E = \{\sin(X) > 0\}$ holds. However if we assign a finitely additive location invariant distribution to X then this tells us nothing about E because sin is not a measurable function under this measure.

For these and other technical reasons great care needs to be exercised when using these probability distributions. Of course when the DM's considered beliefs *genuinely* correspond to these assignments then if we are to follow the Bayesian paradigm then they must be employed. However since the occasions when these distributions are used tend to be ones where a high level of uninformedness is assumed it appears to me that there is a strong case for using an even more general framework of inference, for example one based on belief functions or upper and lower probabilities instead of working within the Bayesian paradigm but with finitely additive probabilities. Note that even from a psychological perspective it is when the DM is most uninformed about some part of her domain that probabilities elicited within a strictly Bayesian framework are most unreliable: see the next section.

For an excellent discussion and a rather different view on these matters see Kadane *et al.* (1986). The close link between certain finitely additive distributions and various classes of uninformative distributions where these are used for the currently fashionable "objective" priors is given in Heath and Sudderth (1989)

4.3 Improving the assessment of prior information

In my view the formal case for recommending that the DM follow a coherent approach and proceed as a Bayesian is very strong when good domain knowledge is available. But this does not mean it can be easily enacted. Over recent years considerable activity has been applied to develop techniques that can elicit probabilities as faithfully as possible. An excellent recent review containing many useful tips can be found in O'Hagan *et al.* (2006), see also Hora (2007) and Curley (2008). Despite its importance to a practising decision analyst, it is impossible to give a comprehensive overview of this important area in this small book. Indeed many of the recommended techniques are specific to the domain of application and so would be distracting in a general text. However I will devote some pages to some of the elicitation issues I myself have found to be critical in elicitation exercises I have been engaged in.

4.3.1 What is feasible?

Coherent analyses need a subjective probability distribution to be elicited. It is important to realise that in some circumstances elicited subjective probabilities cannot be reliably estimated. Some of the major barriers to good elicitation are given below.

- It is unreasonable to expect people who are *innumerate* to produce probabilities on a numerical scale that represent to any real extent their degrees of belief associated with propositions. Experts therefore need to have a level of mathematical training to engage in the process described above.

All scientists are candidates for potentially successful elicitation. However some powerful experts in some professions – for example the law – lack even the most rudimentary skills in numbers. In such domains and with such individuals the elicitation techniques described below tend to be futile.

- *Ignorance* distorts assessments of probabilities, for reasons including some already discussed. Unless the "expert" is informed, studies set in a variety of different scenarios have indicated a strong tendency to over-confidence in judgements and a spuriously high or low probability assigned to events. This is often a result of a lack of ability in imagining the range of different ways the future might unfold. The poorly informed person is also much more prone to the types of biasing effects discussed below.
- In my own experience, an informed numerate genuine expert gives well-elicited probability judgements on a single event and tends to quote probabilities *with errors* at best in the range 0.02 to 0.05 in probability depending on the type of event elicited. By this I mean that different but equivalently good elicitation of probabilities taken on different occasions from the same expert with no change in her underlying expertise still tend to vary within this range. We have seen that whilst these sorts of inherent measurement errors do not usually substantially distort the results of a decision analysis they nevertheless need to be borne in mind. The errors can be particularly influential for events that have an associated small probability but with big impacts on consequence: a scenario often referred to as *risk*.
- Direct estimation of events with very *small probabilities* are hazardous to elicit not only because of the effects of the last bullet but also because the expert may find it difficult to imagine *how* such an event could occur at all.
- People providing subjective probabilities *unaided* can be prone to make big errors. There is now well-documented evidence that in such circumstances they will commonly lean heavily on heuristics. Whilst informative and sometimes effective, these heuristics do not directly translate into probabilities relevant to a decision analysis and can seriously distort the transmission of beliefs: see below. In particular this means that if an analysis is dependent on probability judgements of inaccessible experts then, in turn, the subsequent analysis may well be seriously distorted by mis-specified inputs.
- Although still popular in some quarters it has been shown that any functional links between certain qualitative terms (almost impossible, most unlikely, quite likely, . . .) and probability values *strongly* depends on the person using these terms, the contexts and disciplines that are being drawn upon and the events themselves. Thus the elicitation of probabilities using such scales can be very unreliable.
- People tend to be less accurate at assigning probabilities to events whose truth value is known – e.g. assigning a probability to the event "The length of the Nile is greater than x km" – as opposed to assigning probabilities to events whose truth value is not known for certain by the questioner – e.g. "Will the DM win the future contract?", "Was the suspect at the house?" This is reassuring for a decision analyst whose events are usually in the second category. However many psychological experiments – see below – have the first property. Therefore results of such experiments must be read with a degree of scepticism. It is often helpful to train a DM so she can calibrate her probability forecasts to events about which she is uncertain but whose truth is known. However such training is not perfect because the calibrating events are in the first category.

4.3.2 Typical biases and ways to minimise these

People use a variety of heuristics to judge the probability of an event which help them answer the elicitation questions but also introduce different types of bias. These biases are strongest when the truth values are known by the questioner or the elicitation is not guided by a skilled analyst. Nevertheless significant elicitation errors can also be introduced when the elicited probabilities are of events regularly met in a typical decision analysis and the analysis is performed carefully. Some of these heuristics are outlined below.

Availability is the heuristic where a person tries to gauge her subjective probability by recalling instances of an event of the same type as the event of interest and set these against instances of its complement. This method of evaluating a probability can obviously introduce a bias unless the instances linked to the event in question in the person's mind appear as if at random. We argued in Chapter 1 that an expert witness who deals regularly with parents who abuse their children might well have an inflated probability of any given person being a child abuser if she uses this heuristic. This is a subjective analogue to the well-known "missingness at random" hypothesis Little and Rubin (2002) which if violated can seriously distort a statistical analysis. Obviously random sampling of the domain helps ameliorate this bias, but such sampling is not always a practical option.

A second heuristic which is commonly used is *anchoring*. Here the person starts with a particular value for her probability of the event and then adjusts it in the appropriate direction away from this point. Such an anchor is easily inadvertently introduced by the analyst. The definition of subjective probability as described earlier in this chapter has an anchor at probability $\frac{1}{2}$. The problem with such an anchor is the psychological one that in practice people tend not to adjust away from the anchor enough. Consequently the stated probability is closer to the anchor than it should be. To minimise this bias it is important to order elicitation questions well. However experiemts suggest that effects on elicited quantities – especially under-estimated elicited variances – can persist.

Support theory Koehler *et al.* (2003) describes and tries to explain a third problem: how different descriptions of the same event give rise to different probability statements. For example it has been noted that the more instances included as illustrations of the event whose probability is elicited the higher the quoted subjective probability of this event tends to be. This observed phenomenon could not be consistent with any reasonable definition of rationality as we define it above. If these discrepancies are transferred to the hypothetical market place we discussed in the construction earlier in the chapter the agent will be prone to getting ambiguous instructions about how to act and this can lead to potential incoherences. On the other hand from the practical point of view if this dependence of stated probabilities on the way an event is described is not guarded against, it can seriously corrupt elicited probabilities.

Elicitation of conditional probabilities can be distorted in other ways. We saw one common problem – the prosecutor fallacy – where if event A causes B and you ask for $P(A|B)$ you will often get $P(B|A)$ which can be quite different. It is therefore important as far as possible to elicit conditional probabilities consistently with the order in which they happen. More subtly the extent of the *representativeness* of B given A can replace $P(B|A)$. Because representativeness is associated with similarity rather than the likelihood of mutual occurrence such a heuristic can again distort the elicited conditional probability.

These are some of the many biases discovered by psychologists in their experimental work: for more details see for example Hora (2007); O'Hagan *et al.* (2006); Oaksford and Chater (1998), and Oaksford and Chater (2006). What these results underline is that heuristic and unguided probability forecasts can well be misleading.

4.3.3 General principles for the analyst to follow

Despite the very real pitfalls outlined above, within the domain defined by the bullets of the last section elicitation can be made to work well and robustly enough for most decision analyses I have encountered. Some general pointers that I have found helpful are given below.

(1) As far as possible the analyst should elicit probabilities herself *directly* through conversation with the expert or DM. This way the analyst controls the inputs and can be more aware of any potential biases that might be being introduced.
(2) *Training* in expressing degrees of belief over events as probabilities which can subsequently be observed and then fed back can be very helpful in enabling an expert to calibrate her probabilties to an appropriate scale. Proper scoring rules are a useful training tool in this regard: see below.
(3) In all but the most simple problems it is essential to first *elicit a qualitative framework* for the decision analysis. This framework needs to be based on a *verbal* as opposed to numerical description of the DM's problem. We have already seen one such framework: the decision tree. Other sorts of framework, more suited to larger but more homogeneous decision problems, will be introduced later in this book. It has been found that whilst quoted probability assessments are prone to biases, structure directly reflecting a verbal explanation is much more robust and is much easier to elicit faithfully and reliably. Probabilities in all but the most trivial settings should simply be embellishments of this type of structure.
(4) As far as possible probabilities should be elicited about *transparently important events* so that the DM appreciates why the questions are important and can apply her expertise to the answers in an appropriate way. Notice that once an appropriate framework has been elicited as in the previous point this principle usually follows directly.
(5) It is useful wherever possible to try to *draw away from a given instance* into a more general context. Asking for a probability of a logically equivalent but easier to compare statement of an event may help in this regard, see the first example of the next subsection. If, for example, a probability is elicited concerning an event that a machine will break down in the next 24 hours it is usually helpful to encourage the DM to recall past analogous scenarios and consider what happened then and why so as to inform this judgement. But note that from the third point in the last section such drawing away must be balanced: with instances recalled of the complement of the event as well as the event itself. This point links to the accommodation of data discussed in the next chapter.
(6) Try to elicit probabilities about events that are *currently unknown but will be resolvable in the future* – at least in principle. This helps avoid ambiguity and makes it easier for the person to bring relevant evidence to mind. Predictive probabilities are much more reliably elicited than for example distributions on parameters (which can often be deduced as functions of elicited predictive statements) see e.g. Carlson (1993).

(7) By *restructuring* the analysis appropriately it is often possible for the DM to better bring information to mind. This is particularly useful as a way of avoiding assessing probabilities that will be *close to* 0 *or* 1. See the examples in the following section.
(8) It is useful to *break down the elicitation* of a single event into smaller components in a similar way as when producing a qualitative framework for the problem as a whole. Setting the event in a wider framework and then marginalising encourages the expert not only to extend her vision of possibilities, but the averaging process intrinsic to marginalisation can help smooth out systematic biases in the elicitation process: see e.g. Goodwin and Wright (2003); Kleinmauntz *et al.* (1996); Wright *et al.* (1994).
(9) *Potential biases* like those discussed above need to be kept in mind by the analyst throughout the elicitation process.
(10) An analysis of the *sensitivity* of various assumptions is useful – see e.g. French and Rios Insua (2000). This can be performed numerically or mathematically by the analyst. This information can then be fed back to the DM so that she appreciates the extent different types of her inputs are having on the overall analysis. This helps to identify those aspects of the model where more information or care needs to be applied to the elicited inputs.

4.3.4 Some illustrations of elicitation

We end this section with a few examples to illustrate some of the principles stated above The first – based on a simple example by Larry Phillips – is an example of the 7th and 8th points above where a probability assessment is refined by restating the probability or decomposing it into smaller components.

Example 4.5. Suppose you need to assign a probability to the event A that there have been more than 100 monarchs of England since the Norman conquest in 1066. You could just give a number. However even if you are a non-English national you could translate this problem into one about the average number of years a that monarchs of England might reign. If the year in which you read this is t then for A to be true

$$a \geq \frac{t - 1066}{100}.$$

The point of re-expressing the event in this way is that you might be able to bring relevant information to bear on the average length of reign, for example life expectancy in Europe over that time period and so on. As a bonus note that you can use this to document any reasons for your guess to give to an auditor.

There is another restructuring that uses a nontrivial decomposition of this query into subcomponents and is even more helpful if you have partial knowledge about the kings and queens of England. You could try to list the names of all the monarchs (m of them) and then guess the number of monarchs x_i with that listed name. For example you might remember Henry the eighth so know there must be at least 8 Henries. The total number of monarchs will then be $x = \sum_{i=1}^{m} x_i$. If this number is much greater than 100 you can be fairly certain the event is true and if much less then fairly sure it is untrue. By decomposing in this way you have not only improved the accuracy of your probability forecast but you can determine where the source of your uncertainty lies – for example the number of monarchs with names you have forgotten – and so more accurately quantify this uncertainty. Again the documentation of the way in which you came to this assessment is helpful additional information to give to an auditor who does not know the answer to the question either, making your quoted probability more or less plausible to him.

This example is not ideal as an illustration of an event we might need in a decision analysis because it concerns an event whose truth value is predetermined. So consider now three simplistic examples where the event is one whose value is as yet unknown by anyone.

Example 4.6. You need to elicit the probability of the event B that an accidental emission of above a safe quantity of radioactivity from a particular nuclear plant of a certain type into the atmosphere will occur in the next 5 years. This probability is small and so difficult to elicit faithfully: see points 6 and 7. However the DM tells you that B could occur if and only if the core overheated – event B_1 – the cooling system was dysfunctional when this happened – event B_2 – and the resulting temperature increase caused a breach of the casing of the core – event B_3. Using this structural knowledge provides a qualitative framework around which to decompose the problem – point 8. Thus from the usual rules of probability we have

$$P(B) = P(B_1 \cap B_2 \cap B_3) = P(B_1)P(B_2|B_1)P(B_3|B_1 \cap B_2).$$

Each of the three probabilities on the right-hand side will be much larger than $P(B)$ and therefore more reliable to estimate. Also it is easier for the DM to bring to mind instances like B_1 when the core had overheated but the cooling system was functional at this and similar plants so that no emission resulted. She can therefore specify her probability $P(B_1)$ with reasonable confidence – point 5. Note here we have conditioned on events consistently with the order in which they could happen so as to avoid the DM erroneously reversing the conditioning when quoting probabilities. Although the elicited probability of B using this decomposition can be expected to correspond more faithfully to the DM's beliefs there are still pitfalls associated with assessing the probability of this rare event. These have a tendency to bias the quoted probability so that it is an under-estimate. First it is tempting to set $P(B_2|B_1) = P(B_2)$ assuming independence between these two events – and to relate the probability of $P(B_2|B_1)$ to the proportion of time that, because of random failures, the cooling system is not operative. But circumstances that might cause the core to overheat might also adversely affect the cooling system as well in which case $P(B_2|B_1) > P(B_2)$. The DM needs to be confronted with the possible existence of these "common causes" and if they plausibly exist they need to be brought into a new qualitative framework which will lead to a different decomposition: see Chapter 8 for a description of this process. Second there may be other albeit rare chains of events leading to an accidental release not accounted for in this calculation. Again the existence of such chain will demand a change in the structure of the underlying decomposition.

Example 4.7. The law of total probability often gives a useful decomposition of an elicited probability. For example suppose that there is a danger to the health of an unborn child if the mother is exposed to a substance emitted when she performs a particular task. Here the event B of interest is that an employee performing this task is unknowingly pregnant. To aid the DM to evaluate her subjective probability of this event it is helpful to split up the population of the relevant cohort of such employees into disjoint exhaustive risk groups $\{A_1, A_2, \ldots, A_n\}$. The law of total probability gives that

$$P(B) = \sum_{1 \le i \le n} P(B|A_i)P(A_i).$$

For example the subset A_1 could consist of male employees and female employees outside childbearing age. In this case the DM knows for sure that $P(B|A_1) = 0$. For other risk groups $P(B|A_i)$ could be elicited with reference to publicly available survey data and so the values the DM chooses for these conditional probabilities are justified. Furthermore in this type of breakdown the probabilities $P(A_i)$ will often relate directly to staff records and so are also reliably and auditably assessed. So as well as improving the quality of faithfulness of the elicitation because many components on the right-hand side will have probability much larger than $P(B)$ the decomposition helps the DM to relate assessments to available data.

Example 4.8. Bayes rule can be used to assess a conditional probability where the conditions are stated anticausally. A simple situation is when a mechanic observes a shudder – event B – in a machine and needs to diagnose which of the n possible disjoint causes $\{A_1, A_2, \ldots, A_n\}$ are responsible. Rather than directly eliciting $\{P(A_1|B), P(A_2|B), \ldots, P(A_n|B)\}$ as in the clinical example of Chapter 2 it is usually wiser to elicit $\{P(B|A_1), P(B|A_2), \ldots, P(B|A_n)\}$ and $\{P(A_1), P(A_2), \ldots, P(A_n)\}$ and then use Bayes rule to obtain $\{P(A_1|B), P(A_2|B), \ldots, P(A_n|B)\}$. As in the last example the nature of these probability assignments are also usually easier to justify both as an extrapolation of other related past events and from a scientific standpoint.

Notice from these simple examples that the basic message is that the more the DM is encouraged to address the underlying process and the underlying dependence structure leading to the event of interest the more reliable the analyst can believe the DM's inputs to be. Several new ways of eliciting and exploring such underlying structure with the DM will be encountered in later chapters.

4.4 Calibration and successful probability predictions

4.4.1 Introduction

In previous sections it has been argued that in a wide range of circumstances it is appropriate for a DM to follow the Bayesian paradigm and assign probabilities to events intrinsic to the successful outcome of a decision process. Her beliefs can then be conveyed as a systematic entity whose meaning can be understood and analysed, over a space of events which has unambiguous interpretation. But how can an analyst determine whether the probabilities elicited from the DM or her trusted expert are good? We have already demonstrated how the DM's beliefs evolve as she begins to think more deeply and systematically about the events important to her analysis.

The first principle is the comfort the DM has herself in her own specification. She should be happy that the probability assignments she expresses are a sufficiently honest and precise representation of what she currently believes about the events that matter to the decision analysis. She is therefore willing to use the decomposition and the probabilities to explain to a third party her current position. These probability assignments she communicates are then called *requisite* (Phillips, 1984). She may want to change these assignments in the future either in the light of new evidence or because important issues subsequently occur to her which did not appear in the original evaluation. But for now she is content to express these as her own.

But the DM's – or trusted expert's – own conviction that her beliefs as expressed through the probabilities she assigns are faithful to her current thinking is quite a weak requirement. Even though her probabilities cohere and she is happy with their elicited values, after all she could be totally misguided. Recall that we required in the first point above that subjective probabilities need to be rational in the sense that they can be appreciated as being rational by a third party. If the probability forecaster makes statements that clearly run counter to evidence then they are not adequate.

However there are ways for an auditor to check the broad appropriateness of a DM's probability statements and if someone is initially uncalibrated in this sense then she can be trained to do better. Moreover the DM can check the plausibility of the probabilities proved by a trusted expert. For clarity we will focus here on the second task, although the techniques described below obviously transfer to the first.

4.4.2 The calibrated forecaster

So assume that the DM needs to *adopt* an expert's subjective probability. Can the DM determine, on the basis of their past performance, if an expert is a good probability forecaster, and how can this be measured?

The easiest scenario to consider is when the probability of the event has natural replicates and the expert is believed to be exchangeably competent to assess the probabilities of this sequence of events. We are then in the situation where we can reasonably expect stated probabilities to be broadly consistent with observed frequencies in a sense described below. To focus the terminology and discussion I will illustrate the ideas using a simple example: a weather forecaster who is forecasting precipitation – here referred to as rain. It is important to be explicit here. So we assume that the forecaster states her probability each day about an event that a measurable amount of precipitation falls at a site S over a future time period T (e.g. 24 hrs) during the next day. Henceforth let A_t denote an indicator function taking a value $a_t = 1$ if it rains tomorrow – time t – at the site S over the period T and $a_t = 0$

otherwise. The DM sees the forecaster's decision to quote a probability $Q_t = q_t$ about $A_t = a_t$ and index this by the time t it is made. A typical table of an expert's daily quoted probabilities and their success over a fortnight is given below.

Day t	1	2	3	4	5	6	7	8	9	10	11	12	13	14
Forecast q_t	0.5	0.0	0.6	1.0	0.6	0.8	0.8	0.6	0.8	0.8	0.5	0.8	0.6	0.6
Rain? a_t	1	0	1	1	0	1	0	1	1	1	0	1	1	1

Definition 4.9. A forecaster is said to be *empirically well calibrated* over a set of n time periods, if over the set of $n(q)$ periods (e.g. days) she quotes the probability of rain as q, the proportion $\widehat{q}(q) = \frac{r(q)}{n(q)}$ of those periods it is rainy is equal to q, this being true for all values of q she quotes, where $r(q)$ denotes the number of days it rains when she quotes q.

In the example above it is easily checked that the forecaster is empirically well calibrated over the 14 days of prediction. For example on the 5 days she quotes the probability of rain as 0.8 it rains 4 times: i.e. 80% of the time. For the purposes of calibration it often convenient to aggregate over the day index into bins labelling the forecast made for that day. Such a table for three different probability forecasters predicting rain over 1,000 days is given below.

Stated forecast q	0.0	0.1	0.2	0.3	0.4	0.5	0.6	0.7	0.8	0.9	1.0
It rains $r(q)$ times when P_1 says q	0	0	0	0	32	290	204	0	0	0	0
P_1's frequency $n(q)$ when saying q	0	0	0	0	80	580	340	0	0	0	0
It rains $r(q)$ times when P_2 says q	20	0	0	0	16	20	0	0	0	0	480
P_2's frequency $n(q)$ when saying q	400	0	0	0	40	40	0	0	0	0	520
It rains $r(q)$ times when P_3 says q	0	8	25	36	32	50	36	84	124	81	50
P_3's frequency $n(q)$ when saying q	20	80	125	120	80	100	60	120	155	90	50

Note that P_1 and P_3 are empirically well calibrated. For example over the probabilities P_1

$$\frac{r(.4)}{n(.4)} = \frac{32}{80} = .4, \quad \frac{r(.5)}{n(.5)} = \frac{290}{580}, \quad \frac{r(.6)}{n(.6)} = \frac{204}{340}.$$

On the other hand P_2 is not since for example

$$\frac{r(.4)}{n(.4)} = \frac{20}{400} = 0.05 \neq 0.$$

It can be demonstrated that the DM can legitimately expect the ideal expert who provides a probability forecaster to be *approximately* empirically well calibrated. Thus suppose the DM believes that the observed sequence of values $\{A_t = a_t : t \geq 1\}$ of the indicator variables above was a random draw from a perfectly known probability model. In this sense the forecaster would then be believed to know the generating process defining the conditional probabilities and use this knowledge directly to predict it: i.e. she would be as well informed as she could be about the outcome of A_t. Now consider the sequence of random variables $\{Q_t : t = 1, 2, \ldots\}$ that correspond to the DM's probabilities the forecaster quotes so that

$$Q_1 = P(A_1 = 1)$$

and for $t = 2, 3, \ldots$

$$Q_t(A_1, A_2, \ldots, A_{t-1}) \triangleq P(A_t = 1 | A_1, A_2, \ldots, A_{t-1})$$

where the probabilities on the right-hand side of these equations are the true generating probability function. Here we assume that the forecaster is making her probability forecasts in time order and if necessary taking into account her success or failure in her past forecasts. We will use the usual convention of employing small case letters to denote realisations of the random variables above. The DM now chooses a "test" subsequence of days $I = \{t_1, t_2, \ldots\}$ where $t_i < t_{i+1}$ where her choice of whether or not $t \in I$ is allowed to depend on $\{a_1, a_2, \ldots, a_{t-1}\}$ and $\{q_1, q_2, \ldots, q_t\}$ and anything else known to her before time t but not on $\{a_t, a_{t+1}, \ldots\}$ or $\{q_{t+1}, q_{t+2}, \ldots\}$ i.e. anything occurring after the event predicted. To be fair to the forecaster the DM must obviously not use hindsight to construct her test set: i.e. use a test set that can be constructed as a computable sequences (Dawid, 1982). It would be legitimate for example for her to choose all the days as her test sequence, all weekends or all days that it had rained on the previous day, but not the days on which it has been observed that it has rained.

Let $I_k = \{t \in I : 1 \leq t \leq k\}$, let N_k denote the number of elements in I_k and R_k the number of such elements when it rained, and

$$\overline{Q}_k = N_k^{-1} \sum_{t \in T_k} Q_t.$$

It is now possible to prove a remarkable result.

Theorem 4.10. *With probability one, if T_k is chosen so that $N(T_k) \to \infty$ as $k \to \infty$ then*

$$\frac{R_k}{N_k} - \overline{Q}_k \to 0.$$

Proof A proof is found in Dawid (1982) uses martingale theory and so is beyond the scope of this book. □

Calibration in the sense above can be achieved as a corollary of this result. Thus suppose the forecaster is using the appropriate model and the auditor chooses a test set $I(p)$ where $t \in I(p)$ if and only if $q_t = q$. Let $I_k(q) \triangleq \{t \in I : 1 \leq t \leq k\}$, let $N_k(q)$ denote the number of elements in $I_k(q)$ and $R_k(q)$ the number of such elements when it rained, then $\overline{Q}_k \triangleq Q$. It follows that, in the sense of the theorem above a DM can expect $\widehat{q}_k$

and q to be close to each other when the number of times the expert $n_k(q)$ is large whatever the value of the quoted probability q. Of course it would be unreasonable to expect $\widehat{q}_k$ to be exactly equal to q just approximately so. Reassuringly it has been shown that weather forecasters are actually close to being well calibrated, with least success when q is close to 0 or 1, see e.g. Murphy and Winkler (1977). In fact similar conclusions are also possible when quoted probabilities are rounded to lie in a given interval. Some probability forecasters in other domains such as bookmaking (Dowie, 1976), sport (Yates and Curley, 1985), medicine (McLish and Powell, 1989) and economics (Wilkie and Pollack, 1996) have all demonstrated skills in attaining good levels of calibration although admittedly there are demonstrably many badly calibrated forecasters (especially in the last two areas) as well! It appears that calibration tends to improve when the expert applies an appropriate credence decomposition, Kleinmauntz *et al.* (1996) and Wright *et al.* (1994). There is also now an established theory explaining what can be expected of a probability forecaster making a sequence of forecasts called *prequential analysis:* see for example Dawid (1982); Dawid (1992); Dawid and Vovk (1999) and Cowell *et al.* (1999).

Note that, contrary to popular beliefs, the Bayesian expert who offers his probability judgements, actually puts his head on the block. The predictions he makes can be held up to scrutiny, at least in repetitive contexts like the one above. If the probability judgements are consistently flawed then this become apparent very quickly though comparing his forecasts with what actually happens: for example through calibration tables like those given above.

Calibration is a property that will be exhibited by an excellent probability forecaster. However it is a necessary and not sufficient condition for a forecaster to be good. In fact an approximately well-calibrated forecaster understands well his ability to predict, without necessarily having good domain knowledge. Thus looking at the two calibrated forecasters above, P_3 appears more useful than P_1. Consider which you would choose to decide whether or not to take an umbrella next day. P_1 is not that useful: always giving a probability between 0.4 and 0.6. In fact, on the basis of history you may well prefer to use the uncalibrated forecaster P_2 than either of these two.

4.4.3 Continuous calibration*

There is a slight technical problem above in that, because the event is binary, the sample space of the quoted probability is also 2-dimensional so the joint distribution has a rather complex form. However if the forecaster produces a sequence of forecast distributions $\{Q_n : n = 1, 2, \ldots\}$ of a sequence of real-valued continuous random variables $\{Y_n : n = 1, 2, \ldots\}$ so that

$$Q_n(y) \triangleq P(Y_n | Y_1, Y_2, \ldots, Y_{n-1})$$

and the densities of $\{Q_n : n = 1, 2, \ldots\}$ are all nonzero in their support the situation is a lot easier. Let $U_n = Q_n(Y_n)$. Then if the auditor believes that the forecaster has the appropriate model he can conclude that $\{U_n : n \geq 1\}$ is a sequence of independent uniform random variables on the unit interval. This gives a myriad of different ways for an auditor to check the veracity of a forecaster when she makes a sequence of distributional statements. This was first pointed out in Dawid (1982) and has been subsequently used as a practical tool by a number of authors (see e.g. Cowell *et al.* (1999); Smith (1985)).

Perhaps this is most used to help a probability forecaster to check a sequence of interval estimates and hence become a better probability forecaster. Thus suppose Y_t denotes the maximum temperature on day t and that a forecaster gave a sequence of intervals J_t, $t = 1, 2, \ldots$, where she believed that

$$P(Y_t \in J_t | Y_1, Y_2, \ldots, Y_{t-1}) = p.$$

Then the result quoted above implies that the auditor – here the forecaster herself – can conclude that the indicator on $\{Y_t \in J_t\}, t = 1, 2, \ldots$ will be a sequence of Bernoulli random variables. This can obviously be checked (see e.g. Bedford and Cooke (2001)).

4.5 Scoring forecasters

4.5.1 Proper scores

One simple way of judging whether one probability forecaster is more reliable than another is to score the performance of forecasters over a long period of time using a score function that penalises inappropriate forecasts. The forecaster who receives the lowest penalised aggregate score could then be adjudged to be the most reliable.

Definition 4.11. A loss function $L(a, q) : a = 0, 1$ and $0 \leq q \leq 1$, is called a *scoring rule* if it is used to assess the quality of probability forecasts q.

If we make the heroic decision that the forecaster has a linear utility on money – approximately true if they have a differentiable utility function and the penalties are small – we can expect the EMV forecaster to quote the probability q^*, $0 \leq q \leq 1$, which minimises her expected loss

$$\overline{L}(q|p) = pL(1, q) + (1 - p)L(0, q)$$

where p, $0 \leq p \leq 1$, is her subjective probability of rain tomorrow.

Definition 4.12. Any loss function $L(a, q)$ (uniquely) minimising $\overline{L}(q|p)$ when $q^* = p$ is called a *(strictly) proper scoring rule [(s)psr].*

Psr's encourage honest forecasts from EMV forecasters.

Example 4.13 (The Brier score). $L(a, q) = (a - q)^2$ gives

$$\begin{aligned}\overline{L}(q|p) &= p(1-q)^2 + (1-p)q^2 \\ &= (q-p)^2 + p(1-p)\end{aligned}$$

which for fixed p is clearly uniquely minimised when $q^* = p$. So the Brier score is an spsr. Note that this is a bounded score taking a value between 0 and 1. In fact if the DM has no domain knowledge she can guarantee a score of $\frac{1}{4}$ simply by choosing $p = \frac{1}{2}$.

It is interesting to note that – inspired by De Finetti (1974) – Goldstein and Wooff (2007) have developed a whole system of inference based round the elicitation of the results called previsions under this loss but generalised away from the probability prediction of binary random variables like A above to general ones. The big advantage of doing this is that a coherent system can be built based only on a moderate number of elicited features and nothing else. Analogues of Bayes rule and the law of total probability exist in this system and the methodology also has an associated semi-graphoid – see below – and so a natural measure of the existence of dependence. Moreover the finiteness of the number of elicited quantities in this belief system means that he can address practically important issues much earlier and with much more ease than is possible within the usual Bayesian paradigm. Whether you find this methodology compelling depends on how convincing you find the use of the elicited previsions as primitives that express genuine beliefs. My personal worry about this method is the influence a DM's preference – in normal Bayesian inference reflected by her utility function – might have on the elicited quantities: see Exercise 4.5. However many interesting applications of this fully formal method now appear in the literature: see Goldstein and Wooff (2007) for a recent review of some of these.

Example 4.14 (Logarithmic score). Here $L(1, q) = -\log q$ and $L(0, q) = -\log(1-q)$ gives

$$\overline{L}(q|p) = -\{p \log q + (1-p)\log(1-q)\}.$$

Differentiating and setting to zero now gives, for $0 < p, q < 1$

$$\frac{p}{q} = \frac{(1-p)}{(1-q)} \Longleftrightarrow q = p.$$

By checking the second derivative at $q^* = p$ is positive we can then assert that the logarithmic scoring rule is spsr.

This scoring rule is widely used and has close links with information theory. It can also be used to elicit densities – see e.g. Bernardo and Smith (1996) – and it exhibits many interesting theoretical properties. Several authors make it central to their discussion of the

Bayesian paradigm. However because of its unboundedness the method suffers from the sorts of instability to slight mis-specification discussed in the last section of the previous chapter. I would therefore not recommend it as a practical tool for elicitation.

Example 4.15 (Absolute score). Most loss functions you write down will actually not be proper scoring rules. The simplest scoring rule to illustrate this is the absolute scoring rule $L(a, q) = |a - q|$. We set as an exercise that for this scoring rule the optimal quoted probability $q^* = 1$ if $p > \frac{1}{2}$ and $q^* = 0$ if $q < \frac{1}{2}$. When $p = \frac{1}{2}$ any decision is optimal. Were an analyst to use this elicitation tool then it would be optimal for an EMV DM or expert to pretend to be certain. Other scoring rules are illustrated in the exercises at the end of this chapter.

4.5.2 Empirical checks of a probability forecaster

There is an obvious way to check how well a probability forecaster is performing. An auditor can simply check her wealth determined by her score after the forecaster has predicted rain over a large number of days by the following.

Definition 4.16. Let $\{(a_i, q_i) : 1 \leq i \leq n\}$ denote the pairs of outcome and probability predictions over n periods. Then that forecaster's *empirical score* S_n over those n days is given by

$$S_n = \sum_{i=1}^{n} L(a_i, q_i). \tag{4.7}$$

It can be shown that a forecaster who knew the probabilistic generating mechanism and quoted $q = p$ would, with probability 1, in the limit as $n \to \infty$ obtain at least as low a score as any other forecaster who wasn't clairvoyant: i.e. could not see into the future (Dawid, 1982). We could therefore conclude that a good measure of a forecaster's $P's$ performance, encouraged to be (approximately honest by a proper scoring rule) is his empirical score $S_n(P)$. In particular a forecaster with the lowest empirical score in a long sequence of forecasting periods could reasonably be considered "best" in the sense that they have the highest reward from his forecasts.

Example 4.17. The forecaster P_2 in Section 4.2 performs much better than the other two when the empirical score is calculated for the Brier score, where we can calculate

$$S_n(P_1) = 245.8, \quad S_n(P_2) = 79.0, \quad S_n(P_3) = 169.0.$$

4.5.3 Relating calibration to score under the Brier scoring rule

So good forecasters should be expected to be calibrated and also should be expected to produce a relatively low score under a proper scoring rule. We end this section by demonstrating how these two ideas can be brought together when the scoring rule used is the Brier score. We first need to introduce some notation. Suppose the forecaster quotes only m probabilities $q_1, q_2, q_3, \ldots, q_m$ where

$$0 \leq q_1 < q_2 < q_3 < \cdots < q_m \leq 1$$

and quotes q_i $n_i > 0$ times $1 \leq i \leq m$, so that

$$\sum_{i=1}^{m} n_i = n.$$

Write $\boldsymbol{q} = (q_1, q_2, q_3, \ldots, q_m)$. It will be convenient to index the outcomes in terms of the quoted probabilities. So let $a_i(j)$ be the outcome arising from the jth period, $1 \leq j \leq n_i$, that the forecaster happens to quote the probability q_i, $1 \leq i \leq m$. Finally let $\widehat{q}_i$ denote the proportion of periods it rains when she says q_i, $1 \leq i \leq m$. Thus

$$\widehat{q}_i = n_i^{-1} \sum_{j=1}^{n_i} a_i(j).$$

Theorem 4.18. *A forecaster's empirical score will be at least as low if he replaces his quoted probabilities q_i by $\widehat{q}_i$.*

Proof We first note that

$$S_n(\boldsymbol{q}) = \sum_{i=1}^{m} \sum_{j=1}^{n_i} (a_i(j) - q_i)^2$$

where

$$\begin{aligned}
\sum_{j=1}^{n_i} (a_i(j) - q_i)^2 &= \sum_{j=1}^{n_i} [(a_i(j) - \widehat{q}_i) + (\widehat{q}_i - q_i)]^2 \\
&= \sum_{j=1}^{n_i} (a_i(j) - \widehat{q}_i)^2 + 2(\widehat{q}_i - q_i) \sum_{j=1}^{n_i} (a_i(j) - \widehat{q}_i) + n_i(\widehat{q}_i - q_i)^2 \\
&= \sum_{j=1}^{n_i} (a_i(j) - \widehat{q}_i)^2 + n_i(\widehat{q}_i - q_i)^2
\end{aligned}$$

since the middle term vanishes by the definition of $\widehat{q}_i$. Summing over the index j we therefore have that

$$\begin{aligned}
S_n(\boldsymbol{q}) &= S_n(\widehat{\boldsymbol{q}}) + \sum_{i=1}^{m} n_i(\widehat{q}_i - q_i)^2 \\
&\geq S_n(\widehat{\boldsymbol{q}})
\end{aligned}$$

with strict inequality unless $\boldsymbol{q} = \widehat{\boldsymbol{q}}$. □

It follows that, unless a forecaster, is empirically well calibrated the DM can obtain a better Brier score by (retrospectively) substituting her vector of empirical success rate $\widehat{\boldsymbol{q}}$ for $\boldsymbol{q}$.

Example 4.19. The forecaster P_2 can be recalibrated by letting $\frac{20}{420} = 0.0467$ replace 0 and $\frac{476}{520} = 0.9154$ replace 1. $P_2^{\prime}s$ score can be quickly calculated using the penultimate equation above to have reduced from 79 to 74.32 and she is now empirically well calibrated.

It is a contentious issue whether or not to recalibrate a probability forecaster. On the plus side in the exercises below you will see that if the probability forecaster has a nonlinear utility on reward then she will not state her true probability: i.e. her effective score is not then strictly proper. But in many cases her true probabilities can be retrieved from recalibrating the quoted probability. However these arguments rely on her being naturally a good probability forecaster. If she is still learning to give good forecasts then recalibration can be counter-productive. Furthermore if an expert learns that she will be recalibrated then she may try to compensate for this in her quoted probability, leading the analyst into a very complicated game-playing scenario. Finally in some environments for an analyst to perform such recalibration can be seen as showing a lack of respect and can have a tendency to disengage the expert. So whether or not it is appropriate to recalibrate depends heavily on context: see Bedford and Cooke (2001) and Clemen and Lichtendahl (2002) for further discussion.

For a more general discussion of the relationship between proper scoring rules and other inferential constructs and their theoretical properties see Dawid (2007).

4.6 Summary

We have seen how a DM's subjective probabilities can be unambiguously defined and that, under certain conditions, such a subjective probability is a probability in the usual mathematical sense of the word when restricted to a finite-dimensional event space. Techniques for improving forecasts using credence decompositions to break up the problem into smaller components and then aggregating these up into a composite were illustrated. These ideas will be further elaborated later in the book. Furthermore we showed how an auditor or the DM herself can use calibration and scoring techniques to examine the plausibility either of the DM's own probability forecasting skills or of those of the trusted expert who she chooses to deliver these judgements.

However we have also noted that directly elicited subjective probabilities are susceptible to biases. In the next chapter we discuss how information from samples, experiments and observational studies can be marshalled together to improve probability forecasts and deliver judgements that are supportable by commonly agreed facts. We then apply these techniques to moderately sized homogeneous, decision problems. This methodology will then be extended and elaborated in Chapter 9 so that data accommodation can be applied to decision models of very large-scale problems.

4.7 Exercises

4.1 Show that the reward distributions associated with the called off bet and the reward distribution of the gamble where the events labelling its edges are substituted by the events associated with the respective calibrating betting wheel gambles, are the same. Hence deduce that the rational DM will specify her subjective probabilities such that her conditional probability assignments always satisfy $P(A \cap B) = P(A|B)P(B)$.

4.2 A DM is asked her beliefs about a random variable Z taking values in $[0, \infty)$. She tells you that she believes that the probability of Z lying in any finite length set is 0. Prove that these statements are consistent with her having a coherent probability distribution but not a countably additive one.

4.3 Two contractors C_1 and C_2 bid for work. They quote respective prices X_1 and X_2 and the one submitting the lower price is awarded the contract. If they both submit the same price the award is determined on the toss of a coin. A DM is a regulator who believes the probability C_1 wins the contract is $\frac{1}{2}$. Because she knows nothing about the nature of the contract bid her probability of C_1 winning the contract given she learns the price $X_1 = x_1$ given by C_1 is also $\frac{1}{2}$ regardless of the value of x_1. If she does not believe that the two contractors will quote the same price with probability one, then prove that these beliefs cannot be represented by a countably additive probability distribution on (X_1, X_2), Atwell and Smith (1991); Heath and Sudderth (1989); Hill (1988).

4.4 The binary random variable Y takes values either 0 or 1 and X takes values on the integers. The DM states the finite additive probabilities

$$P(Y = 0, X = x) = 2^{-(1+x)}, \quad P(Y = 1, X = x) = 2^{-(2+x)}$$

for $x = 1, 2, 3, \ldots$ and $P(Y = 1) = 1/2$. Show that for all $x = 1, 2, 3, \ldots$ $P(Y = 1|X = x) = 1/3$. Hence or otherwise show that the law of total probability fails for this distribution. (This is called the *nonconglomerability* property of finitely additive distributions.)

4.5 For the purposes of assessing probabilities given by weather forecasters you choose to use the Brier scoring rule $S_1(a, q)$ where

$$S_1(a, q) = (a - q)^2.$$

You suspect, however, that your this expert's utility function is not linear but is of the form

$$U(S_1) = 1 - S_1^{\lambda}$$

for some value of $\lambda > 0$. If this is the case, prove that the forecaster will quote $q = p$ for all values of p iff $\lambda = 1$. Prove, however, that if $\lambda > \frac{1}{2}$ and you are able to elicit the value of λ, then you are able to express the expert's true probability p as a function of their quoted probability q. Write down this function explicitly and explain in what sense, when $\frac{1}{2} < \lambda < 1$, your client will appear over-confident and if $\lambda > 1$, under-confident in her probability predictions. Finally prove that if $0 \le \lambda \le \frac{1}{2}$, it will only be possible to determine from her quoted probability whether or not she believes that $p \le \frac{1}{2}$.

4.6 The elicitation of probabilities using scoring rules has been criticised on the grounds that the decision maker will not be EMV and her utility will be a function of her current fortune x – whose density before she gambles we will denote by $g(x)$. Suppose that you have elicited this utility $U(x)$ and found that this takes the form

$$U(x) = 1 - e^{-\lambda x} \text{ where } \lambda > 0.$$

Suppose that the probability forecaster will be scored with the logarithmic scoring rule,

$$S(a,q) = -(\log q)^a (\log[1-q])^{1-a}$$

so that her fortune x^* after quoting a probability q and observing $A = a$, $a = 0, 1$ will be $x^* = x - s(a,q)$.

i) Prove that the probability q, quoted by a rational probability forecaster, will not depend on the value of her fortune x before scoring and will be chosen so as to minimise

$$f(q) = \frac{p}{q^\lambda} + \frac{1-p}{(1-q)^\lambda}.$$

ii) Hence or otherwise prove that the rational forecaster will quote a probability q^* where q^* must satisfy

$$\phi(q^*) = \frac{\phi(p)}{1+\lambda}$$

where

$$\phi(y) = \log\left(\frac{y}{1-y}\right).$$

iii) How will this quoted probability differ from the one obtained from the forecaster which assumes that she will ignore her current fortune and has a linear utility function? In particular, how might this affect the forecaster's quoted probability when her true probability p is

a) very small but not zero,

b) very large but not one?

If you could elicit the value of λ accurately, how should you adjust your client's quoted probability to input this into a statistical model?

4.7 Over a period of 100 days a weather forecaster quotes probability forecasts $q = 0.1, 0.3, 0.5, 0.7, 0.9$ on $n[q]$ occasions, it raining on $r[q]$ of those days. Her results are given below.

q	0.1	0.3	0.5	0.7	0.9
$r[q]$	4	9	5	14	15
$n[q]$	20	30	10	20	20

Is she empirically well-calibrated? Without proof describe how you could improve her Brier score by reinterpreting the probabilities she quotes.

4.8 On each of 650 consecutive days, two probability forecasters F_1 and F_2 state their probabilities that rain will occur the next day. Each forecaster only chooses to state one of the probabilities $\{q(1) = 0, q(2) = 0.25, q(3) = 0.5, q(4) = 0.75, q(5) = 1\}$. The results of these forecasts are given below. The unbracketed numbers in the (i, j)th element of the table give the number of times F_1 quoted $q_1(i)$ and F_2 quoted $q_2(j)$ whilst the bracketed numbers in the (i, j)th element give the number of times F_1 quoted $q_1(i)$ and F_2 quoted $q_2(j)$ and it also rained on that day, $1 \leq i, j \leq 5$.

$q_1(i)\backslash q_2(j)$	0.0	0.25	0.5	0.75	1.0
0.0	40, (0)	20, (0)	20, (0)	20, (0)	0, (0)
0.25	20, (0)	20, (0)	20, (0)	20, (5)	20, (20)
0.5	20, (0)	20, (0)	50, (25)	30, (20)	30, (30)
0.75	20, (0)	20, (5)	30, (20)	30, (25)	50, (50)
1.0	0, (0)	20, (20)	30, (30)	50, (50)	50, (50)

Show that neither forecaster is well calibrated. Noting the symmetry of the table above, calculate the empirical Brier score they share. Stating without proof any result you might use, adapt the forecasts of each forecaster so that their empirical Brier score improves, and calculate the extent of this improvement. It is suggested that you could obtain improved forecasts by using the table above to combine the forecasts of the two individual forecasters in some way. Find a probability forecasting formula that has a lower empirical Brier score than the Brier score of either of the individual forecasters.

5
Bayesian inference for decision analysis

5.1 Introduction

In the last section we considered how domain knowledge could be expressed probabilistically to provide the basis for coherent acts. We now turn our attention to how the Bayesian DM can draw into her analyses evidence from other sources and so make it more compelling both to herself and an external auditor.

In many situations factual evidence in the form of data can be collected and drawn on to support the DM's inferences. We have seen several examples already in this book where such evidence might be available. It is important to accommodate such information on two counts. First by using such supporting evidence the DM herself will have more confidence in her probability statements and will be able to explain herself better. We saw in the previous chapter that probabilities can rarely be elicited with total accuracy. By refining these judgements and incorporating evidence from data whose sample distribution can be treated as known the DM can help improve her judgement and minimise unintentional biases she introduces. Second if she supports her judgements by accommodating evidence from well designed experiments and sample surveys, generally accepted as genuinely related to the case in hand, then this will often make her stated inferences more compelling. Although expert judgements about the probability of an event or the distribution of a random variable are often open to question and psychological biases, it is usually possible to treat data from a well-designed experiment as *facts* agreed by the DM and any auditor.

This chapter is about Bayesian parametric inference, how this can be related to a Bayesian decision analysis and where care needs to be exercised in exploiting this relationship. Bayesian parametric inference is a method which uses a full probability model over not only the data collected but also the parameters – which in many of our examples are unknown probabilities – to make inferences about the parameters/probabilities associated with the experiment after the data is observed. We will see that in the Bayesian framework, when data is only seen after the marginal (prior) probability density is specified by the DM then this is entirely automatic. Given the DM's set of beliefs before she sees the evidence and given she really believes the sampling distributions associated with that model the only beliefs she can legitimately hold about her parameters after the experimental evidence is seen are given by Bayes rule. Furthermore what she believes she will see can be calculated from the inputs above using the law of total probability formula.

If an auditor accepts the Bayesian paradigm then to criticise the DM's conclusions she therefore needs either to criticise the DM's assumptions about the sampling scheme, for example the randomness of the sampling in the survey or the modelling assumptions that lie behind the DM's chosen family of sampling distributions, or the DM's prior margin over the parameters of these distributions. Any contentious features of the model can thus be explored and reappraised if necessary. A formal development and illustrations of this process are presented below.

This chapter mainly focuses on how a Bayesian accommodates into her model evidence from sample survey data and simple experimental designs. Using Bayes rule and the law of total probability to do this is called a *prior to posterior analysis*. The implications of such analyses make up the bedrock of Bayesian inference and have now been widely studied. In particular there are many excellent texts on the implications of following this simple algorithm (Bernardo and Smith, 1996; De Groot, 1970; Gelman *et al.*, 1995; O'Hagan and Forster, 2004; Raiffa and Schlaifer, 1961), and also about the outworkings of this procedure as it applies to a myriad of different sampling schemes and experimental structures (Denison *et al.*, 2005; Fruthwirth-Schnatter, 2006; Gelman and Hill, 2007; Lancaster, 2004; Rossi *et al.*, 2005; West and Harrison, 1997). It would be impossible to do justice to all this material in this small book. Because these methodologies are so well documented elsewhere our discussion will be limited. However I will give enough scenarios to support the various illustrations of decision analyses seen in this text. We will also use mixtures of models to demonstrate some basic aspects of Bayesian model selection. Further discussions on learning in more complicated scenarios as it applies to decision analyses is deferred to Chapters 7, 8 and 9.

Bayesian inference is a rather different discipline from Bayesian decision analysis. Bayesian inference will typically focus on the logical implications of a particular set of experiments on inferences about the generating process of that particular data set once it has been seen. Inferences about what will happen in the future usually comprise of posterior or predictive inferences. *Posterior inferences* concern the distribution of the parameters or probabilities associated with the *population* from which the individual sample was drawn. *Predictive inferences* are associated directly with the probability of another *unit drawn at random from the sampled population*.

Of course both these inferences are an important part of a decision analysis. They can also sometimes have a direct relevance to a decision analysis. The DM may believe that the sample of the potentially allergenic shampoo in Chapter 2 is a genuine random sample from the whole population in which case the distribution of the probability of anyone in the population having an adverse reaction or becoming sensitised could be identified as the same probability associated with the sampled units. Similarly the DM needing to forecast the probability that a green fibre would be found at the scene of the crime outlined in Chapter 2 might credibly be based on a large population survey of similar events.

But this direct correspondence tends to be the exception rather than the rule. The parameters or probabilities needed in decision analyses are those associated with events affecting the value of the DM's expected utility under different decision she might make in the *given* scenario she faces. If these do not correspond precisely to a sample proportion or further replicate a previous experiment then more work is needed. We end the chapter with a discussion of this issue.

5.2 The basics of Bayesian inference

Very early in the development of Bayesian inference it was proved that if a sequence of binary random variables $\{Y_i : i = 1, 2, 3, \ldots\}$ was such that when a finite number of the indices were permuted the joint marginal distributions of all the finite subsets of the same cardinality were the same – this is termed the sequence being *exchangeable* – then this was logically the same as saying that $\amalg_{i=1}^{\infty} Y_i|\theta$ where θ, a random variable taking values between zero and one could be interpreted as the probability that any one of the observations took the value one. This is the simplest scenario where the data is the value of the first n observations $\{y_i : i = 1, 2, 3, \ldots, n\}$ in this exchangeable sequence and the interest is in predicting the probability of the next observation Y_{n+1}. There has been enormous activity proving various different variants of De Finetti theorems. A surprising number of problems can be embedded into a sample space that exhibits the types of invariances required for such theorems to hold. Despite this the scope for the use of these elegant ideas, especially in a decision analysis, has therefore been found to be somewhat too limited at least in its unrefined form. For examples of these results about exchangeability see Bernardo and Smith (1996) and Schervish (1995).

However, far more valuable to the decision analyst was the fact that the study of exchangeable systems excited the idea of using hierarchical (or multilevel) structures to model relationships between variables. Thus the idea was spawned that by adding new unobserved random variables a description of dependence between observables could actually be simplified. This not only allowed the accommodation of sample survey data, whose parameters could not be directly associated with the instance of interest into a Bayesian analysis, but also permitted the Bayesian to draw into her analysis relevant observational studies. It will be seen later that intrinsic to these descriptions is the qualitative concept of conditional independence relationships between the variables in the model. These sets of conditional independences help us to explain probabilistic relationships clearly and persuasively and provide the framework for many different types of elegant and feasible inference to be performed.

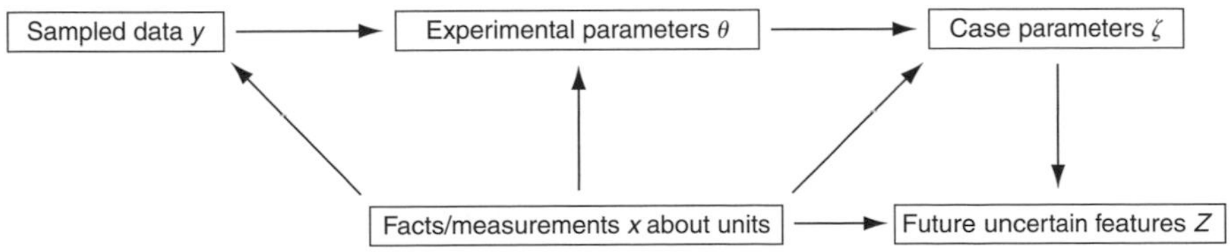

The basic idea of hierarchical Bayesian modelling is to construct a new explanatory random vector $\boldsymbol{\theta}$. This vector is built to capture all those aspects of the past data $\boldsymbol{Y} = \boldsymbol{y} = (y_1, y_2, \ldots, y_n)$ that are relevant to the future uncertain quantities. These will include the distribution of random probabilities and parameters of their distribution of $\boldsymbol{Y}$ given all known facts and measurements $\boldsymbol{X} = \boldsymbol{x}$. The parameters such as the probabilities $\boldsymbol{\zeta}$ of the case in hand together with the actual outcome of features of that analysis $\boldsymbol{Z}$, given $\boldsymbol{X} = \boldsymbol{x}$, are assumed to depend on $\boldsymbol{Y}$ only through $\boldsymbol{\theta}$. In the simplest problems $\boldsymbol{\zeta} = \boldsymbol{\theta}$.

It quickly became apparent that this was a very useful paradigm of inferential model building and admitted a wide range of useful elaborations. For example the extension of this idea prompted much of the development of large dependent structures through the Bayesian network: a topic for a later chapter. We will wait until that chapter to give practical illustrations of how such a vector of explanatory uncertain quantities can be derived from a given context. Instead we begin by describing how we can proceed having first constructed such a vector $\boldsymbol{\theta}$ to act as a conduit for information from sampled units or results of experiments into the random vector of parameters $\boldsymbol{\zeta}$ of interest and then for $\boldsymbol{\zeta}$ to act as a conduit for this information into $\boldsymbol{Z}$, in the situation faced by the DM.

The two technical conditions for this to be so are that $\boldsymbol{\zeta}$ is independent of $\boldsymbol{Y}$ given $\boldsymbol{\theta}$ and $\boldsymbol{X}$ – written $\boldsymbol{\zeta} \amalg \boldsymbol{Y} \,|\, (\boldsymbol{\theta}, \boldsymbol{X})$ – and that $\boldsymbol{Z}$ is independent of $(\boldsymbol{Y}, \boldsymbol{\theta})$ given $\boldsymbol{\zeta}$ and $\boldsymbol{X}$ – written $\boldsymbol{Z} \amalg (\boldsymbol{Y}, \boldsymbol{\theta}) \,|\, (\boldsymbol{\zeta}, \boldsymbol{X})$. Thus in particular, given the known facts $\boldsymbol{X} = \boldsymbol{x}$ everything needed to predict the future $(\boldsymbol{Z}, \boldsymbol{\zeta})$ is conveyed by what we can learn about $\boldsymbol{\theta}$ from $\boldsymbol{Y}$.

These two conditions can be equivalently written in terms of the arguments of conditional densities. Thus

$$p(z, \boldsymbol{\zeta} | \boldsymbol{x}, \mathbf{y}, \boldsymbol{\theta}) \triangleq p(z | \boldsymbol{\zeta}, \mathbf{x}, \mathbf{y}, \boldsymbol{\theta}) p(\boldsymbol{\zeta} | \mathbf{x}, \boldsymbol{y}, \boldsymbol{\theta})$$

where our two conditions imply

$$p(\boldsymbol{\zeta} | \mathbf{x}, \mathbf{y}, \boldsymbol{\theta}) = p(\boldsymbol{\zeta} | \boldsymbol{x}, \boldsymbol{\theta}),$$
$$p(z | \boldsymbol{\zeta}, \boldsymbol{x}, \boldsymbol{y}, \boldsymbol{\theta}) = p(z | \boldsymbol{\zeta}, \boldsymbol{x}).$$

From the usual rules of probability we have that the density $p(\boldsymbol{\zeta} | \boldsymbol{x}, \boldsymbol{y})$ of the future parameter vector $\boldsymbol{\zeta}$ is given by

$$p(\boldsymbol{\zeta} | \boldsymbol{x}, \boldsymbol{y}) \triangleq \int p(\boldsymbol{\zeta} | \boldsymbol{x}, \boldsymbol{y}, \boldsymbol{\theta}) p(\boldsymbol{\theta} | \boldsymbol{x}, \boldsymbol{y}) d\boldsymbol{\theta} = \int p(\boldsymbol{\zeta} | \boldsymbol{\theta}, \boldsymbol{x}) p(\boldsymbol{\theta} | \boldsymbol{x}, \boldsymbol{y}) d\boldsymbol{\theta} \tag{5.1}$$

and the mass function or density $p(z | \boldsymbol{\zeta}, \boldsymbol{x}, \boldsymbol{y})$ of $\boldsymbol{Z}$ given $\boldsymbol{\zeta}, \boldsymbol{X}, \boldsymbol{Y}$ can be calculated using the formula

$$p(z | \boldsymbol{\zeta}, \boldsymbol{x}, \boldsymbol{y}) \triangleq \int p(z | \boldsymbol{\zeta}, \boldsymbol{x}, \boldsymbol{y}, \boldsymbol{\theta}) p(\boldsymbol{\theta} | \boldsymbol{x}, \boldsymbol{y}) d\boldsymbol{\theta} = p(z | \boldsymbol{\zeta}, \boldsymbol{x}). \tag{5.2}$$

It follows from the usual probability fomulae that its mass function $p(z | \boldsymbol{x}, \boldsymbol{y})$ is given by

$$p(z | \boldsymbol{x}, \boldsymbol{y}) \triangleq \int p(z | \boldsymbol{\zeta}, \boldsymbol{x}) p(\boldsymbol{\zeta} | \boldsymbol{x}, \boldsymbol{y}) d\boldsymbol{\zeta}. \tag{5.3}$$

In many simple problems $\boldsymbol{\zeta} = \boldsymbol{\theta}$ in which case $p(\boldsymbol{\zeta}|\boldsymbol{x},\boldsymbol{y}) = p(\boldsymbol{\theta}|\boldsymbol{x},\boldsymbol{y})$. Inference then simply focuses on the joint distribution of $(\boldsymbol{\theta}, \boldsymbol{X}, \boldsymbol{Y})$ and then

$$p(z|\boldsymbol{x},\boldsymbol{y}) \triangleq \int p(z|\boldsymbol{x},\boldsymbol{\theta})p(\boldsymbol{\theta}|\boldsymbol{x},\boldsymbol{y})d\boldsymbol{\theta}.$$

In more complicated scenarios we still need to calculate $p(\boldsymbol{\theta}|\boldsymbol{x},\boldsymbol{y})$ but then use either (5.3) or (5.1) to calculate the appropriate probability distributions.

There is an important point to note about the use of such models in a decision analysis where typically the DM's utility function is usually a function $U_1(\boldsymbol{\zeta}, d(\boldsymbol{x},\boldsymbol{y}))$ of the pair $(\boldsymbol{\zeta}, d(\boldsymbol{x},\boldsymbol{y}))$ or a function $U_2(z, d(\boldsymbol{x},\boldsymbol{y}))$ of $(z, d(\boldsymbol{x},\boldsymbol{y}))$. In both cases the decision analysis can equivalently be formulated as a problem of finding a utility maximising decision albeit using a transformed utility function $U_3(\boldsymbol{\theta}, d(\boldsymbol{x},\boldsymbol{y}))$. Thus by (5.1) the DM's expected utility

$$\begin{aligned}\overline{U}(d(\boldsymbol{x},\boldsymbol{y})) &= \int U_1(\boldsymbol{\zeta}, d(\boldsymbol{x},\boldsymbol{y}))p(\boldsymbol{\zeta}|\boldsymbol{x},\boldsymbol{y})d\boldsymbol{\zeta} \\ &= \int\int U_1(\boldsymbol{\zeta}, d(\boldsymbol{x},\boldsymbol{y}))p(\boldsymbol{\zeta}|\boldsymbol{\theta},\boldsymbol{x})p(\boldsymbol{\theta}|\boldsymbol{x},\boldsymbol{y})d\boldsymbol{\theta}d\boldsymbol{\zeta} \\ &= \int U_3(\boldsymbol{\theta}, d(\boldsymbol{x},\boldsymbol{y}))p(\boldsymbol{\theta}|\boldsymbol{x},\boldsymbol{y})d\boldsymbol{\theta} \qquad (5.4)\end{aligned}$$

where

$$U_3(\boldsymbol{\theta}, d(\boldsymbol{x},\boldsymbol{y})) = \int U_1(\boldsymbol{\zeta}, d(\boldsymbol{x},\boldsymbol{y}))p(\boldsymbol{\zeta}|\boldsymbol{\theta},\boldsymbol{x})d\boldsymbol{\zeta}.$$

So a Bayes decision can be seen implicitly as a utility maximising decision on the parameter vector $\boldsymbol{\theta}$. Similarly by (5.3) when $\mathbb{Z}$ is discrete

$$\begin{aligned}\overline{U}(d(\boldsymbol{x},\boldsymbol{y})) &= \sum_{z\in\mathbb{Z}} U_2(z, d(\boldsymbol{x},\boldsymbol{y}))p(z|\boldsymbol{x},\boldsymbol{y}) \\ &= \sum_{z\in\mathbb{Z}} U_2(z, d(\boldsymbol{x},\boldsymbol{y})) \int p(z|\boldsymbol{\zeta},\boldsymbol{x})p(\boldsymbol{\zeta}|\boldsymbol{x},\boldsymbol{y})d\boldsymbol{\zeta} \\ &= \int U_1(z, d(\boldsymbol{x},\boldsymbol{y}))p(\boldsymbol{\zeta}|\boldsymbol{x},\boldsymbol{y})d\boldsymbol{\zeta}\end{aligned}$$

where

$$U_3(z, d(\boldsymbol{x},\boldsymbol{y})) \triangleq \sum_{z\in\mathbb{Z}} U_2(z, d(\boldsymbol{x},\boldsymbol{y}))p(z|\boldsymbol{\zeta},\boldsymbol{x})$$

which by (5.4) can again be expressed as the expectation over $U_3(\boldsymbol{\theta}, d(\boldsymbol{x},\boldsymbol{y}))$ defined above. Notice that the sum above corresponds to the first operation in a backwards induction step from the leaves of a rollback decision tree.

It important to keep in mind that the utility U_3 is an expectation of a utility. However if $p(\boldsymbol{\zeta}|\boldsymbol{x},\boldsymbol{\theta})$ and if necessary $p(z|\boldsymbol{\zeta},\boldsymbol{x})$ and her utility function have all been elicited from the DM it is *technically* sufficient only to discover $p(\boldsymbol{\theta}|\boldsymbol{x},\boldsymbol{y})$ to determine her Bayes decision. We describe and illustrate how this is done below.

5.3 Prior to posterior analyses

The consequences of using the rules of probability for inference and decision making are most straightforward when the transformation from the DM's prior beliefs about a system before data are seen to her beliefs after the data are accommodated can be conducted in closed form. So in this book I will mainly address this case. Surprisingly there is a rich variety of problems admitting the sort of simple analysis some of which I describe below.

From the comments above, to be able to calculate the expected utilities from (5.2) and (5.4) we simply need to calculate $p(\boldsymbol{\theta}|\boldsymbol{x},\boldsymbol{y})$. Since we will condition on what is known $\boldsymbol{x}$ throughout, to keep notation simple we will suppress the $\boldsymbol{x}$ index of known facts and measurements unless I need to address this vector explicitly. By its definition this vector of parameters $\boldsymbol{\theta} = (\theta_1, \theta_2, \ldots, \theta_r) : \boldsymbol{\theta} \in \Theta \subseteq \mathbb{R}^r$ will embody all information relevant to the problem in hand. Let the DM's beliefs about the quantities of interest $\boldsymbol{\theta}$ in the experiment or sample before she sees any observations be represented by the probability density $p(\boldsymbol{\theta})$ – henceforth called her *prior probability density*. When the data $\boldsymbol{Y}$ has an absolutely continuous density $p(\boldsymbol{y}|\boldsymbol{\theta})$ given each possible value of $\boldsymbol{\theta}$ then this is called the *sampling density* given $\boldsymbol{\theta}$. On the other hand if observations are discrete then their joint probability mass function $p(\boldsymbol{y}|\boldsymbol{\theta})$ is called the *sampling probability mass function* given $\boldsymbol{\theta}$. In either case consider $p(\boldsymbol{y}|\boldsymbol{\theta})$ as a function of $\boldsymbol{\theta}$.

In this book we will focus much of our attention on the analysis of discrete models. Here it will often be necessary for the DM to convince an auditor about the propriety of transparently and appropriately accommodating evidence sampling about *probabilities* into an analysis. In other settings we need to elicit parameters that define this probability distribution rather than a vector of probabilities. Typical examples of this type of setting are illustrated below.

Example 5.1. A crime involved a person throwing a brick from a distance of 5 metres and breaking a window. The court is interested in the probability that the suspect was at the scene of this crime after matching glass fragments were found on the suspect's clothing. To assess the evidence, jurors will need to elicit two probabilities. The first is the probability that someone matching the age group and life style of the suspect chosen at random from this population would have this type of glass on their clothing anyway. Surveys have been conducted that search for glass on randomly selected individuals. The sort of information available in these studies is the number of individuals in a given category and the number of those exhibiting fragments of glass (indexed by type) on their clothes. A second type of evidence corresponds to an experiment conducted by forensic scientists where someone throws a brick at various distances from a window and the number of glass fragments landing on their clothing is counted and recorded when the pane breaks.

Example 5.2. A DM recently employed at a company has prepared a tender for a contract. She then notices that the company has records of its success in similar circumstances to the one she faces. She notices that her company has won y of the N such contracts.

In many circumstances designed experimental data is conducted so that $p(\boldsymbol{y}|\boldsymbol{\theta})$ is known. In the second experiment of the first example above, provided the experiment was properly randomised, standard probability theory tells us that

$$p(y|\theta) = \binom{N}{y}\theta^{y}(1-\theta)^{N-y}$$

$y = 0, 1, 2, \ldots, N$ where θ is the probability and y is the number of experimental units with glass landing on them.

We argued above that the sampling density or mass function from a well-conducted sample survey or from a designed experiment often has a special status because the DM and auditor are likely to agree about its probability distribution. Thus in the example above, provided they both believe that the experiment was properly conducted then given θ the DM and the auditor will share a common mass function $p(y|\theta)$ of $Y|\theta$ over the number of the fragments. It will be shown later in Chapter 9 that drawing on this shared appreciation of evidence tends to pull the beliefs of a Bayesiain DM and auditor closer together even if their respective prior beliefs were initially far apart.

Returning to calculations, recall that any function $l(\boldsymbol{\theta}|\boldsymbol{y})$ in $\boldsymbol{\theta}$ which is proportional to $p(\boldsymbol{y}|\boldsymbol{\theta})$, i.e. which can be written as

$$p(\boldsymbol{y}|\boldsymbol{\theta}) = A(\boldsymbol{y})l(\boldsymbol{\theta}|\boldsymbol{y}) \tag{5.5}$$

where $A(\boldsymbol{y})$ is not a function of $\boldsymbol{\theta}$, is called a *likelihood* of $\boldsymbol{\theta}$ given $\boldsymbol{y}$. Much statistical analysis of evidence is based on the likelihood and many inferential principles were crystallised before Bayesian inference became so fashionable. For example *the strong likelihood principle* states that two experiments giving rise to the same observed likelihood should also give rise to the same inferences about parameters: a principle we will see is automatically obeyed if the Bayesian methodology is used, see below.

The usual rules of probability tell us that

$$p(\boldsymbol{\theta}|\boldsymbol{y})p(\boldsymbol{y}) = p(\boldsymbol{y}|\boldsymbol{\theta})p(\boldsymbol{\theta}). \tag{5.6}$$

In this context the density $p(\boldsymbol{\theta}|\boldsymbol{y})$ represents the DM's revised belief about the probability vector $\boldsymbol{\theta}$ after having observed $\boldsymbol{y}$ and is called her *posterior density* of $\boldsymbol{\theta}$. This is what we need to calculate the DM's expected utility. The density or mass function $p(\boldsymbol{y})$ is often called the *marginal likelihood* of $\boldsymbol{y}$ – and represents the DM's marginal probability density/mass function of what data she had expected to see. Thus after observing $\boldsymbol{y}$, $p(\boldsymbol{y})$ gives a numerical evaluation of the surprise the DM experiences about the value of data she observes in the light of the specification of her prior density, where the lower the value of $p(\mathbf{y})$ the greater the surprise. Note that this term is therefore often very sensitive to the prior density the DM specifies. It has the biggest impact on inference involved in selecting a model: see later.

A good way of specifying a prior density on parameters is to ensure that the predictions about the results of an experiment before they are seen are plausible in the given context: i.e. to calibrate them to a specification of $p(\mathbf{y})$. This is because probability statements embodied in $p(\mathbf{y})$ are often more tangible than statements about $p(\boldsymbol{\theta})$ itself and so more easy to elicit accurately: see the review in the previous chapter. It is certainly always wise, even if there is a case for directly eliciting $p(\boldsymbol{\theta})$, to double check that the prior settings give plausible predictions about the results of an experiment that might be observed, even if it is necessary to simply hypothesise an experiment rather than conduct it. Examples of how this can be done in a particular practical scenario can be seen in Liverani *et al.* (2008); Wakefield *et al.* (2003).

Again using the usual rules of probability $p(\mathbf{y})$ can be calculated from $p(\mathbf{y}|\boldsymbol{\theta})$ and our prior density $p(\boldsymbol{\theta})$, from the continuous version of the law of total probability

$$p(\mathbf{y}) = \int_{\Theta} p(\mathbf{y}|\boldsymbol{\theta})p(\boldsymbol{\theta})d\boldsymbol{\theta}. \tag{5.7}$$

This integral can be difficult to evaluate. Indeed it may not be possible to write it in closed form. However from equation (5.5) and reading equation (5.6) as a function of $\boldsymbol{\theta}$ we can also see that

$$p(\boldsymbol{\theta}|\mathbf{y}) \propto p(\mathbf{y}|\boldsymbol{\theta})p(\boldsymbol{\theta}) \propto l(\boldsymbol{\theta}|\mathbf{y})p(\boldsymbol{\theta}). \tag{5.8}$$

This is a critical equation, because it represents what the DM believes about the parameters of her model – expressed in terms of a new density – now that she has seen the values of the observations $\mathbf{y}$. Note that it is often possible to use this equation alone to identify $p(\boldsymbol{\theta}|\mathbf{y})$, because the proportionality constant can be found indirectly from the fact that, because $p(\boldsymbol{\theta}|\mathbf{y})$ is a density

$$\int_{\Theta} p(\boldsymbol{\theta}|\mathbf{y})d\boldsymbol{\theta} = 1. \tag{5.9}$$

Examples of this are given later. If the proportionality constant cannot be calculated in closed form then the integral (5.7) will need to be calculated numerically. There are many cases when no closed form analysis is possible. However over the last 25 years a wide variety of ways of calculating good numerical approximations have been developed and in many instances free software is available to perform these tasks. Most of these methods find ways of drawing an approximate random sample of massive size from the given posterior to estimate the joint sample distribution for the theoretical one of interest. There is now a vast literature on this topic which has excited many researchers interested in probabilistic approximation techniques and computation. These numerical methods are now extensively researched and of a rather technical nature and so outside the scope of this small volume.

Introductions to these techniques are given in Gelman *et al.* (1995); Robert and Casella (2004) and O'Hagan and Forster (2004) with more detail for example in Chen *et al.* (2000); Evans and Swartz (2000); Gammerman (2006); Marin and Robert (2007) and Robert (2001).

Finally note that by taking logs of (5.8), provided no term with $\boldsymbol{\theta}$ as an argument is zero,

$$\log p(\boldsymbol{\theta}|\mathbf{y}) = \log l(\boldsymbol{\theta}|\mathbf{y}) + \log p(\boldsymbol{\theta}) + a(\mathbf{y}) \tag{5.10}$$

where $a(\mathbf{y})$ is not a function of $\boldsymbol{\theta}$. This equation is sometimes more useful than (5.8), not only because it is linear but also because the logarithm of many common densities and likelihoods have a particularly simple algebraic form: see below.

5.4 Distributions which are closed under sampling

When moving to a decision analysis of large-scale problems it is important to try to keep the analysis as transparent and fast as possible. Of course this is not always an option and then numerical methods need to be used the calculate posterior densities and marginal likelihoods. However there are a surprising number of scenarios when an analysis can be performed when the effect of data is easy to determine because the posterior density is in the same family as the prior.

Definition 5.3. A family of prior distributions $\mathcal{P} = \{P(\boldsymbol{\theta}|\boldsymbol{\alpha}) : \boldsymbol{\alpha} \in A \subseteq \mathbb{R}^m\}$ is said to be *closed under sampling* for a likelihood $l(\boldsymbol{\theta}|\mathbf{y})$, if for any prior $P(\boldsymbol{\theta}) \in \mathcal{P}$, the corresponding posterior distribution $P(\boldsymbol{\theta}|\mathbf{y}) \in \mathcal{P}$.

Of course for a given problem there may exist no nontrivial family closed under sampling with an algebraic form. Furthermore it may not be possible to represent a DM's beliefs faithfully using such a family because the parametric form of the prior density forces her to make probabilistic statements she could not entertain as plausible. But if such a family does exist and also contains a density faithful to the DM's prior beliefs then this is a big advantage. It is then not only easy and quick to compute the posterior density – very useful when problems are scaled up in size: see later – but also to understand how and why the results of the experiment modified the beliefs expressed in the prior to those now expressed in the posterior density. Whilst it is often now possible to approximate a DM's posterior density numerically very well by sampling, this method does not tend to provide the basis for a narrative to support *why* her beliefs have adjusted in the way they have. In the context of decision modelling such methods should therefore be avoided if possible.

Happily there are several important examples of structured multivariate distributions like those that can be described by trees or the Bayesian networks we discuss later where such closure to sampling often exists. Furthermore if one family, closed under sampling, is overly restrictive in this sense, then richer families, also closed under sampling, can be built through mixing: see Section 5.7.

One important family of prior distributions $\mathcal{P} = \{P(\boldsymbol{\theta}|\boldsymbol{\alpha}^0) : \alpha^0 \in A \subseteq \mathbb{R}^m\}$ over a vector $\boldsymbol{\theta}$ of parameters is one for which the density $p(\boldsymbol{\theta}|\boldsymbol{\alpha})$ associated with $P(\boldsymbol{\theta}|\boldsymbol{\alpha})$ takes the form

$$p(\boldsymbol{\theta}|\boldsymbol{\alpha}^0) = \exp\{\alpha^0.\psi'(\boldsymbol{\theta}) + k_1(\boldsymbol{\alpha}^0)\}$$

where α^0 and $\psi(\boldsymbol{\theta})$ are both vectors of length m, $\psi'(\boldsymbol{\theta})$ is the transpose of $\psi(\boldsymbol{\theta})$ and $k_1(\boldsymbol{\alpha})$ a function of $\boldsymbol{\alpha}$ but not $\boldsymbol{\theta}$ ensuring that $p(\boldsymbol{\theta}|\boldsymbol{\alpha})$ integrates to unity. It is quite common for experiments to be designed so that the sample distribution has a likelihood $l(\boldsymbol{\theta}|\mathbf{y})$ that can be written

$$l(\boldsymbol{\theta}|\mathbf{y}) = \exp\{r(\mathbf{y}).\psi'(\boldsymbol{\theta}) + k_2(\mathbf{y})\}. \tag{5.11}$$

Equation (5.10) then gives us that

$$p(\boldsymbol{\theta}|\boldsymbol{\alpha},\mathbf{y}) = \exp\{\alpha^+(\mathbf{y}).\psi'(\boldsymbol{\theta}) + k_1(\boldsymbol{\alpha}_+)\}$$

where

$$\alpha^+(\mathbf{y}) = \alpha^0 + r(\mathbf{y})$$

which lies in $\mathcal{P}$. Provided that $\alpha^+(\mathbf{y}) \in A$ it follows that this family – sometimes called the *conjugate family* – is closed under sampling.

5.5 Posterior densities for absolutely continuous parameters

5.5.1 Updating probabilities using evidence about a single probability

For a Bayesian analysis, the first task is to elicit a prior density. This density can be based on logical arguments. We will illustrate this technique by considering the simplest possible scenario where the DM wants to back up a choice of probability. So return to Example 4.15 above. The DM – here a forensic scientist – needs to describe how glass might innocently be present on clothing in the case of the first probability or the physics of how glass flies out of a broken window in the second. A convenient family of prior distributions on a variable like a probability taking values in the interval [0, 1] is the beta family.

5.5.1.1 The beta density and its elicitation

The beta $Be(\alpha, \beta)$ density $p(\theta|\alpha, \beta)$ of a probability θ, $0 \leq \theta \leq 1$, is given by

$$p(\theta|\alpha, \beta) = \frac{\Gamma(\alpha + \beta)}{\Gamma(\alpha).\Gamma(\beta)}\theta^{\alpha-1}(1-\theta)^{\beta-1} \tag{5.12}$$

where $\alpha, \beta > 0$ and $\Gamma(y) = \int_0^\infty x^{y-1}e^{-x}dx$. This means in particular that for $y = 1, 2, \ldots, \Gamma(y) = (y-1)!$. The density $p(\theta|\alpha, \beta)$ is unimodal when $\alpha, \beta > 1$, with its mode at $\frac{\alpha-1}{\alpha+\beta-2}$. Its mean μ and variance σ^2 are given by

$$\begin{aligned} \mu &= \alpha(\alpha+\beta)^{-1}, \\ \sigma^2 &= \mu(1-\mu)\,(\alpha+\beta+1)^{-1}. \end{aligned} \tag{5.13}$$

Note that, since, for $0 \leq \mu \leq 1, 0 \leq \mu(1-\mu) \leq \frac{1}{4}$,

$$\lim_{\alpha+\beta\to\infty} \sigma^2 = 0$$

so that as $\alpha+\beta$ becomes large, $p(\theta|\alpha,\beta)$ concentrates its mass very close to μ. The density $p(\theta|\alpha,\beta)$ is symmetric if and only if $\alpha=\beta$ and is uniform if $\alpha=\beta=1$.

A variety of software is now available that displays plots of the beta density together with its bounds. Suppose the expert – in the example given above this would be the forensic scientist – is prepared to state her beliefs about the probability θ with a beta density $Be(\alpha^0,\beta^0)$. Then the mean $\mu^0 = \alpha^0(\alpha^0+\beta^0)^{-1}$ is the prior probability that the analyst might elicit using the types of techniques discussed in the last section, based on logic or experience. To fully specify her prior density as described by the pair of hyperparameters (α^0,β^0) once she gives her value of μ^0 she just need to specify $\alpha^0+\beta^0$. This can be done using the graphics above with credibility intervals, e.g. by specifying an interval inside which she is 90% certain that an infinitely sampled proportion would lie, from which the software can calculate the corresponding value of $\alpha^0+\beta^0$. Alternatively we will see below that $\alpha^0+\beta^0$ can be thought of as the equivalent sample size her prior information would be worth. In my experience for many applications it is rare for this sum to take a value greater than 10 so that the prior density is usually quite flat. From the above properties of the beta a uniform density is obtained if $\mu^0 = 0.5$ and $\alpha^0+\beta^0 = 2$. Note that if the probability forecaster is concerned that none of the beta densities she is offered accurately fits her beliefs – for example if she believes that the distribution is bimodal with modes lying in $(0,1)$ – then she can extend her options and use mixtures of beta densities which can express a much richer family of shapes: see below.

5.5.1.2 A beta prior to posterior analysis

Now suppose, as in our forensic example above, we take a random sample of size N from a population sharing the same value of covariate x – describing what happened in the brick-throwing incident and whose probability of success is θ. Then standard statistical arguments tell us that the number y of successes – in our example the number of individuals where fragments of glass fell on the thrower – we observe has a binomial mass function, with nonzero masses on $y = 0,1,2,\ldots,N$ given by

$$p(y|\theta) = A(y)\theta^y(1-\theta)^{N-y}$$

where $A(y) = \binom{N}{y}$. Thus using equation (5.8),

$$p(\theta|\alpha^0, \beta^0, y) \propto \frac{\Gamma(\alpha^0 + \beta^0)}{\Gamma(\alpha_0).\Gamma(\beta_0)} \theta^{\alpha^0 - 1}(1-\theta)^{\beta^0 - 1}.\theta^x(1-\theta)^{N-x}$$
$$\propto \theta^{\alpha^+ - 1}(1-\theta)^{\beta^+ - 1}$$

where $\alpha^+ = \alpha^0 + y$ and $\beta^+ = \beta^0 + (N - y)$. So in particular this posterior density is proportional to a beta $Be(\alpha^+, \beta^+)$ density. But since any density must integrate to 1, this implies that

$$p(\theta|\alpha^0, \beta^0, y) = p(\theta|\alpha^+, \beta^+) \sim Be(\alpha^+, \beta^+).$$

So in particular, the beta prior family is closed under binomial sampling. The effect of this sampling is to map y

$$\alpha^0 \to \alpha^+ = \alpha^0 + y,$$
$$\beta^0 \to \beta^+ = \beta^0 + N - y.$$

Note here that the beta family is conjugate where $l(\theta|\mathbf{y})$ given in (5.11) where $r(\mathbf{y}) = (y, N - y)$ and $\psi = (\log\theta, \log(1-\theta))$.

To understand how the probabilistic framework determines how the Bayesian's beliefs change in the light of this sampling information, first consider how her posterior mean μ^+ responds as a function of her prior mean μ_0 and the data.

$$\mu^+ = \alpha^+(\alpha^+ + \beta^+)^{-1} = \rho\frac{y}{N} + (1-\rho)\mu^0$$

where $\rho = (\alpha_0 + \beta_0 + N)^{-1}N$, which is a weighted average of her mean of the probability of success before she saw the data and the sample proportion of successes she subsequently observed. The weight ρ given to the sample proportion equals $\frac{1}{2}$ when $\alpha^0 + \beta^0 = N$. This is why $\alpha^0 + \beta^0$ is sometimes called the equivalent sample size. When she has little relevant available evidence and N is very small then she weights her prior information more highly and her posterior variance remains large. On the other hand when the sample size N large compared with $\alpha^0 + \beta^0$ then

$$\rho \simeq 1 \Rightarrow \mu^+ \simeq \frac{y}{N}$$

so the posterior mean is almost identical to the sample proportion and she effectively jettisons her prior. Furthermore, for large N, because $\alpha^+ + \beta^+ = \alpha^0 + \beta^0 + N$ by (5.13) the forecaster's posterior variance

$$Var(\theta|y) = \mu^+(1-\mu^+)(\alpha^+ + \beta^+ + 1)^{-1}$$

will be very small, i.e. she will be very certain that her posterior mean is about right. Note these ways of accommodating the data are all very plausible even to someone who was not insisting that a Bayesian approach be used.

From (5.6) it is easy to calculate the marginal likelihood of the number of individuals in the sample of N having glass on their clothes takes positive values only on $y = 0, 1, \ldots, N$ when $p(y)$ takes the form

$$p(y) = \frac{N!\Gamma(\alpha^0 + \beta^0)\Gamma(\alpha^0 + y)\Gamma(\beta^0 + N - y)}{\Gamma(\alpha^0 + \beta^0 + N)y!(N - y)!\Gamma(\alpha^0).\Gamma(\beta^0)}.$$

Note that with a uniform prior on θ when $\alpha^0 = \beta^0 = 1$ this reduces to a uniform distribution on $y = 0, 1, \ldots, N$. In our running example of course we would normally expect a higher prior probability of lower numbers of glass fragments to be found in the sample than is provided by this mass function. However α^0, β^0 can then be chosen so that they calibrate appropriately to her predictive mass function $p(y)$.

Another quantity of interest is the predictive mass function of the random variable of interest in the study. In the running example, Z would be the indicator on whether or not the suspect would have glass on his clothing given he threw the brick. This can be calculated from the formula

$$p(\theta|y, z)p(z|y) = p(z|\theta, y)p(\theta|y) = p(z|\theta)p(\theta|y)$$

since, because we are assuming that the next individual is drawn at random from this population Z is independent of Z given θ. It follows that – as we might have predicted –

$$P(Z = 1|y) = \alpha^+(\alpha^+ + \beta^+)^{-1} = \mu^+$$

the posterior mean of θ.

5.5.2 Revising beliefs on a vector of parameters

Of course most interesting problems have many variables describing them and this makes the problem of how the DM's beliefs should be adjusted in the light of the data subsequently collected somewhat harder. However the principles are exactly the same and we use exactly the same fomulae. We will see later that the challenge for high-dimensional problems is to break them up coherently into small pieces – like joint distributions on low-dimensional margins. A composite model can then often be formed as a composite, utilising elicited structural information like conditional independences discussed late in this book.

5.5.2.1 The Dirichlet joint density

So next consider the forensic scenario like the one described in Example 4.15 but where the *number* of fragments of matching glass is counted, not just whether or not some is found. Suppose that the scientist believes that up to $r - 1$ fragments could be found where $r \geq 2$.

The prior density then needs to assign a distribution on r probabilities – the example above the possible outcomes of finding $0, 1, 2, \ldots, r-1$ fragments on an individual – so this vector of probabilities $\boldsymbol{\theta} = (\theta_1, \theta_2, \ldots, \theta_r)$ had components that were positive and summed to one: for which $\boldsymbol{\theta}$ took values in the simplex

$$\mathcal{S} = \left\{ \boldsymbol{\theta} : \sum_{i=1}^{r} \theta_i = 1, \theta_i \geq 0, 1 \leq i \leq r \right\}.$$

The most common choice for this multivariate density is the Dirichlet joint density $D(\boldsymbol{\alpha})$ where $\boldsymbol{\alpha} = (\alpha_1, \alpha_2, \ldots, \alpha_r)$ has density $p(\boldsymbol{\theta}|\boldsymbol{\alpha})$, given by

$$p(\boldsymbol{\theta}|\boldsymbol{\alpha}) = \frac{\Gamma(\alpha_1 + \alpha_2 + \cdots + \alpha_r)}{\Gamma(\alpha_1)\Gamma(\alpha_2)\ldots\Gamma(\alpha_r)} \theta_1^{\alpha_1 - 1} \theta_2^{\alpha_2 - 1} \ldots \theta_r^{\alpha_r - 1} \tag{5.14}$$

when $\boldsymbol{\theta} \in \mathcal{S}$ and zero elsewhere, where $\alpha_i > 0$, $1 \leq i \leq r$. Note that when $r = 2$, setting $\theta_1 = \theta, \theta_2 = 1 - \theta, \alpha_1 = \alpha$, and $\alpha_2 = \beta$ we have that $\theta|\ \boldsymbol{\alpha} \sim Be(\alpha, \beta)$. So the Dirichlet is in this sense just a generalisation of the beta to cases when the observation can take more than two levels. Henceforth we will consider the beta analysis given above as a special case of the Dirichlet prior to posterior analysis given below.

The Dirichlet distribution is useful when we are estimating probabilities of discrete random variables which are finite but not binary. Letting $\alpha_. = \sum_{i=1}^{r} \alpha_i$ and denoting the mean $\mathbf{E}(\theta_i|\boldsymbol{\alpha}) = \mu_i$ and the variance $Var(\theta_i|\boldsymbol{\alpha})$ then it is easily checked that

$$\mu_i = \alpha_i \alpha_.^{-1} \quad \text{and} \quad Var(\theta_i|\boldsymbol{\alpha}) = \mu_i(1 - \mu_i)\,(\alpha_. + 1)^{-1}.$$

As well as its property of closure under multinomial sampling discussed below, the Dirichlet family has many convenient properties. For example for its relationship to the Gamma density see the exercises below.

As for the beta, to elicit the vector of hyperparameters $\boldsymbol{\alpha}^0$ of a Dirichlet $D(\boldsymbol{\alpha}^0)$ density it is usual to elicit the vector $\boldsymbol{\mu}^0 = (\mu_1^0, \ldots, \mu_r^0)$ of prior means, which just leaves the equivalent sample size parameter $\alpha_.$ to specify. The larger this parameter is chosen the more confidence is shown in the accuracy of the prior. Just as for the beta its precise value can be chosen for example either to ensure $Var(\theta_1|\boldsymbol{\alpha})$ exhibits the right order of credibility interval or with regard to strength in data points equivalence.

When the number of categories is large it is sometimes expedient to use a simple distribution of $\boldsymbol{\mu}_0$ based on some suitable random hypotheses. For example the DM might believe that glass fragments landing on clothing might arrive approximately randomly at a certain rate. The DM might then believe the qualitative hypothesis that prior cell counts would be approximately Poisson distributed with a rate λ say so that

$$\mu_i = \begin{cases} \dfrac{e^{-\lambda}\lambda^i}{i!}, & i = 1, \ldots, r-1 \\ 1 - \displaystyle\sum_{i=0}^{r-1} \mu_i, & i = r. \end{cases}$$

If this were so then after eliciting her prior expectation $\mu_1 = -\log \lambda$ of the number of fragments being found she could find the putative values of $\mu_2, \ldots, \mu_r$. There are also more systematic ways of inputting information using hierarchical models, as we will discuss in Chapter 9.

5.5.3 A prior to posterior analysis for multinomial data

Now suppose that N randomly selected units are sampled and the number of pieces of glass counted, a configuration $\boldsymbol{y} = (y_1, y_2, \ldots, y_r)$ where we observe y_i experimental units with s_i fragments of glass $1 \leq i \leq r$, where $\sum_{i=1}^{r} y_i = N$. Standard probability theory then tells us that $\boldsymbol{Y}$ has a multinomial $M(N, \boldsymbol{\theta})$ distribution conditional on the values of the vector $\boldsymbol{\theta}$ of cell probabilities whose probability mass function $p(\boldsymbol{y}|N, \boldsymbol{\theta})$ is given by

$$p(\boldsymbol{y}|N, \boldsymbol{\theta}) = \frac{N!}{y_1! y_2! \ldots y_r!} \theta_1^{y_1} \theta_2^{y_2} \ldots \theta_r^{y_r}. \tag{5.15}$$

If the DM's, prior density over $\boldsymbol{\theta}$ is $D(\boldsymbol{\alpha}_0)$, using equation(5.8) and writing $\boldsymbol{\alpha}_0 = (\alpha_{0,1}, \alpha_{0,2}, \ldots, \alpha_{0,r})$, we then have that for $\boldsymbol{\theta} \in \mathcal{S}$,

$$\begin{aligned} p(\boldsymbol{\theta}|\boldsymbol{y}) &\propto \theta_1^{\alpha_1^0 - 1} \theta_2^{\alpha_2^0 - 1} \ldots \theta_r^{\alpha_r^0 - 1} \theta_1^{y_1} \theta_2^{y_2} \ldots \theta_r^{y_r} \\ &= \theta_1^{\alpha_1^+ - 1} \theta_2^{\alpha_2^+ - 1} \ldots \theta_r^{\alpha_r^+ - 1} \end{aligned}$$

where, for $1 \leq i \leq r$,

$$\alpha_i^+ = \alpha_i^0 + y_i.$$

We note that this joint density is proportional to (and so equal to) a $D(\boldsymbol{\alpha}^+)$ density where $\boldsymbol{\alpha}^+ = (\alpha_1^+, \alpha_2^+, \ldots, \alpha_r^+)$.

It is easy to check that the probability forecaster's posterior mean of the ith category is given by

$$\mathbb{E}(\theta_i|\boldsymbol{y}) = \mu_i^+ = \rho \left(\frac{y_i}{N}\right) + (1 - \rho)\mu_i^0$$

where $\rho = N(\alpha^0 + N)^{-1}$. So again as the sample size N increases relative to α^0, $\mu_i^+ \to N^{-1} y_i$ and $Var(\theta_i|\boldsymbol{y}) \to 0$, $1 \leq i \leq r$.

5.6 Some standard inferences using conjugate families

The example given above is one of many situations where a simple closed form prior to posterior analysis is possible provided the prior family is chosen carefully. Usually in these cases a location hyperparameter – often the posterior mean – can be expressed as a weighted average of the prior mean and a function of the data. As illustrated in the examples above, the posterior density of the parameters usually has a variance which decreases as the sample size becomes large. We will demonstrate this below. The actual derivation of these results follows the illustrations above. They are technical but straightforward and more appropriate to a text on Bayesian inference and are carefully laid out and discussed elsewhere: see for example Bernardo and Smith (1996); De Groot (1970); O'Hagan and Forster (2004); Raiffa and Schlaifer (1961). We therefore limit our discussion to a few important cases we elude to elsewhere and leave their proof as an exercise. Throughout we denote the prior vector of hyperparameter with a 0 superscript and the posterior with a + superscript just as in the example above.

5.6.1 The gamma prior distribution

The gamma distribution $G(\alpha, \beta)$ is strictly positive only on $\theta \in [0, \infty)$ where it is given by

$$p(\theta) = \frac{\alpha^{\beta}}{\Gamma(\alpha)} \theta^{\alpha-1} \exp(-\beta\theta). \tag{5.16}$$

It is unimodal, has a mean $\mu = \alpha/\beta$ and variance $\sigma^2 = \mu/\beta$. This is particularly useful as a prior for the *rate* θ of a process. It is closed under sampling to a likelihood of the form

$$l(\theta|\mathbf{y}) \propto \theta^{s} \exp(-t\theta)$$

when the prior to posterior updating equations of the hyperparameters (α, β) are

$$\alpha^{+} = \alpha^{0} + s, \quad \beta^{+} = \beta^{0} + t.$$

This form of likelihood and conjugate analysis is often met. For example from standard texts or by direct calculation, you will see that when $\{Y_i : i = 1, 2, \ldots, n\}$ are independent each with an exponential density $p(y|\theta)$ given by

$$p(y|\theta) = \theta \exp(-\theta y)$$

then

$$s = n \quad \text{and} \quad t = y_1 + y_2 + \cdots + y_n.$$

Suppose we are interested in the distribution of the next observation in this sample. The log marginal likelihood $\log p(\mathbf{y})$ and predictive density $p(z|\mathbf{y})$ of the next observation in the sequence – a *power law* or *Pareto density* are given by

$$\log p(\mathbf{y}) = \beta^{0} \log \alpha^{0} - \beta^{+} \log \alpha^{+} - \log \Gamma(\alpha^{0}) + \log \Gamma(\alpha^{+}),$$

$$p(z|\mathbf{y}) = \frac{\alpha^{+}(\beta^{+})^{\alpha_{+}}}{(\beta^{+} + z)^{\alpha^{+}+1}} \text{ when } z > 0.$$

When $\{Y_i : i = 1, 2, \ldots, n\}$ are independent each with a Poisson mass function $p(y|\theta)$ on support $y = 0, 1, 2, \ldots$ given by

$$p(y|\theta) = \frac{\theta^y}{y!} \exp(-\theta)$$

then

$$s = y_1 + y_2 + \cdots + y_n \quad \text{and} \quad t = n.$$

The log marginal likelihood $\log p(\mathbf{y})$ and predictive mass function $p(z|\mathbf{y})$ – a negative binomial – of the next observation in the sequence are given by

$$\log p(\mathbf{y}) = \beta^0 \log \alpha^0 - \beta^+ \log \alpha^+ - \log \Gamma(\alpha^0) + \log \Gamma(\alpha^+) + \sum_{i=1}^{n} \log(y_i!),$$

$$p(z|\mathbf{y}) = \frac{\Gamma(\alpha^+ + z)(\beta^+)^{\alpha+}}{\Gamma(\alpha^+)(\beta^+ + 1)^{\alpha^+ + z}} \text{ when } z = 0, 1, 2, \ldots,$$

respectively. Notice that this predictive density is very similar to the conditional density which is Poisson with its rate parameter substituted by its posterior mean. However the negative binomial above is somewhat more spread out and has thicker tail probabilities, and so automatically accommodates the uncertainty associated with the best rate estimate. This sort of sensible and automatic combination of different sources of information – here that associated with sampling variation and uncertainty associated with the estimation process – which is critically important when samples are small, is one reason why Bayesian methods have recently and rightly become so popular.

When $\{Y_i : i = 1, 2, \ldots, n\}$ are independent each with a normal density $p(y|\theta)$ with zero mean and variance θ^{-1}given by

$$p(y|\theta) = (2\pi)^{-1/2} \theta^{1/2} \exp\left(-\frac{1}{2}\theta y^2\right)$$

then

$$s = n/2 \quad \text{and} \quad t = (y_1^2 + y_2^2 + \cdots + y_n^2)/2.$$

Note in all these cases that the posterior means tend to the sample proportion and variances tends to zero with probability one as the number n of independent observations tends to zero.

5.6.2 Normal-inverse gamma prior

The Normal-inverse gamma is the most common of the prior distributions used when we observe a random sample of normal random variables $\{Y_i : i = 1, 2, \ldots, n\}$. Let $\amalg_{i=1}^{n} Y_i|\boldsymbol{\theta}$ where $\boldsymbol{\theta} = (\theta_1, \theta_2)$ and each $Y_i|\boldsymbol{\theta}$ is normally distributed $N(\theta_1, \theta_2^{-1})$ with the same mean

θ_1 and precision (the reciprocal of the variance) θ_2 and so has density $p(y_i|\boldsymbol{\theta})$, $\boldsymbol{\theta} = (\theta_1, \theta_2)$ is given by

$$p(y_i|\boldsymbol{\theta}) = (2\pi)^{-1/2}\,\theta_2^{1/2}\exp\left(-\frac{1}{2}\theta_2\,(y_i-\theta_1)^2\right)$$

$i = 1, 2, \ldots, n$. We need to specify a prior density $p(\boldsymbol{\theta}) = p(\theta_1|\theta_2)p(\theta_2)$. A convenient choice for $p(\theta_2)$ the prior on the marginal density of the precision is to give this a gamma $G(\alpha^0, \beta^0)$ density given by (5.16) whilst setting $p(\theta_1|\theta_2)$ to be normally distributed $N(\mu^0, (n^0\theta_2)^{-1})$, i.e. with mean μ^0 and precision $\lambda^0 = n^0\theta_2$. Here the parameter n^0 can be thought of as the strength of prior information measured by comparative numbers of data much as we set $\alpha^0 + \beta^0$ for the beta prior in Section 5.5. If the prior density above accurately expresses the DM's beliefs then this prior is closed under sampling with hyperparameters of the prior updating thus:

$$\begin{aligned}
\mu^+ &= \rho\bar{y} + (1-\rho)\mu^0, \\
\lambda^+ &= (n^0+n)\theta_2, \\
\alpha^+ &= \alpha^0 + n/2, \\
\beta^+ &= \beta^0 + \frac{1}{2}\left\{\sum_{i=1}^{n}(y_i-\bar{y})^2 + n^0\rho\left(\mu^0-\bar{y}\right)^2\right\}
\end{aligned}$$

where $\bar{y} = n^{-1}\sum_{i=1}^{n} y_i$ $\rho = n\left(n^0+n\right)^{-1}$. Note in particular that the posterior mean is a weighted average of the prior mean and the sample proportion which tends to the sample proportion as n becomes large. The predictive density $p(z|\mathbf{y})$ – a Student t density – of the next observation in the sample is given by

$$p(z|\mathbf{y}) = \frac{\Gamma\left(\alpha^+ + \frac{1}{2}\right)\varsigma^{1/2}}{\Gamma(\alpha^+)\left(2\alpha^+\pi\right)^{1/2}\left(1+\frac{\varsigma}{2\alpha^+}\left(x-\mu^+\right)^2\right)^{-\left(\alpha^+ + \frac{1}{2}\right)}} \tag{5.17}$$

where

$$\varsigma = \frac{\left(n^0+n\right)\alpha^+}{\left(n^0+n+1\right)\beta^+}.$$

This predictive density is symmetric and looks very like the normal when α^+ is large and has mean μ^+ if $\alpha^+ > \frac{1}{2}$. However it has thicker tails than the normal and so is more prone to exhibiting outlying observations. This again reflects the uncertainty induced by having to estimate parameters – here especially the unknown sample variance which by chance, albeit with a small probability, might be a lot larger than the sample variance of the observations.

It is worth pointing out that this is not the only prior that is closed under sampling. For example we also obtain closure using a gamma on the variance rather than a gamma prior on the inverse of the variance. This makes the predictive distributions have tighter tails and is quite useful in modelling clustering problems. The fact that the priors on the mean and variance of the sample distribution are not independent of each other, whilst being defended by e.g. De Groot (1970) has also been criticised as being inappropriate for many practical scenarios: see for example O'Hagan and Forster (2004). If these parameters are assumed to be independent conjugacy is lost and we obtain a posterior density which is the product of a Student t density and a normal density: see Exercise 5.7. This can give rise to a predictive density which is bimodal if the prior mean and sample mean of the data are moderately far apart. In this sense when the data is in conflict with the prior then we believe that one or other is likely to be mis-specified and if using a bounded loss function will select a decision either close to the sample mean or the prior mean. This contrasts with the conjugate analysis which always forces symmetry and unimodality on the predictive density and hence forces us to compromise with any symmetric loss function even when there is dissonance between data and prior location. This is an example where there is a case for not using the usual conjugate forms; but see the next section on mixtures which gives a solution to this problem using mixture distributions. Other examples of this type of posterior bimodality resulting from simple common scenarios can be found in Lui and Hodges (2003); Smith (1977) and Andrade and O'Hagan (2006).

5.6.3 Multivariate normal-inverse gamma regression prior

Linear models are a very well used statistical tool and there is a conjugate model for these too. Here we need to explicitly condition on the values $\boldsymbol{x}_i$ of the covariates $\boldsymbol{X}_i$ of the ith unit. We observe a vector of random variables $\boldsymbol{Y}(\boldsymbol{x}) = (Y_1(\boldsymbol{x}_1), Y_2(\boldsymbol{x}_2), \ldots, Y_n(\boldsymbol{x}_n))$ indexed by a vector of known functions of covariates. Here $\boldsymbol{x}_i$ is a vector of length q indexing various not necessarily independent known features or facts about the ith unit.

Example 5.4. The n random variables $\boldsymbol{Y}(\boldsymbol{x}) = (Y_1(\boldsymbol{x}_1), Y_2(\boldsymbol{x}_2), \ldots, Y_n(\boldsymbol{x}_n))$ represent a random sample of the log of the breaking point of a steel cable. The vector of covariates $\boldsymbol{x}_i = (x_{i1}, x_{i2}, x_{i3}, x_{i4}, x_{i5}, x_{i6})$ indexes each sample of cable $i = 1, 2, \ldots, n$. Here $x_{.1}$ is always set to one, $x_{.2}$ is the log diameter of the cable, $x_{.3}$ the log percentage of carbon, $x_{.4}$ a measure of its level of rusting, $x_{.5} = x_{.3}x_{.4}$ and $x_{.6}$ is an indicator of whether the cable is bolted or clamped. Before he sees the result of the experiment the expert has good scientific reasons to believe that, given its covariates, each random variable Y_i is normally distributed with a mean

$$\psi_i(\boldsymbol{x}_i) = \theta_1 + \sum_{j=1}^{5} \theta_{j+1} x_{ij} \tag{5.18}$$

and a variance θ_7^{-1}. The DM's problem requires her to give a distribution of the mean breaking point $\zeta(\boldsymbol{x}_z)$ of the cable used in her safety device informed by the experiment above and its vector of covariates $\boldsymbol{x}_z$ which will be a function of the decision made by the DM. The DM and auditor is assured by the expert that the cable being used can be considered as a further replicate in the experiment above. So in particular the distribution of the actual breaking strain $Z|\boldsymbol{\theta}, \boldsymbol{x}_z$ conditional on the parameters of the model on the covariates of that cable is given as above.

This is a simple example of the linear model which has been studied for over 100 years and admits a straightforward Bayesian analysis which is closed under sampling. The class lies at the very foundation of the theory of design of experiments and many econometric models. One of its advantages is that it gives a simple representation in terms of a small number of parameters (here 7) of an infinite variety of types of model as characterised by the particular values of their covariates. Of course some assumptions have been made to achieve this: for example the randomness, normality and the structure of the mean function has the type of linear form given above. However for the purpose of this example assume these are justified. Here our purpose is simply to demonstrate how to perform a prior to posterior analysis of this model class when this is the case.

So let A denote the design matrix of this experiment which is $n \times q$ and whose ith row is $\boldsymbol{x}_i$. The normal *linear model* or *regression model* assumes that $\amalg_{i=1}^{n} Y_i(\boldsymbol{x}_i)|\boldsymbol{\theta}, A$ where $\boldsymbol{\theta} = \left(\boldsymbol{\theta}_1, \theta_{q+1}\right)$ and that $Y_i(\boldsymbol{x}_i)|\boldsymbol{\theta}, \boldsymbol{x}_i \sim N(\psi_i(\boldsymbol{\theta}_1, A), \theta_{q+1}^{-1})$ with density normal random variables $\{Y_i(\boldsymbol{x}) : i = 1, 2, \ldots, n\}$ with mean ψ_i given by

$$p_i(y_i|\boldsymbol{\theta}, \boldsymbol{x}_i) = (2\pi)^{-1/2}\, \theta_{q+1}^{1/2} \exp\left(-\frac{1}{2}\theta_{q+1}\, (y_i - \psi_i(\boldsymbol{\theta}_1, \boldsymbol{x}_i))^2\right)$$

where if $\boldsymbol{\psi} = (\psi_1, \psi_2, \ldots, \psi_q)$ then

$$\boldsymbol{\psi} = A\boldsymbol{\theta}_1$$

It follows that a likelihood of this data can be written as

$$l(\boldsymbol{\theta}|\boldsymbol{y}, A) = \theta_{q+1}^{n/2} \exp\left(-\frac{1}{2}\theta_{q+1}\, (\boldsymbol{y} - A\boldsymbol{\theta}_1)^T\, (\boldsymbol{y} - A\boldsymbol{\theta}_1)\right). \tag{5.19}$$

The most usual choice of model which is closed under sampling sets the prior $p_0(\boldsymbol{\theta}|\boldsymbol{x}) = p_0(\boldsymbol{\theta}_1|\theta_{q+1}, \boldsymbol{x})p_0(\theta_{q+1}|\boldsymbol{x})$. As in the simple normal analysis the prior density $p_0(\theta_{q+1}|\boldsymbol{x})$ of the precision is set to be gamma $G(\alpha^0, \beta^0)$ given by (5.16). Finally the density of the regression coefficients $\boldsymbol{\theta}_1|\theta_{q+1}, \boldsymbol{x}$ is set to have a multivariate normal distribution $N(\mu^0, (R^0\theta_{q+1})^{-1})$ where μ^0 is the q-dimensional mean vector of $\boldsymbol{\theta}_1|\theta_{q+1}, \boldsymbol{x}$ and R^0 is some symmetric positive definite $q \times q$ matrix so that its density is given by

$$p_0(\boldsymbol{\theta}_1|\theta_{q+1}, \boldsymbol{x}) = \left(\frac{\theta_{q+1}\, |R^0|}{(2\pi)^p}\right)^{1/2} \exp\left(-\frac{1}{2}\theta_{q+1}(\boldsymbol{\theta}_1 - \mu^0)^T R^0 (\boldsymbol{\theta}_1 - \mu^0)\right).$$

Note here that R^0 is the inverse of the covariance matrix (or the *precision matrix*) of $\boldsymbol{\theta}_1|\theta_2, \boldsymbol{x}$ divided by the precision of each observation and $|R^0|$ its determinant. The conditional density of $\boldsymbol{\theta}_1|\theta_2,\boldsymbol{x}$ is still multivariate normal with new mean vector μ^+ and the scaled precision matrices R^+ given by

$$\mu^+ = \Psi\widehat{\mu}(y) + (I_q - \Psi)\mu^+,$$
$$R^{\dagger} = R^0 + A^T A$$

where Ψ is the $q \times q$ matrix

$$\Psi = (R^0 + A^T A)^{-1} A^T A$$

and $\widehat{\mu}(\boldsymbol{y})$ is the usual maximum likelihood estimate $\left(A^T A\right)^{-1} A\boldsymbol{y}$ of $\boldsymbol{\theta}_1$. It can be shown that as $n \to \infty$, Ψ tends term by term to the identity matrix provided $A^T A$ is full rank. So, as with the univariate analysis, the posterior mean is a sort of weighted average of the prior mean and the maximum likelihood estimate and gets ever closer to the maximum likelihood estimate as n gets large. However until a moderate amount of data is collected the DM or expert will shrink her estimate by a significant extent to what she and he believed before sampling.

Because of closure under sampling we find that the posterior marginal densty of the precision is still gamma $G(\alpha^+, \beta^+)$ with

$$\alpha^+ = \alpha^0 + n/2$$
$$\beta^+ = \beta^0 + \frac{1}{2}(\mu^{0T} R^0 \mu^0 + \boldsymbol{y}^T \boldsymbol{y} - \mu^{+T} R^+ \mu^+).$$

The predictive density is a multivariate Student t and has a simple closed form analogous to its univariate analogue given above. For a careful recent discussion of the theory and varied uses of this model class see for example Denison *et al.* (2005) and O'Hagan and Forster (2004).

5.7 Non-conjugate inference*

5.7.1 Introduction

It is often a practical necessity to deviate from a prior to posterior analysis which is not closed under sampling. In this case the relationships between the prior hyperparameters, sample statistics and the posterior density can no longer be expressed through simple algebraic relationships, so it is somewhat harder to judge the sensitivity to mis-specification of various components of the model or to produce a narrative to the DM to explain why the data is directing the inference in the way it does. On the other hand, for many classes of models these posterior distributions admit quick and good numerical approximations of their distributions, using a variety of techniques, so in particular the distribution of $\boldsymbol{\theta}|\boldsymbol{x},\boldsymbol{y}$ we need can very often be specified to an arbitrarily high level of accuracy. The demands of faithfulness to the underlying science often make such numerical analyses necessary.

Again to do proper justice to this kind of model would require a much more detailed development than I have space for but is well-documented elsewhere. However it is very important to understand how flexible these techniques make a Bayesian analysis and the wide variety of possible structures they can be used to analyse. So I will briefly discuss one class of models, closely linked to some of the discrete models illustrated above, that can be used to demonstrate some of the advantages and a few disadvantages of these techniques.

5.7.2 Logistic regression

Consider the following example.

Example 5.5. The prosecution case asserts that a suspect was standing $x_{z2} = 12\ m$ from a window of pane thickness $x_{z3} = 5\ mm$ with a large rock $x_{z4} = 1$ (rather than a small one). Her interest is in the probability $\zeta(\boldsymbol{x}_z)$ where $\boldsymbol{x}_z = (1, x_{z2}, x_{z3}, x_{z4})$ that glass will land on the suspect's coat. To answer questions like this an expert takes a random sample of 500 experiments. On the ith experiment a unit throws a small ($x_{i4} = 0$) or large ($x_{i4} = 1$) rock standing at various distances x_{i2} from panes of different thicknesses x_{i3} and records whether ($y_i = 1$) or not ($y_i = 0$) glass landed on the coat of the thrower.

This setting looks very similar to the one described by the linear model. However there are two differences between this type of model and the normal linear model. First the parameter of interest ζ is a probability. It therefore must lie in the interval [0, 1]. So it is unreasonable to assume that its mean respects a linear form like (5.18). Second the data collected from each unit in the sample is not the value of a normally distributed random variable but an indicator variable Y indicating whether or not a glass fragment landed on the thrower.

A way to close the first difference is to reparameterise the probability ζ to a new parameter β that takes values on the whole real line. We argued in Chapter 1 that log odds were a useful and directly interpretable reparameterisation of a probability. So one way of reparameterising the problem so that it is more like a linear model is to specify it in terms of the *logistic link function* $\beta(\boldsymbol{x}_i) \triangleq \log(\zeta(\boldsymbol{x}_i)) - \log(1 - \zeta(\boldsymbol{x}_i))$ where we assume that this probability is given by the linear equation in the explanatory parameter vector $\boldsymbol{\theta}$ as

$$\beta_i(\boldsymbol{x}_i) = \theta_1 + \sum_{j=1}^{3} \theta_{j+1} x_{ij}$$

$i = 1, 2, \ldots, n, z$. A prior distribution on $\boldsymbol{\theta}$ can then be chosen exactly as we did for the normal linear model, for example using a conjugate normal inverse gamma prior. The prior distribution for $\beta_i(\boldsymbol{x}_i)$ can then be calculated just as before and transformation of variable techniques used to find the prior distribution of $\zeta(\boldsymbol{x}_i)$.

Of course what the DM needs is the distribution of ζ *posterior* to the sampling information $\boldsymbol{y}(\boldsymbol{x}) = (y_1(\boldsymbol{x}_1), y_2(\boldsymbol{x}_2), \ldots, y_n(\boldsymbol{x}_n))$. Were this a random normal sample with prior variance as set in the way suggested for the linear model then this would be straightforward. We would obtain the same prior to posterior analysis as in the example where a posteriori the distribution of $\boldsymbol{\theta}|\boldsymbol{y}(\boldsymbol{x})$ is multivariate Student t with hyperparameters linked to the prior hyperparameters in the way described above. From this we could calculate the distribution of $\zeta(\boldsymbol{x}_i)|\boldsymbol{y}(\boldsymbol{x})$.

Unfortunately the data is not of this form. However the likelihood of this sample in the case above is simply

$$l(\boldsymbol{\theta}|\boldsymbol{y}(\boldsymbol{x})) = \prod_{i=1}^{n} \zeta(\boldsymbol{\theta},\boldsymbol{x}_i)^{y_i}(1-\zeta(\boldsymbol{\theta},\boldsymbol{x}_i))^{y_i}.$$

To obtain the posterior density $p(\boldsymbol{\theta}|\boldsymbol{y}(\boldsymbol{x}))$ of $\boldsymbol{\theta}|\boldsymbol{y}(\boldsymbol{x})$ all we need do is to multiply $l(\boldsymbol{\theta}|\boldsymbol{y}(\boldsymbol{x}))$ by the chosen prior and then renormalise this product and multiply it by a proportionality constant so this integrates to one. The proportionality function cannot be calculated directly. However the density $p(\boldsymbol{\theta}|\boldsymbol{y}(\boldsymbol{x}))$ can be approximated numerically by using the functional forms of the prior and likelihood to draw a random sample whose values can be formally proved to have this density. Because of the convexity of the logarithm of the likelihood most standard numerical methods of this type work well for this problem. By performing this analysis the expert can provide, usually to drawing accuracy, his posterior density for $\boldsymbol{\theta}|\boldsymbol{y}(\boldsymbol{x})$ which the DM can then adopt as her own.

These techniques can obviously be used in a wide range of problems of unbounded complexity, provided it can be proved that the numerical methods used provide good estimates of the required posterior. Furthermore, because it makes little difference to the numerical algorithms whether or not the priors are chosen from a particular family a wide range of alternative priors can be used in this context and the methodology sketched above will still apply.

Most useful methods like this one also have their downsides. First the implications of an expert's hypothesis of the linear model on a transformed scale like β, called a *generalized linear model,* are usually more difficult to appreciate fully than in the pure linear model. The logistic model described above is slightly easier in this regard because linearity of response can be linked to certain families of conditional independence statements: see Lauritzen (1996); Whittaker (1990) and O'Hagan and Forster (2004). It is often possible to explain the implications of the model class to the DM in terms of hypotheses about how different variables in the problem might or might not influence each other. Nevertheless such explanations of even the logistic models are not always available.

The setting of appropriate priors is also more difficult in this general setting because the roles of the hyperparameters can be transformed by the link function. For example the Student t prior from the usual conjugate analysis translates into one which typically exhibits up to three modes and always two modes in the prior density for ζ suggesting that the expert believes either the probability is very close to zero or one or alternatively somewhere in between. The variance of β is therefore linked to bifurcation rather than uncertainty especially when the expectation of this variance is greater than 2. Furthermore very high prior variances on $\boldsymbol{\theta}$ mean that the expert believes that observations will take the same value with probability very close to one. So high variances in the prior parameters in the model do not translate on to a statement of prior uncertainly about the observations as it does in the normal linear model. In fact if only small samples are taken and all observations take the same value then because of this prior the probability of the next data point being different will be infinitesimally small. This will not usually correspond to conclusions one likes to make in most applications. Great care needs to be exercised in the setting of priors in these models especially when there are many probabilities to estimate each with only moderate numbers of observations.

But with these caveats these methods provide a very flexible toolkit for addressing data accommodation in a wide range of interesting problems. They are currently widely used.

5.8 Discrete mixtures and model selection

5.8.1 Mixtures are closed under sampling

Sometimes we are in the situation where the DM's elicited prior does not have a particularly simple form. For example it might be bimodal or highly skewed in some direction. The following result is then very useful. Take a family of prior distributions $\mathcal{P} = \{P(\boldsymbol{\theta}|\boldsymbol{\alpha}) : \alpha \in A \subseteq \mathbb{R}^n\}$ of absolutely continuous distributions with densities $p(\boldsymbol{\theta}|\boldsymbol{\alpha},\boldsymbol{\pi})$. Write

$$\Big\{\overline{\mathcal{P}} = P(\boldsymbol{\theta}|\boldsymbol{\alpha},\boldsymbol{\pi}) : \{P(\boldsymbol{\theta}|\boldsymbol{\alpha},\boldsymbol{\pi}) = \sum_{i=1}^{p} \pi_i P(\boldsymbol{\theta}|\boldsymbol{\alpha}_i) :$$

$$P(\boldsymbol{\theta}|\boldsymbol{\alpha}_i) \in \mathcal{P}, \alpha_i \in A \subseteq \mathbb{R}^n, 1 \le i \le p, \boldsymbol{\pi} \in \mathbb{S}^p\Big\}$$

where $\mathbb{S}^p \subset \mathbb{R}^p$ is the simplex of probability mass functions on p values: i.e. contains all $\boldsymbol{\pi} = (\pi_1, \pi_2, \ldots, \pi_p)$ such that $\pi_i \geq 0$, $1 \leq i \leq p$ and $\sum_{i=1}^{p} \pi_i = 1$, where by abuse of notation we allow p to be possibly infinite. It follows that any prior $P_0(\boldsymbol{\theta}|\boldsymbol{\alpha}_i) \in \overline{\mathcal{P}}$ has density $p_0(\boldsymbol{\theta}|\boldsymbol{\alpha}_i)$ where

$$p_0(\boldsymbol{\theta}|\boldsymbol{\alpha}_i) = \sum_{i=1}^{p} \pi_i p_0(\boldsymbol{\theta}|\boldsymbol{\alpha}_i).$$

Therefore the posterior density $p(\boldsymbol{\theta}|\boldsymbol{\alpha},\boldsymbol{\pi},\boldsymbol{y})$ associated with this mixture prior satisfies

$$\begin{aligned} p(\boldsymbol{\theta}|\boldsymbol{\alpha},\boldsymbol{\pi},\boldsymbol{y})p(\boldsymbol{y}) &= \sum_{i=1}^{p} \pi_i p_0(\boldsymbol{\theta}|\boldsymbol{\alpha}_i)p(\boldsymbol{y}|\boldsymbol{\theta}) \\ &= \sum_{i=1}^{p} \pi_i p_i(\boldsymbol{y}|\boldsymbol{\alpha}_i)p_0(\boldsymbol{\theta}|\boldsymbol{\alpha}_i,\boldsymbol{y}). \end{aligned}$$

Thus

$$p(\boldsymbol{\theta}|\boldsymbol{\alpha},\boldsymbol{\pi},\boldsymbol{y}) = \sum_{i=1}^{p} \pi_i^+ p_0(\boldsymbol{\theta}|\boldsymbol{\alpha}_i,\boldsymbol{y}) \tag{5.20}$$

where

$$\pi_i^+ = \frac{\pi_i p_i(\boldsymbol{y}|\boldsymbol{\alpha}_i)}{\sum_{j=1}^{p} \pi_j p_j(\boldsymbol{y}|\boldsymbol{\alpha}_j)}. \tag{5.21}$$

Note in particular that the log odds $o_{i,j}^+ = \log \pi_i^+ - \log \pi_j^+$of these posterior probabilities link to the prior odds $o_{i,j}^0 = \log \pi_i - \log \pi_j$ and log marginal likelihood ratios $\lambda_{i,j} = \log p_i(\boldsymbol{y}|\boldsymbol{\alpha}_i) - \log p_j(\boldsymbol{y}|\boldsymbol{\alpha}_j)$ by the by now familiar equation

$$o_{i,j}^+ = o_{i,j}^0 + \lambda_{i,j}$$

where $\lambda_{i,j}$ is a function not only of the data but also of the hyperparameters in the two components being compared. It follows in particular that if $\mathcal{P} = \{P(\boldsymbol{\theta}|\boldsymbol{\alpha}) : \alpha \in A \subseteq \mathbb{R}^n\}$ is closed under sampling for a likelihood $l(\boldsymbol{\theta}|\boldsymbol{y})$, then so is $\overline{\mathcal{P}}$.

This property of closure of mixtures under sampling is useful for a number of reasons. The first we explore is the potential flexibility this gives us in choosing a prior density closed under mixing.

5.8.2 Mixing to improve the representation of a prior

Typically it is possible to take the standard family – like the beta for binomial sampling and the prior for normal sampling – and use mixtures of these to approximate to an arbitrary degree of accuracy any prior density elicited from a client. When approximations of this kind lead to approximately the same inferences and decisions it is possible at least in principle to perform an excellent decision analysis of many problems where data are sampled from a standard distribution and an arbitrary prior density is used. This continuity under approximation – with a couple of caveats – does in fact hold as we will demonstrate later in this chapter. Fast exact calculations based on this closure are therefore possible in a surprisingly wide range of applications; see e.g. Heard *et al.* (2006); Smith *et al.* (2008a). These approximations are now embedded in various software tools where densities can be drawn and these are then automatically approximated by a suitable mixture. Consider the following simple example of how learning under mixed priors can differ from a standard analysis.

Example 5.6. A random variable Y has a binomial $Bi(N,\theta)$. A DM is interested in the proportion θ of particles that contain a particular chemical C and she takes a random sample of N such particles. If the particles have not been contaminated she believes that θ will have a $Be(2, 98)$ beta distribution. However she believes that there is a probability 0.05 that the sample is contaminated and given this she believes that θ will have a $Be(2, 8)$ distribution. In this case her prior density must take the form

$$p_0(\theta) = 0.05p(\theta|2, 8) + 0.95p(\theta|2, 98)$$

where $p(\theta|\alpha, \beta)$ is a beta $Be(\alpha, \beta)$. From the above

$$p_0(\theta) = \pi^+ p(\theta|2 + x, 8 + N - x) + (1 - \pi^+)p(\theta|2 + x, 98 + N - x)$$

where the posterior odds o^+ are related to the prior odds o by the equation

$$o^+ = -\log 20 + \log \left\{ \frac{\Gamma(10)\Gamma(98)}{\Gamma(8)\Gamma(100)} \frac{\Gamma(100 + N)\Gamma(8 + N - x)}{\Gamma(98 + N - x)\Gamma(10 + N)} \right\}.$$

The last term in this expression increase dramatically as y/N increases. Thus suppose that $N = x = 3$ and the first three of the samples we observe are positive. Then some simple calculations give us that $\pi^+ = 0.97$ suggesting the sample is almost certainly contaminated: the prior expected level $\mu_0 = 0.029$ to a posterior mean $\mu_+ = 0.37$ is almost identical to the mean we would have obtained initially had we started with the hypothesis that the sample was contaminated. Had we used a beta prior with the same mean and variance as the mixture above the posterior mean would be much smaller and, because it did not explicitly model the possibility of contamination, would not have accommodated this unexpected information nearly so quickly.

There is now a considerable literature on the useful theoretical properties of mixture distributions and these methods have now been applied in a wide range of applications. See Marin *et al.* (2004) and Fruthwirth-Schnatter (2006) for extensive reviews of this important area.

5.8.3 Mixing for model selection

What happens when an analyst is faced with several different explanations of what is happening and has to choose between these? To consider this issue it is first helpful to consider the following artificial scenario. Suppose the analyst knows that a sequence of random vectors $\boldsymbol{Y} = \{\boldsymbol{Y}_t : t = 1, 2, \ldots, T\}$ is produced – via transformations of outputs from perfect random number generators – to be the output of one of p possible data generating mechanisms $M_1, M_2, \ldots, M_p$: the analyst's prior probability of M_i generating $\boldsymbol{Y}$ being π_i^0, $1 \leq i \leq p$. If the observed data were $\boldsymbol{y}$ and the density value of $\boldsymbol{y}$ given M_i were true is denoted by $p(\boldsymbol{y}|M_i)$, $1 \leq i \leq p$ – where we have marginalised over any parameters in the model – then it is clear how the Bayesian analyst should proceed. Following the same argument leading to equation (5.21) the analyst simply updates her prior probabilities using the formula for her posterior probabilities $\boldsymbol{\pi}^+ = (\pi_1^+, \pi_2^+, \ldots, \pi_p^+)$

$$\pi_i^+ = \frac{\pi_i p_i(\mathbf{y}|M_i)}{\sum_{j=1}^{p} \pi_j p_j(\mathbf{y}|M_j)}. \tag{5.22}$$

A model M_i with the largest value of π_i^+ is called the *maximum a posteriori (MAP)* model and if the analyst must choose a model and obtains strictly positive gain only if she chooses the right one then she should choose a MAP model. Notice that to identify a MAP model we need only calculate the posterior log odds

$$s_i = \log \pi_i^+ - \log \pi_i^+ = \left(\log \pi_i^+ - \log \pi_i^+\right) + (\log p_i(\mathbf{y}|M_i) - \log p_1(\mathbf{y}|M_1))$$

choosing M_i, $i = 2, 3, \ldots, n$ maximising this expression if $\log \pi_i^+ - \log \pi_i^+ \geq 0$ and otherwise choosing M_1. Also note that if the DM believed all models were a priori equally probable then the best choice of model is one for which $\log p_i(\mathbf{y}|M_i)$ is maximised. Alternatively to forecast the next observation $\mathbf{y}_{T+1}$ she should use the mixture predictive density/mass function.

When $\boldsymbol{Y}$ is a random sample and data is drawn from from M_i then it can be proved that this method of model selection has a nice asymptotic property. Thus provided that $\pi_i > 0$ and $j \neq i$

$$\int \left(\log p_i(\mathbf{y}_t|M_i) - \log p_j(\mathbf{y}|M_j)\right) p(\mathbf{y}_t|M_i) d\boldsymbol{x}_t > 0$$

i.e. that no other model M_j gives exactly the same probability over observables as M_i then $\pi_i^+ \to 1$ almost surely as $T \to \infty$; see Bernardo and Smith (1996) or French and Rios Insua (2000). So in the long run the analyst can expect to select the right model. Furthermore, even if none of the models is the generating model the method will choose the model within the class which is closest to the generating model in Kulback–Liebler distance; see e.g. Bernardo and Smith (1996).

Because of their simplicity such model selection methods are now widely used in practical inferential problems, see for example Bernardo and Smith (1996); Denison *et al.* (2005), especially when the number of models p searched over is huge. Of course, as argued in Bernardo and Smith (1996) the scenario above where this method is fully justified is hardly ever met. Usually the DM will need to choose appropriate priors over the parameters of the different models in her model class. To do this accurately will be a practical impossibility unless some sort of homogeneity is assumed. But one critical issue we eluded to earlier is that marginal likelihoods, and therefore these selection methods, are heavily dependent on the prior density used on the parameters within a putative model. So it is important to carefully calibrate these priors to aspects the DM genuinely believes she will see in the data. There are ways of doing this in various different contexts – see Chapter 9 for an example.

However this does mean that these selection techniques are necessarily linked to a particular DM's prior beliefs about an underlying context.

One application of model selection in a decision analysis is when a DM has several experts giving different predictions of sequences of events of interest. If she believes that one of the experts is right but is uncertain which one then she should use the mixture of their forecasts given above and update the prior probabilities π_i of the ith expert being the right one using the formula above. After a few training observations the DM will often then find that all but one expert has a very small posterior probability of being correct. Care needs to be exercised in using this method however. This process can be distorted by the assumption that there is at least one expert that gives best probability forecasts in all scenarios about everything associated with the future variables of interest. It can discard good experts simply because they are uncalibrated over events about which the DM is almost indifferent.

5.9 How a decision analysis can use Bayesian inferences

5.9.1 Statistics and their impact on a decision analysis

In the preceding section we illustrated how to help the DM to build a full joint model over past data and the future variables conditional on known facts. An elegant use of the rules of probability then allows her formally to deduce her probability distribution that takes account of the data and the facts. As well as using probabilistic judgements that are likely to be shared by an auditor, such a procedure has the advantage that it helps the DM avoid the sorts of systematic biases discussed in the previous chapter. However this formal process is not without its own pitfalls. When planning to accommodate data into a decision analysis it is important to ask: "Are the probabilities/parameters associated with the experiment about events the DM believes can be logically related to events with the same probabilities/parameters in the instance at hand, considered by the DM? In particular is the instance we consider simply another replicate of the experimental evidence in hand or do we have more instance-specific information which also needs to be accommodated?"

In the former case we can assume $\boldsymbol{\zeta} = \boldsymbol{\theta}$ and once the DM's utility function has been elicited, analyses like those above give sufficient information for the decision analysis. However it is often impossible to answer this question positively. For instance consider the forensic example where the DM needs to assign the probability ζ that the suspect has innocently acquired glass on his clothing. The original survey that was intended to inform this assignment assesses – through a random sample of the population – the probability θ that someone has different numbers of glass fragments on their clothes. However a typical suspect – who were often young men who spent a lot of time on the streets – had orders of magnitude more glass on their clothing that the average person in the general population. This is a simple example where the answer to the question in the last paragraph is "No".

Therefore in many cases the probability $p(\boldsymbol{\zeta}|\boldsymbol{\theta}, \boldsymbol{x})$ is not degenerate and has to be elicited from the DM. Furthermore there is usually no data supporting this probability so it *must* be reflect a subjective judgement. In this sense the majority of Bayesian decision analyses need to seriously engage in the elicitation of subjective assessments regardless of the difficulties this might bring if their conclusions are not going to mislead. Incidentally there is an advantage when $p(\boldsymbol{\zeta}|\boldsymbol{\theta},\boldsymbol{x})$ is not degenerate. We will see in the last section of this book that the DM's inferences about ζ will then be much more robust to mis-specification of a trusted expert's prior on $\boldsymbol{\theta}$.

Now instead of eliciting $p(\boldsymbol{\zeta}|\boldsymbol{\theta},\boldsymbol{x})$ and using probability calculus to find $p(\boldsymbol{\zeta}|\boldsymbol{x},\boldsymbol{y})$ we could alternatively simply display summaries of results $\boldsymbol{y}$ and then encourage the DM to specify these densities directly. From a practical perspective this may well be the best option when $p(\boldsymbol{\zeta}|\boldsymbol{\theta},\boldsymbol{x})$ is an unnatural construct for the DM to elicit, for example when $\boldsymbol{\zeta}|\boldsymbol{\theta},\boldsymbol{x}$ reflects an anticausal conditioning. However well-designed experiments will often produce good benchmark cases – e.g. information about parameters $\boldsymbol{\theta}$ in an ideal or ordinary setting from which to measure beliefs about $\boldsymbol{\zeta}$ – exhibiting some form of distortion from the ideal or ordinary setting – associated with the current instance at hand. In the example above for example it is not unnatural to consider how much more glass a person like the suspect has on their clothing given they spend an above average time on the streets.

My practical advice to the DM is to use a full Bayesian model when incorporating data from well-designed experiments or observational studies which match the case in point well, but to be aware that the inferences made in this way are approximate and may be prone to error, including this explicitly through elicitation of $p(\boldsymbol{\zeta}|\boldsymbol{\theta},\boldsymbol{x})$, carefully documenting the reasoning behind these probability assignments. The positive advantages of the implicit credence decomposition in reducing psychological biases as well as the transparency of the inference tend to support this formal incorporation of data. On the other hand when great leaps of belief are necessary to link the experimental evidence and/or survey information it is often better simply to directly present this information in support of the chosen probability distributions of the analysis. In either case the DM and auditor should be cogniscent of the fact that the presented distributions are just one honest subjective and balanced interpretation of the evidence at hand. There is no panacea here: different contexts demand different protocols.

5.9.2 *Other's prior judgements*

The DM often need a trusted expert to provide her with his posterior density $p(\boldsymbol{\theta}|\boldsymbol{x},\boldsymbol{y})$ so she can proceed with her analysis. In some circumstances, the expert might be unwilling to specify a subjective prior $p(\boldsymbol{\theta}|\boldsymbol{x})$ but to strive to make this as "objective" as possible and reflective as possible of the likelihood to "let the data speak for itself".

If an expert has a genuine prior then he should be encouraged to use it when conveying his posterior probability. If he has a likelihood and the DM has sufficient domain knowledge and time then she should provide her own prior. Otherwise the DM has just to go with what she is given. Some software tries to address this issue by suggesting default prior settings which give proper but diffuse prior distributions. When data sets give rise to a likelihood which is very spiked – and this would normally be the case for such studies – it is shown in Chapter 8 that with certain caveats any analysis will not be sensitive to the expert's choice. So if the expert has used this sort of software in deriving his posterior then the analysis is likely to be more robust than if he blindly uses some vague prior. If the expert is determined to use improper prior settings my recommendation would be to encourage him to use appropriate reference priors developed and discussed in Bernardo and Smith (1996) which in a particular sense are least formally least informative and so give a $p(\boldsymbol{\theta}|\boldsymbol{x})$ which makes the most conservative evaluation of the the strength of evidence in his data.

Fortunately this type of poor procedure as necessary compromise usually has only a small effect on the effectiveness of a Bayesian decision analysis. This is because, from both a practical and theoretical viewpoint the levels of uncertainty on the conditional $p(\boldsymbol{\zeta}|\boldsymbol{\theta},\boldsymbol{x})$ is usually much higher than that in $p(\boldsymbol{\theta}|\boldsymbol{x})$. This means that in fact although inferences about $\boldsymbol{\theta}$ can be badly distorted by the use of improper priors, unless $\boldsymbol{\zeta} = \boldsymbol{\theta}$ inferences about $\boldsymbol{\zeta}$ rarely are.

*5.9.3 Accommodating data previously seen**

A related issue is the credence the DM can legitimately give to her own or an expert's specification $p(\boldsymbol{\theta}|\boldsymbol{x},\boldsymbol{y})$. So far we have assumed that the person using Bayes rule to calculate this had not seen the data $\boldsymbol{y}$ before she chose her prior density $p(\boldsymbol{\theta}|\boldsymbol{x})$ and her sampling mass function $p(\boldsymbol{y}|\boldsymbol{\theta},\boldsymbol{x})$. Bayes rule tells her what she then planned to do when she observed $\boldsymbol{y}$.

However it is not unusual for her to have seen and made a preliminary study of her data set *before* applying Bayes rule. Indeed it may be unavoidable that she has this information before she makes her choices. The problem here is that a prior to posterior analysis is only formally valid as a provisional plan of what she expects to believe in the future were she to see certain data. If she chooses either $p(\boldsymbol{\theta}|\boldsymbol{x})$ or $p(\boldsymbol{y}|\boldsymbol{\theta},\boldsymbol{x})$ so that they concur with the observed data $\boldsymbol{y}$ then she is guilty of double counting and her analysis is faulty. Even the most committed DM will find it difficult to cast her mind back honestly to what she would have thought before she knew what she now knows.

For data from good sample surveys or designed experiments of the type discussed above, this is not such a problem because then usually the sampling distribution will be defined by the *type* of sample survey designed or experiment conducted and class of dependences reasonable to envisage. So unless the observed data calls into question the validity of the survey or experimental design the DM and auditor will usually agree on this distribution both then and now. The specification of the prior $p(\boldsymbol{\theta}|\boldsymbol{x})$ may be more difficult to provide honestly and accurately. But a sensitivity analysis will then usually demonstrate that its specification will not have a big effect.

On the other hand if the data is undesigned and observational and the family of sampling distributions chosen in the light of what is subsequently learned then the results of the prior to posterior analysis should be treated with extreme caution, especially if this is provided by a remote expert.

5.9.4 Counterfactuals*

An even more delicate problem the analyst may need to face is when the DM is forced to think back to what she might legitimately have thought had something different happened in the past than actually did. This type of issue is routinely addressed in liability claims. For example a person A was exposed to a certain amount of nuclear radiation. Then fifteen years later she develops a cancer of the thyroid. One question that needs to be asked is whether and to what extent the appearance of the cancer can be attributed to the past exposure to radiation. This requires the court to consider the event of the appearance and timing of the cancer had the person not been exposed to radiation – an event that is known not to have happened and so is *counterfactual*.

One way for a Bayesian to think about this question is to try to retrospectively construct what she would have thought before she learned A had developed the cancer, drawing on evidence about the development and existence of this cancer in people similar to A. She can then condition on the event she knows to be true – here that A had cancer. From this she can deduce what her posterior probability would have been had she held this model.

It is well known that this sort of inference is perilous for a number of reasons. First we have the definition of exactly how to define the population "similar" to A. This is very like what we need to do in a standard decision analysis. However the difference is that this needs to be done in the knowledge of what happened to A. It is very difficult to force this fact to be forgotten and not allow it to influence the choice of population, the covariates we use to classify A or the choice of situations in any event tree we draw. For Bayes rule to be valid these covariates need to be the ones we would have selected not knowing what had subsequently happened to A after the exposure. In particular we need to imagine all the different types of ways in which someone just like A might have developed different possibly fatal disease classes A is known not to have developed over the fifteen years, possible exposure to different sources of radiation he might have been exposed to but was not, and so on. This is a very challenging mind experiment to perform even for a philosopher.

Second, note that there is no diagnostic available to check whether the backcasting described above is plausible. In particular a model which says that A was inevitably going to develop a cancer fifteen years after the event with probability one has a higher marginal likelihood than any model that accepted that this development was uncertain. These sorts of issues also impinge on Bayesian model selection and relate to such phenomena as Lindley's paradox (Bernardo and Smith, 1996; Denison *et al.*, 2005). Third, when variables are continuous the construction of a backcast model is ambiguous unless the method of measuring the evidence is set up beforehand. Consider the following simple hypothetical setting discussed in detail by Kadane *et al.* (1986). You have learned that two continuous variables X and Y are equal but you have not seen the value of X. Doing a backcasting mind experiment you decide that you would have believed that $X \amalg Y$ were both normally distributed random variables with mean zero and variance one before you had seen that they were equal. Now you have a problem. For you notice that what you have seen could be expressed as having observed that $Z_1 \triangleq X - Y = 0$ or by the logically equivalent event $Z_2 \triangleq X/Y = 1$. But the rules of probability tell you that the conditional density $p(x|z_1 = 0) \neq p(x|z_2 = 1)$. So which are you going to use to represent your current beliefs?

Of course it is sometimes impossible to avoid a counterfactual analysis. Rubin (Imbens and Rubin, 1997; Rubin, 1978) has carefully defined a methodology for addressing counterfactual issues by treating counterfactual events as missing data and using various collections of conditional independence statements – assumed to have an enduring validity – to devise formal ways of addressing these problems. In the terminology of Chapter 2 he posits many parallel situations across a population of people with different partially matching histories and uses these hypotheses to make the necessary inferences. This methodology has not gone uncriticised however (see e.g. Dawid (2000); Robins (1997) and Kurth *et al.* (2006)). These issues also link to ideas of causality as Pearl uses this term and which we address more fully in a later chapter.

5.10 Summary

In simple scenarios there are straightforward and elegant ways of formally including data from experiments, surveys and observational studies into the probabilistic evaluation of features of the problem related to the distribution of a utility function under various decisions. Often, by choosing a prior closed under sampling, these computations can be made to be quick and to provide a transparent narrative enabling the DM to explain the effects of the data she has used on her current probabilistic beliefs. We will discuss these issues further as they apply to more complicated scenarios in a later chapter. Even when conjugate analyses are impossible, sampling and other methods make a wide variety of statistical analyses possible which can be automatically fed into the decision analysis and at least for moderate sized problems usually provide good approximate evaluations.

The main problems we encounter therefore tend to centre on the DM's ability faithfully to represent the relationships between her information sources, her model of process and her utilities. It is the appropriate structuring of the problem which is the necessary prerequisite for effective decision making. Once we move to the study of large problems, producing effective frameworks for such structuring therefore becomes critical. These are the issues which form the basis of the second part of the book, where we use the theoretical underpinning and practical implementation techniques here described for smaller routine forms of decision analysis in the study of much larger systems.

5.11 Exercises

5.1 Hits at a website are observed over a short interval of time over an evening and the inter arrival times Y_i between the $(i-1)$ th and ith hit are recorded, $i = 1, 2, \ldots, n$. Suppose a DM believes that $\amalg_{i=1}^{n} Y_i|\theta$ have a density with support the positive real line which takes the form $p_i(y_i|\theta) = \theta e^{-\theta x}$ where $\theta > 0$, whatever the index i. A colleague just observes that n observations occurred until time T – the last one occurring exactly at the time of your last hit. She believes this has a Poisson density with rate $n\theta$. If you both believe that the prior density $p_0(\theta)$ on θ is Gamma $G(\alpha_0, \beta_0)$, prove that the DM's posterior density and that of her colleague are identical.

5.2 You have been told that access to regions of a website crashes on average about once every 10 hours with a variance over different regions of about 0.01. You are accessing a particular region and so far the times between a crash have been 3.2, 12.7, 20.6, 7.9, 10.2. Assuming that these times have an exponential density with mean parameter θ and your prior density over θ is Gamma distributed, find the probability distribution before the next crash.

5.3 Let $Y_1, Y_2, \ldots, Y_{n+1}$ be independent random variables conditional on θ with a uniform distribution on the interval $[0, \theta]$ so that inside this interval the density $p_i(y_i) = \theta^{-1}$, $i = 1, 2, \ldots, n$. Your prior density $p(\theta|\alpha, \beta)$ on θ – where $\alpha, \beta > 0$ is given by a Pareto density $Pa(\alpha, \beta)$ – is

$$p(\theta|\alpha, \beta) = \begin{cases} \alpha\beta^{\alpha}\theta^{-(\alpha+1)} & \text{when } \theta > \beta \\ 0 & \text{otherwise.} \end{cases} \tag{5.23}$$

Find the posterior density of θ given $y_1, y_2, \ldots, y_n$ and the predictive density of $Z = Y_{n+1}$ given $y_1, y_2, \ldots, y_n$. Graph this predictive density.

5.4* i) Suppose the random variables $\phi_1, \phi_2, \ldots, \phi_r$ are mutually independent and that ϕ_i is Gamma $G(\alpha_i, \beta)$ distributed. Use change of variables techniques to prove that $\phi_. \triangleq \phi_1 + \phi_2 + \cdots + \phi_r$ is $G(\alpha_., \beta)$ distributed where $\alpha_. \triangleq \alpha_1 + \alpha_2 + \cdots + \alpha_r$ and is independent of $\boldsymbol{\theta}$ where $\boldsymbol{\theta} = (\theta_1, \theta_2, \ldots, \theta_r)$ where $\theta_i \triangleq (\phi_.)^{-1}\phi_i$, $i = 1, 2, \ldots, r$. Also prove that $\boldsymbol{\theta}$ has a Dirichlet $D(\boldsymbol{\alpha})$ distribution where $\boldsymbol{\alpha} = (\alpha_1, \alpha_2, \ldots, \alpha_r)$ where these components are defined above.

ii) Now suppose you observe mutually independent Poisson random variables Y_i with rate ϕ_i, $i = 1, 2, \ldots, n$. Find the posterior distribution of the vector of probabilties $\boldsymbol{\theta}$ defined above. Show that this is the same posterior distribution you would have obtained had $(y_1, y_2, \ldots, y_r)$ been the observations from a multinomial $M(N, \boldsymbol{\theta})$ random vector where $N = y_1 + y_2 + \cdots + y_r$.

5.5 Prove the results about prior to posterior analyses and the marginal likelihoods and predictive distributions for the normal inverse gamma analysis above.

5.6 You have a single observation y which conditional on its median θ has a Cauchy density

$$p(y|\theta) = \pi^{-1}(1 + (\theta - y)^2)^{-1}.$$

Your prior density $p(\theta)$ is also Cauchy but with median μ and so is written

$$p(\theta) = \pi(1 + (\theta - \mu)^2)^{-1}.$$

Let $m = 1/2(y + \mu)$ and $\Delta = \frac{1}{4}(y - \mu)^2$. Show that

$$p(\theta|y) \propto (1 + 2\{(\theta - m)^2 + \Delta\} + 4\{(\theta - m)^4 - \Delta\}^2)^{-1}.$$

Hence or otherwise prove that $p(\theta|y)$ is always symmetric with median $m = 1/2(y + \mu)$. Hence or otherwise show that this is unimodal with mode at m if $|y - \mu| \leq 1$ but otherwise has an antimode at zero and two modes, one between μ and m and the other between y and m. Interpret this result.

5.7* Normally distributed random variables $Y_1, Y_2, \ldots, Y_n$ are independent given their mean θ_1 and variance θ_2^{-1}. You assume that a priori $\theta_1 \amalg \theta_2$ where θ_1 has a normal distribution with known mean μ and variance σ^2, whilst θ_2 is gamma distributed $G(\alpha, \beta)$. Show that the marginal posterior density of θ_1 can be written as proportional to a normal and Student t density. Hence or otherwise show that the posterior density can have two modes for certain values of the hyperparameters.

5.8 $Y_1, Y_2, \ldots, Y_n$ are independent random variables conditional on θ uniformly distributed on the interval $[0, \theta]$. A DM has a prior density $p(\theta)$ given by

$$p(\theta) = \pi_1 p_1(\theta|\alpha, \beta_1) + \pi_2 p(\theta|\alpha, \beta_2)$$

where $\pi_1, \pi_2 > 0, \pi_1 + \pi_2 = 1$ and $p_i(\theta|\alpha, \beta_i)$ are Pareto densities $Pa(\alpha, \beta_i), i = 1, 2$ whose form is given in (5.23) where $\beta_1 < \beta_2$. What sort of beliefs would this prior represent? Show that the posterior distribution of θ can be written in the form

$$p(\theta) = \pi_1^+ p_1(\theta|\alpha^+, \beta_1^+) + \pi_2^+ p(\theta|\alpha^+, \beta_2^+)$$

where $p_i(\theta|\alpha^+, \beta_i^+)$ are also Pareto densities $Pa(\alpha^+, \beta_i^+), i = 1, 2$ and give the values of $\frac{\pi_1^+}{\pi_2^+}, \alpha^+, \beta_1^+, \beta_2^+$ as a function of the prior hyperparameters and the observed values $y_1, y_2, \ldots, y_n$.

5.9 Counts $\boldsymbol{x} = (x_1, x_2, x_3)$ of N units each lying in one of three categories are taken. The sample mass function $p(\boldsymbol{x}|\boldsymbol{\theta})$ of $\boldsymbol{X}$ is multinomial so that

$$p(\boldsymbol{x}|\boldsymbol{\theta}) = \frac{N!}{x_1!x_2!x_3!}\theta_1^{x_1}\theta_2^{x_2}\theta_3^{x_3}$$

where $x_1 + x_2 + x_3 = N, \boldsymbol{\theta} \in \Theta$ where

$$\Theta = \{\boldsymbol{\theta} = (\theta_1, \theta_2, \theta_3), \theta_1, \theta_2, \theta_3 > 0, \theta_1 + \theta_2 + \theta_3 = 1\}.$$

The 3-dimensional Dirichlet $D(\boldsymbol{\alpha})$ density $\pi(\boldsymbol{\theta}|\alpha)$, $\boldsymbol{\alpha} = (\alpha_1, \alpha_2, \alpha_{,3}), \alpha_1, \alpha_2, \alpha_3 > 0$ on the vector of probabilities $\boldsymbol{\theta} \in \Theta$ is given by

$$\pi(\boldsymbol{\theta}|\alpha^0) = \begin{cases} \Delta(\alpha)\theta_1^{\alpha_1 - 1}\theta_2^{\alpha_2 - 1}\theta_3^{\alpha_3 - 1} & \text{when } \boldsymbol{\theta} \in \Theta \\ 0 & \text{otherwise} \end{cases}$$

where the proportionality constant

$$\Delta(\alpha) = \frac{\Gamma(\alpha_1 + \alpha_2 + \alpha_3)}{\Gamma(\alpha_1)\Gamma(\alpha_2)\Gamma(\alpha_3)}$$

and $\Gamma(\alpha) = \int_0^\infty u^{\alpha-1}e^{-u}du$, $\alpha > 0$ is the Gamma function with the property that $\Gamma(\alpha) = (\alpha - 1)\Gamma(\alpha - 1)$, $\Gamma(1) = 1$. Suppose the decision maker believes a priori that the vector of probabilities $\boldsymbol{\theta}$ has a Dirichlet $D(\boldsymbol{\alpha}^0)$ distribution.

Three experts a, b, c have different Dirichlet prior densities. Expert a sets his prior $\pi_a(\boldsymbol{\theta}|\alpha_a^0)$ such that $\boldsymbol{\alpha}_a^0 = (1, 8, 1)$, b sets his prior $\pi_b(\boldsymbol{\theta}|\alpha_b^0)$ so that $\boldsymbol{\alpha}_b^0 = (4, 2, 4)$ and c sets her prior $\pi_c(\boldsymbol{\theta}|\alpha_c^0)$ such that $\boldsymbol{\alpha}_c^0 = (2.5, 5, 2.5)$. A fourth expert d had no prior information of her own but believed that expert a was right with probability 0.5 and expert b with probability 0.5 and so set her prior density $\pi_d(\boldsymbol{\theta})$ so that

$$\pi_d(\boldsymbol{\theta}) = 0.5\pi_a(\boldsymbol{\theta}|\alpha_a^0) + 0.5\pi_b(\boldsymbol{\theta}|\alpha_b^0).$$

Prove that experts c and d have the same prior mean. You now observe $\boldsymbol{x} = (0, 5, 0)$. Calculate the posterior mean for the experts a, b, c. Let

$$p_a(\boldsymbol{x}) = \frac{p(\boldsymbol{x}|\boldsymbol{\theta})\pi_a(\boldsymbol{\theta}|\alpha_a^0)}{\pi_a(\boldsymbol{\theta}|\alpha_a^+)},$$

$$p_b(\boldsymbol{x}) = \frac{p(\boldsymbol{x}|\boldsymbol{\theta})\pi_b(\boldsymbol{\theta}|\alpha_b^0)}{\pi_b(\boldsymbol{\theta}|\alpha_b^+)}$$

where α_a^+ and α_b^+ denote, respectively, the vector of hyperparameters of a and b's posterior density. Note that $p_a(\boldsymbol{x})$ and $p_b(\boldsymbol{x})$ are the probabilities of the observed data predicted a priori by a and b respectively. Prove that

$$\frac{p_a(\boldsymbol{x})}{p_b(\boldsymbol{x})} = \frac{\Delta(\alpha_a^0)\Delta(\alpha_b^+)}{\Delta(\alpha_a^+)\Delta(\alpha_b^0)}$$

and calculate this ratio explicitly for the example above. Hence or otherwise calculate d's posterior mean. How does this differ from expert c's posterior mean?

Part II

Multidimensional Decision Modelling

6
Multiattribute utility theory

6.1 Introduction

So far this book has given a systematic methodology that can be used to address and solve some simple decision problems. However some of the most interesting and challenging real decision problems can have many facets. It is therefore necessary to extend the Bayesian methodology described earlier in the book so that it is a genuinely operational tool for addressing the types of to complex decision problems regularly encountered. Even for moderately sized problems we have seen the advantages of disaggregating a problem into smaller components and then using the rules of probability and expectation within a suitable qualitative framework to draw the different features of a problem into a coherent whole. Although the appropriate decomposition to use depends on the problem addressed there are nevertheless some well-studied decomposition methods that are appropriate for a wide range of decision problem which the analyst is likely to encounter frequently. The remainder of this book will focus on the justification, description and enaction of some of these different methodologies.

When addressing the formal development of simpler models we began by developing a methodology constructing a justifiable articulation and quantification of a DM's preferences. In particular in Chapter 3 a formal rationale was developed describing when and why a DM should be guided into choosing a utility maximising decision rule. But techniques are needed to apply these methods effectively when the vector of attributes of the DM's utility function is moderately large. This chapter begins by discussing how a utility function with more than one attribute can be elicited using appropriate assumptions about various decompositions the DM's utility function might exhibit. These techniques are then demonstrated on a series of examples. The encoding of beliefs over many variables and hence the elicitation and estimation of reward distributions will follow in later chapters.

Recall from Chapter 3 that a utility function can be elicited from a DM by eliciting the values of $\alpha(\boldsymbol{r})$ for each possible vector of values r of attributes in $\mathcal{R}$ where $\alpha(\boldsymbol{r})$ is defined to be the probability making

$$\boldsymbol{r} \sim B(\boldsymbol{r}^0, \boldsymbol{r}^*, \alpha(\boldsymbol{r}))$$

where $b(\boldsymbol{r}^0, \boldsymbol{r}^*, \alpha)$ is the betting scheme which gives the best possible value $\boldsymbol{r}^*$ of attributes with probability α and otherwise gives the worst $\boldsymbol{r}^0$. It was demonstrated in Chapter 3 that it was relatively straightforward to elicit this utility function by using techniques like the midpoint method combined with linear interpolation or an appropriate characterisation when the DM had a single attribute to her utility function. The major practical problem with this approach was the psychological difficulty associated with accurately eliciting indifferences between rewards with certainty and a betting scheme with two consequences extremely different in their desirability: simplifying heuristics then tend to dominate the DM's thinking in a way that makes a set of coherent and faithful specifications fragile.

As the dimension of the attribute space increases so do the number of very different scenarios the DM needs to compare and this difficulty becomes more acute. In particular the DM will often find comparing gambles between consequences different in more that one component attribute very confusing. Thus in the court case example assume the two attributes of the prosecution are whether or not the suspect is convicted and the cost of pursuing the case. The DM can be expected to find it easier to compare cost preferences given a successful prosecution and the cost preferences given an unsuccessful prosecution than to use the reference pair $(\boldsymbol{r}^0, \boldsymbol{r}^*)$ which in this context considers gambles between a high-cost unsuccessful prosecution and a low-cost successful prosecution.

Two distinct methodologies, both used with considerable success by their respective proponents, have been developed to address these issues. The first is a value-based approach. These methods first elicit the contours – sometimes called the *indifference curves* – of values of the vector of attributes which are *preferentially equivalent* to the DM or have the same *preferential value* to her. The vector of attributes is thus reduced to a one-dimensional preferential value attribute. A simple comparator is then selected from each equivalence class of vectors of values of attributes with the same value. One-dimensional techniques to fix the utility of each value of the preferential value can then be used. One of the advantages of this approach is that it leaves the actual quantification of the utility, via betting preferences, to the end of the elicitation process. All it uses early in the process is the DM's preference order. The disadvantage is the difficulty of specifying parametric forms for the indifference sets that are both easy to handle and clearly interpretable by the DM. Advocates of this type of method include Keeney, Bedford and Cooke and excellent articles describing these methods can be found in Bedford and Cooke (2001); Keeney (1992) and Keeney (2007).

A second approach described below supplements the utility axioms with some additional "independence" assumptions concerning the attributes. These utility independence axioms assume certain types of invariance between components of the DM's objectives and allow the analyst to elicit the DM's full utility function via comparing her preferences over gambles. These either compare certain combinations of best possible and worst possible scenarios associated with the different attributes or compare gambles where all but one of the attributes in the gambles is known to share the same value. This methodology leads to an elegant framework where the parameters in the forms of utility implied are simple to understand both by the DM herself and any auditor. This is because the independence assumptions can be specified qualitatively in terms of preferences being the same in different scenarios.

One drawback of this methodology is that it is quite often necessary for the analyst to spend considerable effort helping the DM to transform the attributes she initially specifies so that these exhibit – at least approximately – the types of utility independence that are required for the formalism to hold.

6.2 Utility independence

Let $\boldsymbol{R} = (R_1, R_2, R_3, \ldots, R_n)$ denote a random vector of attributes defining the DM's reward space. In Chapter 3 we saw that each potential decision $d \in D$ she could make will have associated with it a distribution $P_d(\boldsymbol{r})$ over $\boldsymbol{R}$, and to be rational she needs to choose a decision $d \in D$ to maximise the expectation of her utility function $U(\boldsymbol{r})$ with respect to $P_d(\boldsymbol{r})$.

Note that when the number n of attributes is even moderately large to elicit the real-valued function $U(\boldsymbol{r})$, for all combinations of values of $\boldsymbol{r} = (r_1, r_2, r_3, \ldots, r_n)$ would be extremely time consuming. Thus eliciting $U(\boldsymbol{r})$ over a lattice of points taking 5 putative values from each attribute we have to evaluate U at 5^n points: a significant task even when n is as small as 3. Furthermore for psychological reasons it is over-ambitious to expect a DM to be able to balance gambles between extreme alternatives varying in many components. So direct elicitation of U using the definitions of utility directly are hazardous.

However if the DM has a utility function satisfying some additional properties then it is much easier to faithfully elicit its form. The required additional assumptions about the form of the utility function demand a different sorts of separation and "independence" of its attributes. Note that these types of independence are quite different from more familiar uses of the term "independence" used in probability and statistics.

Let $\boldsymbol{R}_S$ denote the subset of the component attributes whose indices lie in a set $S \subseteq \{1, 2, \ldots, n\}$ and let A and B partition the set of indices $\{1, 2, \ldots, n\}$ into the set of indices A of the attributes we are addressing and the indices B of the rest.

Definition 6.1. The set of component attributes $\boldsymbol{R}_A$ is said to be *preferentially independent* of the other attributes $\boldsymbol{R}_B$ if the preference *order* for achieving different attribute scores $\boldsymbol{r}_A$ with certainty does not depend on the levels of $\boldsymbol{r}_B$ to which $\boldsymbol{R}_B$ might be fixed.

To demand preferential independence of a given set of attributes $\boldsymbol{R}_A$ is one that only acts on preferences over known values of attributes. As such it is relatively easy to at least spot check its validity. In the methods discussed below the following is always assumed.

Axiom 6.2 (Simple preferential independence). All sets $\boldsymbol{R}_A = R_i$ for which $A = \{i\}$, $1 \leq i \leq n$, are preferentially independent.

This axiom is a type of monotonicity requirement. It demands that if a value of certain $r_i(1) \preceq r_i(2)$ when the rest of the attributes take one value for certain then this preference remains true whatever values the other attributes take. In this sense $r_i(2)$ is at least as preferable as $r_i(1)$ independent of the consequences described by the other rewards. A simple example of two attributes that for me are not preferentially independent are the reward I receive because I spend time with my partner and the reward I receive when I go to the bar or stay at home. If I receive the reward of being with her then I would prefer to do this at home rather than in the bar. On the other hand if I am not going to receive the reward of spending time with her I find being in the the bar more preferable than being at home alone. The efficacy of the two attributes are therefore intertwined. They need to be redefined if they are to represent genuinely independent measures of my preference: here perhaps simply combined together into a single attribute labelled "intimacy with another" – where intimacy with my partner scores higher than company I enjoy in the bar.

An important implication of this axiom is that the space of attributes R must form a product space $R = R_1 \times R_2 \times \ldots \times R_n$. This is because the axiom could only hold if r_i is a feasible value for attribute R_i to take for one value of its complement then it must be a feasible value for all values of its complement. It follows that attributes can be defined so that they do not logically constrain one another. The first problem an analyst can encounter after having elicited the DM's attributes is that these attributes are not variationally independent. If such variational dependence does exist then, to use the methods described below the analyst will need to help the DM to re-express her problem so that this is not so. Sometimes this will simply be a case of removing redundancies from the description of the problem, for example two attributes that are actually measuring the same consequences, but measured in slightly different ways. However sometimes a subtler transformation of the attributes is necessary. The appropriate procedures for the systematic transformation of initial attributes are rather problem dependent and will be illustrated only later in the chapter.

Once the problem has been reparameterised so that it has no such *logical* constraints it will nevertheless be quite likely that the analyst might still have to perform further transformations before the preferential independence assumption is satisfied. Sign switching of attributes given the values of others is sometimes necessary. It is often necessary to invent hypothetical scenarios corresponding to different extreme values in the hypercube used in the elicitation – as defined by the product space above – and so embed the problem in a richer class of possibilities.

However if the attribute space can be transformed into a product space so that these attributes exhibit simple preferential independence then this has big benefits. In particular it will then follow immediately from preferential independence that the worst possible option $\boldsymbol{r}^0 = (r_1^0, r_2^0, \ldots, r_n^0)$ and the best possible outcome $\boldsymbol{r}^* = (r_1^*, r_2^*, \ldots, r_n^*)$ where all the pairs (r_i^0, r_i^*) of attribute r_i, $i = 1, 2, \ldots, n$, define respectively the worst and best attribute for R_i. The DM's utility function U can then be referenced to $(\boldsymbol{r}^0, \boldsymbol{r}^*)$ so that for her worst scenario, $U(\boldsymbol{r}^0) = 0$ and for her best scenario $U(\boldsymbol{r}^*) = 1$. We will see later that the utilities of particular configurations of rewards are especially important. For $i = 1, 2, \ldots, n$, let the *criterion weight* k_i be defined by

$$k_i = U(r_1^0, r_2^0, \ldots, r_{i-1}^0, r_i^*, r_{i+1}^0, \ldots, r_n^0).$$

Thus k_i denotes the DM's utility of getting the worst outcome for all attributes other than the ith attribute where she obtains the best possible outcome. Clearly by definition $0 \leq k_i \leq 1, 1 \leq i \leq n$.

Definition 6.3. A set of attributes $\boldsymbol{R}_A$ is said to be *utility independent* of the rest of the attributes $\boldsymbol{R}_B$ for DM if her utility function $U(\boldsymbol{r})$ can be written in the form

$$U(\boldsymbol{r}) = a_A(\boldsymbol{r}_B) + b_A(\boldsymbol{r}_B)U_A(\boldsymbol{r}_A) \tag{6.1}$$

where $a_A, b_A > 0$ can be written as functions of components only in B and U_A can be written only as a function of attributes in the set A.

This definition may look a little obscure but it has a simple preferential interpretation. We have already noted in Chapter 3 that utility functions can be identified if one utility function is a strictly increasing linear transformation of the other. Thus suppose $\boldsymbol{R}_A$ is utility independent of $\boldsymbol{R}_B$. Then take any two distributions $P_1(\boldsymbol{r}_B)$ and $P_2(\boldsymbol{r}_B)$ over attributes $\boldsymbol{R}$ where the attributes in $\boldsymbol{R}_B$ are *known* to take the value $\boldsymbol{r}_B$ are such that $\overline{U}(P_1(\boldsymbol{r}_B)) \leq \overline{U}(P_2(\boldsymbol{r}_B))$ – so that $P_1(\boldsymbol{r}_B) \preceq P_2(\boldsymbol{r}_B)$. Suppose that $P_1(\boldsymbol{r}'_B)$ and $P_2(\boldsymbol{r}'_B)$ over attributes $\boldsymbol{R}$ are such that the attributes in $\boldsymbol{R}_B$ *are known* to take the value $\boldsymbol{r}'_B$ but share with $P_1(\boldsymbol{r}_B)$ and $P_2(\boldsymbol{r}_B)$ the same conditional distribution of $\boldsymbol{R}_A|\boldsymbol{R}_B$. It then follows that $\overline{U}(P_1(\boldsymbol{r}'_B)) \leq \overline{U}(P_2(\boldsymbol{r}'_B))$ – so that $P_1(\boldsymbol{r}'_B) \preceq P_2(\boldsymbol{r}'_B)$. So whatever values the remaining attributes $\boldsymbol{r}_B$ take, with the identification above, our preferences between gambles over $\boldsymbol{R}_A$ remain unchanged. In this sense the evaluation of attributes $\boldsymbol{R}_A$ are "independent" of $\boldsymbol{R}_B$ because they are measuring aspects of the circumstances which are – in the sense above – orthogonal to $\boldsymbol{R}_B$.

One of the simpler assumptions using utility independence a DM can make is that utility independence holds for each set $\boldsymbol{R}_A$ consisting of a single attribute.

Definition 6.4. We say that the utility function $U(\boldsymbol{R})$ has *singly utility independent attributes* (*suia*) if all subsets $\boldsymbol{R}_A = \{R_i\}, i = 1, 2, \ldots, n$ are utility independent.

Mathematically, by assuming U has suia implies that U must be *multilinear* in components $(U_1, U_2, \ldots, U_n)$ (Hoffman and Kunze, 1971). This simply means that U must be expressible as a polynomial whose arguments are the set of components $\{U_1, U_2, \ldots, U_n\}$ and furthermore no component in this polynomial can have degree greater than one. Letting $\mathcal{A}(n)$ denote the set of all subsets of $\{1, 2, \ldots, n\}$ other than the empty set we have the following.

Theorem 6.5. *If a utility function $U(\boldsymbol{r})$ has n suia attributes then it must take the form*

$$U(\boldsymbol{r}) = \sum_{A \in \mathcal{A}(n)} l_A U_A(\boldsymbol{r}_A) \tag{6.2}$$

where

$$U_A(\boldsymbol{r}_A) = \prod_{r_i \in A} U_i(r_i)$$

$0 = U_i(r_i^0) \leq U_i(r_i) \leq U_i(r_i^*) = 1$ *are increasing functions of* r_i, $i = 1, 2$.

To ensure simple preferential independence further conditions on the coefficients l_A are required when $n > 2$ – see for example French and Rios Insua (2000) for an explicit statement of these constraints and a proof of the result above.

Sadly, assuming the property of suia is not that useful unless n is small, since it still requires the specification of $2^n - 2$ functionally independent coefficients $\{l_A : A \in \mathcal{A}(n)\}$, in addition to the n functions $U_i(r_i)$. So unless the DM has $n \leq 4$ elicitation is still extremely time consuming unless more assumptions about the DM's preference structure are made. On the other hand when a utility function has just two attributes then it is relatively straightforward to study the implication of these being suia and to discuss their elicitation. Since this simpler scenario is not uncommon it is helpful to consider it first.

6.2.1 The two-attribute case

In the following example the utility of a DM working in a clinic performing radical surgery has just two attributes. The first R_1 is the number of years a patient survives after treatment which here for the sake of simplicity we will assume to lie in the interval $0 \leq R_1 \leq 10$. The second R_2 is an index of the quality of life of the patient after treatment:

$r_2 = 0$ signifies no significant brain function,
$r_2 = 1$ signifies normal brain function but life in a wheelchair,
$r_2 = 2$ signifies normal life.

For attribute R_1 to be utility independent of R_2 would require that – setting $A = \{1\}$ and $B = \{2\}$ – $U(\boldsymbol{r})$ can be written in the form:

$$U(\boldsymbol{r}) = a_1(r_2) + k_1(r_2)U_1(r_1).$$

Let $T_1(2)$ be (a possibly hypothetical) treatment ensuring survival for almost exactly 2 years and a quality of life r_2 and let treatment $T_2(\alpha, r_2)$ ensure survival for a further 10 years with probability α and a quality of life r_2 but risks immediate death and a quality of life r_2 with probability $(1 - \alpha)$, $x_2 = 0, 1, 2$. Then under the form of utility above, the expected utility, for $r_2, r_2' = 0, 1, 2$,

$$\overline{U}(T_1(2, r_2)) \geq \overline{U}(T_2(\alpha, r_2)) \Rightarrow \overline{U}(T_1(2, r_2')) \geq \overline{U}(T_2(\alpha, r_2')).$$

So in this sense we can think of $T_1(2)$ being as least as preferable as $T_2(\alpha)$ "independently of" attribute R_2. In particular to find the break-even point $\alpha^*(2)$ of α making $T_1(2)$ and

$T_2(\alpha^*)$ equally preferable we can simply fix r_2 to some arbitrary chosen value $\overline{r}_2$ (say) and increase α from 0 until $\alpha = \alpha^*(2)$ where

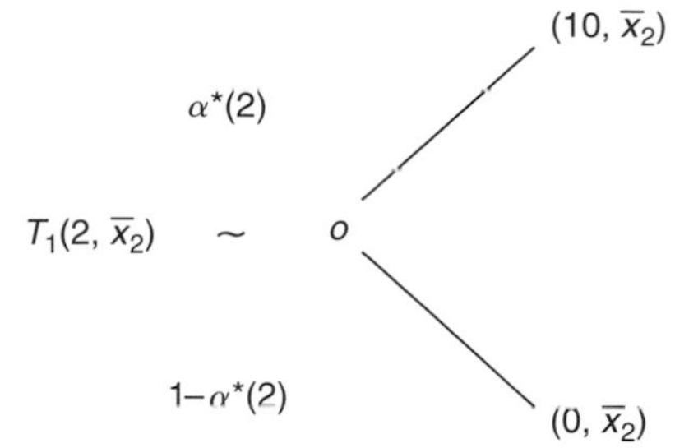

knowing that $\alpha^*(2)$ is not a function of the value of $\overline{r}_2$ but only r_1. To check for suia it is also necessary to check the utility independence of R_2. But this is simply performed by reversing the role of R_1 and R_2 in the above and repeating the elicitation check for independence.

The issue of whether the DM's utility function really does have suia is clearly dependent on her objectives and the context she addresses. Here, if $\alpha^*(2)$ were large when $r_2 = 0$ then certain DM's might be uncomfortable with the assumption. She might judge that certain brain death of a patient may make her less willing to risk surgery and consequent long life than the prospect of a future normal life if the surgery were successful. If she held this opinion then, under the obvious extension of the notation

$$\alpha^*(2,0) > \alpha^*(2,2).$$

One of the skills the decision analyst develops is be able to redefine attributes when the DM is unhappy with the independence assumptions so that they exhibit the suitable degree of independence. For example, for the unhappy DM discussed above she might find the new attributes R_1' – the number of additional years of life with brain fully functional – to be utility independent of the new attribute R_2' – an indicator on whether or not the patient has to be in a wheelchair – utility independent. But often several reparameterisations need to be tried before an appropriate one is found.

Note that whether each single attribute is utility independent can be addressed simply by questioning whether the value the other attributes score could make a difference to the DM's preferences about the attribute focused on. So the plausibility of this assumption as judged by both the DM and a possible auditor is *qualitative* and can be argued about in common language. In this sense if it is appropriate it is stable to differing views about just how much one option is preferred to another. It is therefore a good starting point for subsequently refining an analysis and provides a structure that often captures a common appreciation of the logical consistency and ethical propriety of a particular preferential model. This is one reason I like this method.

Of course occasionally it will be necessary in the light of further elaborations to revisit and perhaps redefine attributes in the larger description of the problem because the independence no longer appears plausible to the DM. But this process of reflection and adjustment, as we have discussed earlier in the book, is an intrinsic aspect of all stages of a decision analysis.

We now return to the specific problem where a utility function has exactly two attributes and where the DM, after the sorts of iterations discussed above, is content to assume that both R_1 and R_2 are utility independent. When a DM has a utility function with two utility independent attributes then it must take a multilinear form. Thus we have the following.

Theorem 6.6. *If a utility function $U(\boldsymbol{r})$ with $U(\boldsymbol{r}^0) = 0$ and $U(\boldsymbol{r}^*) = 1$ with two attributes has suia and exhibits simple preference independence then it must take the form*

$$U(\boldsymbol{r}) = k_1 U_1(r_1) + k_2 U_2(r_2) + k k_1 k_2 U_1(r_1) U_2(r_2) \tag{6.3}$$

where $0 \leq k_i \leq 1$ and $0 = U_i(r_i^0) \leq U_i(r_i) \leq U_i(r_i^*) = 1$ are increasing functions of r_i, $i = 1, 2$ and k is the (unique) solution of

$$1 + k = (1 + kk_1)(1 + kk_2) \Leftrightarrow \tag{6.4}$$

$$k = \frac{1 - (k_1 + k_2)}{k_1 k_2}. \tag{6.5}$$

Note that we can substitute for k in (6.3) when it can be rewritten as

$$U(\boldsymbol{r}) = k_1 U_1(r_1) + k_2 U_2(r_2) + [1 - (k_1 + k_2)] U_1(r_1) U_2(r_2).$$

It is left to an exercise to rescale the representation in (6.2) to prove the result above.

This utility function clearly reduces to a linear function of the functions $U_i(r_i)$, $i = 1, 2$ when $k_1 + k_2 = 1$. The *conditional utility functions* $U_i(r_i) = U_{A_i}(A_i(\boldsymbol{r}))$ where $A_i = \{R_i\}$, $i = 1, 2$ are those utilities obtained by fixing the other attribute to some reference value. The *criterion weights* k_i, $i = 1, 2$ also have a simple interpretation since

$$k_1 = U(r^*, r^0) \quad \text{and} \quad k_2 = U(r^0, r^*)$$

are respectively the utility of getting the best possible return on the first attribute but the worst possible in the second and the utility of getting the best possible return on the second attribute but the worst possible in the first. Note that when k_1 and k_2 are both close to one – so in particular $k_1 + k_2 > 1$ – then the DM is extremely pleased if she obtains a high score in *either one* of the attributes. On the other hand, if both k_1 and k_2 are close to zero the DM needs to achieve a high score in *both* attributes to be satisfied with an outcome.

Thus once these criterion weights have been elicited the analyst has two univariate utility functions to elicit. The effort eliciting a utility function of the form $U(\boldsymbol{r})$ has approximately doubled compared with the single-attribute case rather than increased with the square. So although the elicitation task has grown, it has grown manageably. A discussion of how this elicitation can be performed will be delayed until we have addressed problems with more than two attributes. Many examples and illustrations of two-attribute problems for which $k_1 + k_2 \neq 1$ are given in Keeney and Raiffa (1976) and in Chapter 6 of Keeney (1992).

6.3 Some general characterisation results

Typically when a DM has a utility function with three or more attributes some extra assumptions help further speed up the elicitation process.

Definition 6.7. We say that the utility function $U(\boldsymbol{R})$ has *mutually utility independent attributes (muia)* if all subsets $A(\boldsymbol{R})$ are utility independent.

Definition 6.8. We say that the utility function $U(\boldsymbol{R})$ is *pair preferentially R_1 utility equivalent attributes*, if all the pairs $\{R_1, R_i\}$, $i = 2, 3, \ldots, n$ are preferentially independent of their complements and R_1 is utility independent.

The conditions of the theorem of the last section imply suia and also muia because there are only two attributes and so only two non-trivial sets $\{R_1\}$, $\{R_2\}$ that can generate a partition of the required form. Note that, at least in principle, all these conditions can be spot-checked for their validity from the DM as in the examples of the last section. We now have the following characterisation.

Theorem 6.9. *If a utility function $U(\boldsymbol{r})$ has mutually utility independent attributes or alternatively pair preferentially R_1 utility equivalent attributes then it must be able to be written in either the form*

$$U(\boldsymbol{r}) = \sum_{i=1}^{n} k_i U_i(r_i) \text{ where } \sum_{i=1}^{n} k_i = 1 \tag{6.6}$$

or

$$U(\boldsymbol{r}) = k^{-1} \left\{ \prod_{i=1}^{n} (1 + k k_i U_i(r_i)) - 1 \right\} \text{ where } \sum_{i=1}^{n} k_i \neq 1 \tag{6.7}$$

where, in either case, $0 < k_i < 1$ and $0 = U_i(r_i^0) \leq U_i(r_i) \leq U_i(r_i^*) = 1$ are increasing functions of r_i, $1 \leq i \leq n$. and k is the (unique) solution of

$$1 + k = \prod_{i=1}^{n} (1 + k k_i). \tag{6.8}$$

For the proof of this result for muia see e.g. Keeney and Raiffa (1976) p289 and Keeney (1974). Alternative characterisations where different sets of axioms lead to the same result are given in Pollack (1967) and Meyer (1970). The fact that there are two different forms here is actually an illusion: after a little algebra and noting from the above equation that $\sum_{i=1}^{n} k_i = 1 \Rightarrow k = 0$ it is easy to check that either form can be written as

$$U(\boldsymbol{r}) = \sum_{i=1}^{n} k_i U_i(r_i) + k \sum_{1 \le i < j}^{n} k_i k_j U_i(r_i) U_j(r_j) \tag{6.9}$$

$$+ k^2 \sum_{1 \le i < j}^{n} k_i k_j k_l U_i(r_i) U_j(r_j) U_l(r_l) + \cdots$$

$$+ k^n k_1 k_2 \ldots k_n U_1(r_1) U_2(r_2) \ldots U_n(r_n).$$

When $\sum_{i=1}^{n} k_i > 1$ the DM will be more pleased to obtain a good utility score on a few attributes than moderate scores on all. On the other hand if $\sum_{i=1}^{n} k_i < 1$ then she will be happier to obtain moderate scores on all attributes rather than a good utility score on some and a bad score on others. It can be shown that the scaling function k is such that when $\sum_{i=1}^{n} k_i < 1$ then $0 < k$, $\sum_{i=1}^{n} k_i > 1$ then $-1 < k < 0$. So from the expansion above $\sum_{i=1}^{n} k_i \to 1$ as $k \to 0$. In all cases the larger k_i the more she values obtaining high scores in the ith attribute.

6.4 Eliciting a utility function

When a utility function has muia it is relatively easy to elicit. It is necessary only to elicit the utility weights and the univariate conditional utility functions. Let

$$\boldsymbol{r}_i^* \triangleq (r_1^0, r_2^0, \ldots, r_{i-1}^0, r_i^*, r_{i+1}^0, \ldots, r_n^0)$$

and recall that $B(\boldsymbol{r}^0, \boldsymbol{r}^*, \alpha)$ denotes a lottery giving $\boldsymbol{r}^*$ with probability α and $\boldsymbol{r}^0$ with probability $1-\alpha$. Then directly from its definition k_i is the value of α such that $\boldsymbol{r}_i^* \backsim B(\boldsymbol{r}^0, \boldsymbol{r}^*, k_i)$. Note that the form of $U(\boldsymbol{x})$ will depend on whether or not $\sum_{i=1}^{n} k_i = 1$.

To elicit $U_i(r_i)$ just set all the other attributes $\{r_j = \overline{r}_j : 1 \le j \ne i \le n\}$ – the choice of this value is theoretically arbitrary but is often set to some ordinary/typical vector of values so that it is more easy to think about – and, for each r_i where compare $r_i^0 < r_i < r_i^*$

$$\overline{\boldsymbol{r}}_i \triangleq (\overline{r}_1, \overline{r}_2, \ldots, \overline{r}_{i-1}, r_i, \overline{r}_{i+1}, \ldots, \overline{r}_n),$$

$$\overline{\boldsymbol{r}}_i^0 \triangleq (\overline{r}_1, \overline{r}_2, \ldots, \overline{r}_{i-1}, r_i^0, \overline{r}_{i+1}, \ldots, \overline{r}_n),$$

$$\overline{\boldsymbol{r}}_i^* \triangleq (\overline{r}_1, \overline{r}_2, \ldots, \overline{r}_{i-1}, r_i^*, \overline{r}_{i+1}, \ldots, \overline{r}_n)$$

with $B(\overline{\boldsymbol{r}}_i^0, \overline{\boldsymbol{r}}_i^*, \alpha)$ defined to give the outcome $\overline{\boldsymbol{r}}_i^*$ with probability α and the outcome $\overline{\boldsymbol{r}}_i^0$ with probability $1-\alpha$. The value of the conditional utility function $U_i(r_i)$ is the value of α such that $\overline{\boldsymbol{r}}_i \sim B(\overline{\boldsymbol{r}}_i^0, \overline{\boldsymbol{r}}_i^*, U_i(r_i))$.

A good way to elicit $U_i(r_i)$ in practice is to use the midpoint method. Thus we find the value $r_i[0.5]$ such that $\overline{\boldsymbol{r}}_i \sim B(\overline{\boldsymbol{r}}_i^0, \overline{\boldsymbol{r}}_i^*, 0.5)$. We then elicit $r_i[0.25]$ from a gamble between r_i^0 and $r_i[0.5]$, as used in the midpoint method for a utility function with a single attribute, but here fixing the other attributes to their typical value.

There are a couple of refinements to the direct method of eliciting the criterion weights. First ask the DM for the attribute she thinks is the most important one and label this as the first attribute R_1 with criterion weight k_1. This criterion weight is normally greater than about 0.2 so is associated with a gamble whose probability is not too close to zero or one. Then to elicit the criterion weight k_2 of the next most important attribute r_2 find the value e_2 such that $\boldsymbol{r}_2^* \backsim B(\boldsymbol{r}_1^0, \boldsymbol{r}_1^*, e_2)$. Continue in this way successively finding the values e_i defining the indifference gamble $\boldsymbol{r}_i^* \backsim B(\boldsymbol{r}^0, \boldsymbol{r}_{i-1}^*, e_i)$, $i = 2, 3, \ldots, n$, between achieving the maximum reward only on the ith attribute with certainty compared with obtaining the better reward of only the $(i-1)$th attribute with a given probability. It is easy to check that $k_i = e_i k_{i-1}, i = 2, 3, \ldots, n$. The reason why this method can be more reliable is that it tends to compare gambles with more comparable consequences than the direct method.

Occasionally – when it is very important to the DM that she achieves high scores on at least two attributes – you might find that even k_1 is small. Instead of eliciting the first criterion weight directly it is then sometimes better to elicit the value of α for which $\boldsymbol{r}_i^0 \sim B(\boldsymbol{r}^0, \boldsymbol{r}^*, \alpha)$ where

$$\boldsymbol{r}_i^0 \triangleq (r_1^*, r_2^*, \ldots, r_{i-1}^*, r_i^0, r_{i+1}^*, \ldots, r_n^*)$$

is a maximal reward for all but the ith where the DM receives the worse reward. A little algebra then shows that

$$k_1 = \frac{1-\alpha}{1+k\alpha}.$$

The main practical issue is to devise plausible actions that might have the distributions above. The more plausible these are the more likely you are to faithfully represent the DM's preferences. This process is illustrated below.

Example 6.10. Consider the court case example and suppose you discover that the DM's utility has three muia. The first r_1 is the financial saving of the proposed action, the second r_2 is whether or not the suspect is found guilty and the third attribute r_3 measures the extent to which the public perceive the police as appropriately pursuing crimes against vulnerable victims. To elicit the DM's criterion weights using the methods above it is first necessary to construct acts associated with the two reference gambles $\boldsymbol{r}^0$ and $\boldsymbol{r}^*$. A worst scenario $\boldsymbol{r}^0$ is one that is most costly. This is the one corresponding to spending the most on the forensic investigations, the suspect is found not guilty and the public see the police as totally inappropriately vigorous – perhaps because the conduct of the police in their investigation is heavily criticised by the judge. A best scenario $\boldsymbol{r}^*$ would be one where they have to spend nothing on further forensic investigation – perhaps because they are able to persuade the Home Office to underwrite any expenditure because of it being so high profile, the suspect is found guilty and because the forensic evidence produced clinched the case the police are commended by the judge for the vigour with which they pursued the case. DM's criterion weight k_1 is the value that makes the DM indifferent between a gamble between obtaining the scenario $\boldsymbol{r}^*$ with probability k_1 and otherwise worst possible combination of attributes as described above. This would be to spend nothing – so not pursue the case further – the suspect not to be found guilty and the police to be seen as expending no effort: this happening with certainty. This latter state of affairs actually happens to be the direct consequence of one of the considered acts, so the DM should find this scenario particularly easy to consider. To elicit k_2 the DM needs to compare the worse and best scenario with an option where, for certain, the service spends the most possible and the suspect is found guilty, but the judge heavily criticises the police. The final certain comparator is the scenario where the police spend the maximum, the suspect is found guilty and the judge praises the police for doing all possible to resolve the case. In this example the three conditional utilities are of different types. The first associated with financial saving can be elicited fixing the two attributes to (say) the success of the prosecution and a typical sort of effect on public perception of police and then, for example, using the midpoint method described in Chapter 3. Since the second attribute is binary no elicitation is necessary to calculate U_2. Finally to find U_3 utilities associated with public perception, the DM will be encouraged to consider all possible scenarios – including those that could arise from any root to leaf path in the tree – and to rank these in terms of how well they will be received by the public. Trade-off associated with these for some fixed average cost and suspect conviction then allows us to construct this conditional utility function. A more detailed description of how this type of construction is performed is given in Clive's decision problem below.

There are several points to draw from this example. Clearly it takes some time to perform the types of elicitation above. Moreover some imagination on the part of the analyst is needed to construct scenarios corresponding to reference gambles. But there are many benefits to this process that compensate for this effort. First, by using the utility scores so elicited for each consequence described at a leaf of her decision tree, with appropriate probabilities added to its edges the DM can *calculate her best option* for the case in hand. Second the sort of discussions used to formulate her attributes so that they are utility independent and also the subsequent quantifications outlined above can be used to *annotate her reasons* for finding one consequence better than another and the extent of this preference. This can then be used not only for her own reference but also be made available to anyone who might legitimately appraise her. This therefore provides a *framework for further adjustments* to the preferences to be articulated and possibly implemented. Third, it will often be possible to use the analysis as *a template for future similar problems*. For example in the illustration above, if police/prosecution are faced with a similar crime in the future then although some of the evaluations below might be different many will be the same. So any subsequent elicitation is usually much quicker: much of the structure, such as the definition of independent attributes and some of the equivalent gambles will remain the same.

6.5 Value independent attributes

The simplest and the most widely used assumption is that the DM's utility function has value independent attributes.

Definition 6.11. A utility with attribute vector $\boldsymbol{R}$ is said to have *value independent attributes* (*via*) if two distributions of reward are equally preferred whenever they have identical marginal distributions to each other on all the individual attributes R_i, $1 \leq i \leq n$.

Theorem 6.12. *U has value independent attributes if and only if U has the linear form given in (6.6).*

For a proof of this result see for example Keeney and Raiffa (1976). One reason this form of utility function is so well used is that it is such a familiar way of scoring different attributes a DM might have experienced in the marking of academic programmes she has attended. Each attribute she obtains corresponds to the mark she achieved in a particular module of her course. The corresponding conditional utility is the possibly nonlinear rescaling of the mark to adjust it to reflect the candidate's ability: taking into account the overall difficulty of the module and quality of the teaching and so on. Finally the criterion weights are the percentage of the final aggregate corresponding to a particular module, reflecting its length and quantity of material. Change the score to run from 0 to 100 instead of from zero to one and the analogy is complete!

Another analogy is useful for the elicitation process. The idea is that the DM's utility is her total wealth: albeit measured not just by money. The conditional utilities are elicited as before but the *exchange rate method* is used to find the criterion weights. This method often elicits the DM's utility more faithfully not only because of its transparency but also because it does not require the DM to compare gambles with widely divergent outcomes. It uses the fact that if U has via and two options $d[1]$ and $d[2]$ differ only in the conditional utility score they give in the ith and $(i+1)$th attribute, $i = 1, 2, \ldots, n-1$ then

$$\begin{aligned} 0 &= \overline{U}(d[1]) - \overline{U}(d[2]) \\ &= k_i \left\{ \overline{U}_i(d[1]) - \overline{U}_i(d[2]) \right\} + k_{i+1} \left\{ \overline{U}_{i+1}(d[1]) - \overline{U}_{i+1}(d[2]) \right\} \Leftrightarrow \\ \Delta_i &= \left\{ \overline{U}_i(d[1]) - \overline{U}_i(d[2]) \right\} = e_i \left\{ \overline{U}_{i+1}(d[1]) - \overline{U}_{i+1}(d[2]) \right\} = e_i \Delta_{i+1} \end{aligned}$$

where $e_i = \frac{k_{i+1}}{k_i}$. It follows in particular that the DM should be prepared to trade options ensuring an increase of $e_i \Delta_{i+1}$ in the utility for attribute i for an option ensuring an increase of Δ_{i+1} in the the utility of attributes $i+1$. Here e_i can be thought of as the exchange rate between r_i and r_{i+1} being the amount of utility gain in marginal utility on r_i a unit increase in the marginal utility on r_{i+1} will buy. Note that this is even true when the unit Δ_{i+1} measure is small. So the values of $\{e_i : i = 1, 2, \ldots, n-1\}$ can be elicited from preferences on sure outcomes and furthermore the conseqences leading to these reward outcomes are not too different. This similarity can be further enhanced by listing attributes in the order of their value with the most valuable given first index. Consequently if a DM has via then this method is usually more effective and less prone to elicitation error than the more general one defined in the last section. Because $\sum_{i=1}^{n} k_i = 1$, the criterion weights $\{k_i : i = 1, 2, \ldots, n\}$ can be written, after a little algebra, in terms of the elicited exchange rates $\{e_i : i = 1, 2, \ldots, n-1\}$ between adjacent attributes using the formula

$$k_i = \prod_{j=1}^{i-1} e_j \left(\sum_{i=1}^{n} \prod_{j=1}^{i-1} e_j \right)^{-1}$$

where by abuse of notation we have let $\prod_{j=1}^{0} e_j \triangleq 1$.

Decision analysis can be used to address problems of different scales of difficulty. For the less critical small-scale problem you will find that an analysis assuming via will often give a good answer even when assumptions are only approximately valid. However much more care is usually needed in larger scale or more critical analyses. The next example adapted from a problem analysed by Larry Phillips is typical of a small-scale problem, but with several attributes where the solution, annotation and adaptation took about 2 hours elicitation time.

Example 6.13. Clive sees an advertisement for a new computer design system that purports to be able to perform a one-off task he is contracted to do over the next week. He knows he could do the task adequately well using his old methodology but this will be labour intensive, inflexible, costly, and would not have such a pretty presentation. However on reading the specification of the new system he is certain that if this program could perform the task he needs to do then he could produce his own system which would be equally glossy and flexible as the advertised system and be customised to his particular use. He has various options he could take. He could try to write his own system and if this is not successful then he could use his old method. Alternatively he could try the advertised system. If this fails he has just time to try to write his own system. If this also fails he will revert to his old method. Alternatively after the advertised system failed he could revert to the old method or finally he could simply stick with his old method. How should he act to balance the potential time wasted in the search of an improved product against the advantages of the new options if they are successful?

In small problems like this drawing a decision tree is often helpful. It enables Clive to check that the analyst understands all the options correctly and the tree also provides a framework for the analysis. The tree of the problem above is given below where a represents testing the advertised system, c testing the customised system, and o using the old system, w means that a system works and f that it fails.

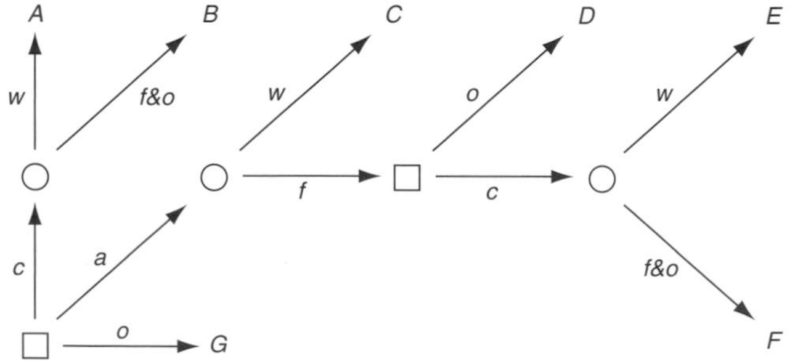

Note here that there are only seven possible consequences: labelled above as A, B, C, D, E, F, G.

Clive now needs to be helped to specify a utility value for each of these consequences. There are four attributes that he articulated as important: R_1 the speed of results, R_2 flexibility, R_3 the appearance of the presentation and R_4 the cost. Going through the seven possible consequences with respect to each attribute he quickly asserts that

$$U_1(F) < U_1(B) < U_1(D) < U_1(G) < U_1(E) < U_1(A) < U_1(C),$$
$$U_2(F) = U_2(B) = U_2(D) = U_2(G) < U_2(E) = U_2(A) = U_2(C),$$
$$U_3(F) = U_3(B) = U_3(D) = U_3(G) < U_3(E) = U_3(A) < U_3(C),$$
$$U_4(F) < U_4(B) < U_4(D) < U_4(G) < U_4(E) < U_4(A) < U_4(C).$$

Here we notice that consequence F is always the worst whilst consequence C is always the best. So we are in the fortunate position that for comparisons needed in the elicitation of criterion weights with no further ado we can assign these the labels $\boldsymbol{r}^0$ and $\boldsymbol{r}^*$ respectively. Assume – as is often the case in practice – that one attribute is judged more important than the others – here speed – and that that the weights turn out to be

$$(k_1, k_2, k_3, k_4) = (0.70, 0.15, 0.05, 0.10).$$

Now the conditional utilities need to be calculated. Having determined the preference orders here helps. Tell Clive that he will be able to check later that provided the numbers are in the right ball park it is not usually particularly critical in such examples to get their values precisely right: similar utilities usually lead to similar solutions. Note that for R_2 no elicitation is needed here since the worst cases must take the value 0 by definition and the best the value 1. Similarly only $U_3(E) = U_3(A)$ can take a value that is not 0 or 1. The elicited values of the four conditional utilities of the seven options are given below.

Consequence	U_1	U_2	U_3	U_4	U
A	.9	1	.9	.9	.915
B	.2	0	0	.2	.160
C	1	1	1	1	1
D	.3	0	0	.4	.250
E	.6	1	.9	.8	.695
F	0	0	0	0	0
G	.4	0	0	.4	.320
Weights	.70	.15	.05	.10	

The utilities in the last column can now be added to the leaves of the decision tree.

There are two probabilities that need to be elicited from Clive before an analysis can be completed. These are the probability that the advertised system works on his problem – he believed this to be 0.8, and the probability his customised method would work given that the advertised method did not work – he sets this at 0.5. From these he can calculate the probability that his customised method would work is

$$0.8 + 0.2 \times 0.5 = 0.9.$$

His decision problem can now be solved using backwards induction as described in Chapter 2. There are now many pieces of software to code such a tree and calculate the decision with the highest expected utility, but in a simple example like this it is almost as easy to calculate this by hand.

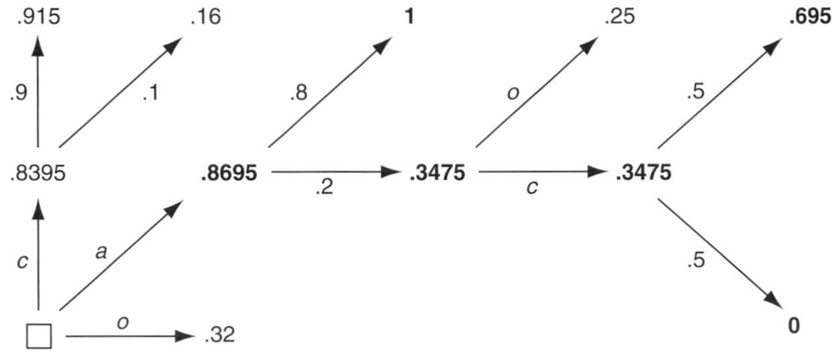

The tree above demonstrates that his best course of action is to first try the advertised system and if this does not work then to try to build a customised system. It can be demonstrated to Clive that by plugging different values into the software that change the specifications of his attribute weights, conditional utilities or his probabilities a little does not change this solution. So the analysis gives him confidence in a course of action as well as providing him with an explanation of why this actually is best.

It is not unusual that after an analysis like this the DM is still uncomfortable and may still be inclined to take another course. If the DM is still uneasy it is usually a sign to the analyst that the problem might not have been fully specified. In this example Clive may still angle to immediately try to build his own system. When the analyst asks him why this is so he might reply that he would enjoy the challenge of building his own system. If this conversation takes place it is clear that he has only partially specified his utility: job satisfaction should have been included as one of his attributes. By re-eliciting criterion weights and marginal utilities to include job satisfaction immediate building may well be found to be optimal. He then has a coherent explanation of what he would like to do and why it is sensible to do it.

6.5.1 Hierarchical utility elicitation and utility trees.

For medium and large scale problems, to elicit the criterion weights for via it is often useful to elicit a tree of attributes. Thus broad attributes are first elicited and label the edges emanating from the root of the tree. The vertices reached by these trees are then further expressed in terms of component attributes and so on.

There are two advantages of proceeding in this way. First if the DM is a body of different people that need to come to a consensus then it is often easiest to obtain a quick consensus as to those broad consequences that are most important. Discussion can then proceed by addressing the refinement of each of the broad consequences in turn. Second and more subtly it is not unusual for a DM body to be layered so that tactical preferences of the strategical DM are delegated to a subordinate team at the local level. The DM is often happy to adopt as her own both the conditional utility evaluations and the relative weights of components leading to the evaluation of the expected utility of each broad consequence. However she will want to have the final say on the relative emphasis she puts on each broad consequence. Such instances are illustrated below. The delegated tactical preferences are represented by the weights given on attributes associated with the leaves on the tree whilst the strategic weights will correspond to the weights given to its interior edges.

If the attributes of the DM's utility function are value independent then the linear form of the tree means that this class is closed under extensions and modifications of the tree. So a DM with via can find this tree – which is then called a *value tree* – particularly useful. This tree is best described through a worked example. Here, for reasons of confidentiality I have adapted the type of analysis described in von Winterfeldt and Edwards (1986) and one I myself led into a hypothetical case study. Notice that in this type of analysis any uncertainty is accommodated only indirectly through the DM's choice of criterion weights and conditional utilities.

Example 6.14. A council is given a grant to rent a property in the city for 10 years with the purpose of supporting young people suffering with heroine addiction. They have a number of possible properties to buy as well as a team of ten personnel to staff the centre. The issue is to find a property to lease. Presceening has suggested that there are several options. Supported by a decision analyst, an executive determines that the choice of property should be determined by three broad criteria: the suitability of the space for client/staff interaction $\overline{r}_1$, availability and attractiveness to the client group $\overline{r}_2$ and quality of the conditions for the staff $\overline{r}_3$. Surveys were then taken to determine those features which might enhance each of these three. It was discovered from interface staff that the three main determinants of $\overline{r}_1$ were the quality and quantity of counselling rooms r_1, quality and quantity of conference rooms r_2 and r_3 – the suitability of waiting areas. The reward associated with the client group $\overline{r}_2$ could be split into the average distance r_4 from the homes of clients, the availability of public transport measured by the distance r_5 from a bus stop and the external appearance of the building r_6, graded 1 as ugly, 2 as average and 3 as attractive. The quality of the property for the staff was measured by the proportion of excellent sized offices r_7, the attractiveness r_8 of the offices on a scale of $1, 2, 3$, average distance of commute for staff r_9 and whether or not r_{10} parking was available for all staff within 200 m of the building. The value tree of this problem is given below.

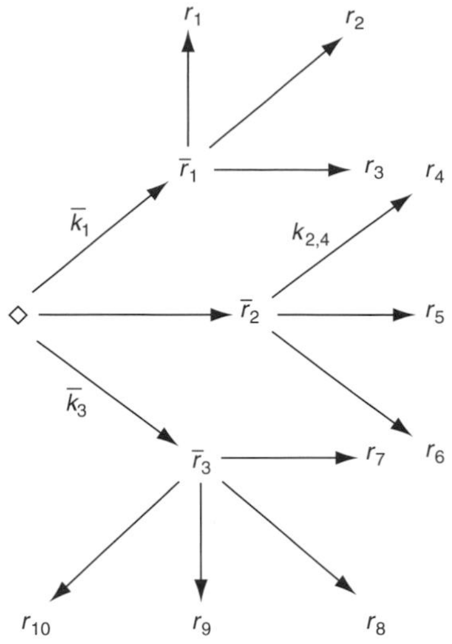

In the problem above if the DM believes that the rewards $\{r_1, r_2, \ldots, r_{10}\}$ are via there is a clear way forward. The DM needs to provide the criterion weights $\{k_1, k_2, \ldots, k_{10}\}$ and the conditional utilities $\{U_1, U_2, \ldots, U_{10}\}$ and choose the property d with the highest score $\overline{U}(d)$ where

$$\overline{U}(d) = \sum_{i=1}^{10} k_i U_i(r_i(d)).$$

However it is very helpful to use the tree to guide the elicitation above. For note that we can write

$$\overline{U}(d) = \sum_{i=1}^{3} \overline{k}_i \overline{U}_i(\overline{r}_i(d))$$

where $\overline{k}_1 = k_1 + k_2 + k_3, \overline{k}_2 = k_4 + k_5 + k_6, \overline{k}_3 = k_7 + k_8 + k_9 + k_{10}$ - so that $\overline{k}_1 + \overline{k}_2 + \overline{k}_3 = 1$ – and

$$\overline{U}_1(\overline{r}_1(d)) = k_{1,1}U_1(r_1(d)) + k_{1,2}U_2(r_2(d)) + k_{1,3}U_3(r_3(d)),$$
$$\overline{U}_2(\overline{r}_2(d)) = k_{2,4}U_4(r_4(d)) + k_{2,5}U_5(r_5(d)) + k_{2,6}U_6(r_6(d)),$$
$$\overline{U}_3(\overline{r}_3(d)) = k_{3,7}U_7(r_7(d)) + k_{3,8}U_8(r_8(d)) + k_{3,9}U_9(r_9(d)),$$

where $k_{i,j} = \overline{k}_i^{-1} k_j$ for $i = 1, 2, 3$ and $j = 1, 2, \ldots, 10$ so that

$$k_{1,1} + k_{1,2} + k_{1,3} = k_{2,4} + k_{2,5} + k_{2,6} = k_{3,7} + k_{3,8} + k_{3,9} = 1,$$

The advantage of this decomposition is that the executive may well be happy to delegate their assessment of each of the benefits reflected in the three scores $\overline{U}_1(\overline{r}_1(d)), \overline{U}_2(\overline{r}_2(d)), \overline{U}_3(\overline{r}_3(d))$ to their respective parties. For example the measure of the benefit to the staff attributable to a certain size of office should surely be theirs, as the attractiveness of a building is to the benefit of the client group. However the weights $(\overline{k}_1, \overline{k}_2, \overline{k}_3)$ should reflect an executive assessment because these balance the important elements they identified in their problem. In an analogous model Keeny describes how the staff group chose through a questionnaire weights that when aggregated put more weight on staff comfort than anything else: giving a large value of $\overline{k}_3$. The executive used the tree to simply down-weight this emphasis whilst keeping the relative weights $k_{i,j}$ intact.

There is of course some elicitation still to do. In practice in problems like this, when there are many attributes, it can be shown that the choice of $U_i(r_i(d))$ is often not critical and is therefore often chosen to be linear although this is obviously not always appropriate.

It should be mentioned that some organisations like to brainstorm random collections of many attributes. After removing logically redundant ones and supervising the transformation of these attributes to value independent ones, it is still often very helpful to group attributes together into broad classes: a bottom-up approach. Again this clarifies executive objectives and helps in the elicitation process exactly as described above. This simple and powerful method of deciding between options in medium sized decision problems is now well used and supported by software.

6.6 Decision conferencing and utility elicitation

6.6.1 Decision conferences

Sometimes decision problems are far more sensitive and important even than ones like that described above. In such circumstances a decision conference is often needed to elicit a group's utility function. A decision conference typically brings together many different people with specific skills. The conference is designed to draw these people into planning to act as a single coherent DM through eliciting a utility for this group. In the example below the form of the elicited utility function – the one with via – would be used to guide the design of a real-time decision support system customised to the needs and priorities of the users. Every facilitator has their own style see e.g. French and Rios Insua (2000); Phillips (2007): I give my own below.

The plan of conferences need to be flexible. Participants are told that discussions about for example the way attributes were measured, the exact values of criterion weights and forms of conditional utility could all be revisited and may well be refined at a later date. Ideally such a conference would have no more than fifteen participants (although the ones described below had rather more) all of whom have some area of expertise pertinent to the problem at hand. They usually last a few days – often an afternoon, a day and a morning (or alternatively a full day and a morning) – and are designed to encourage relaxed wide-ranging discussion about the main issues at hand and not consist simply of formal presentations through a chairperson. Ambience is extremely important to the success of such a conference. For example the conference should take place in a large pleasant room away from the workplace with a semicircular seating plan so that everyone can see each other. Any fears the participants might have about being cajoled into making commitments to policies with which they are not in full agreement need to be allayed as much as possible: see the comments above. So in particular it is important that each participant is briefed about the nature of the conference so they knew what to expect. Furthermore although ideally draft reports of the proceedings should be produced within days of the conference to be circulated to participants, confidentiality clauses will bind the participants until such a time that unanimity about the content of any report is reached, when more general distribution of the proceedings would be facilitated.

The plan of the type of decision conference described below is to elicit the attributes of the group's utility function. A conference of this type often needs three analysts to run it: a *facilitator* who directs the course of the discussion, ensuring that no participant dominates and drawing the discussion away from culs-de-sac; a *recorder* who records the main contents of the discussion and feeds this back to the participants and facilitator at regular intervals, pointing out general shared themes, contradictions and ambiguities to inform future discussion; and a *domain expert*, drawn from the scientific domain of the conference who understands many of the issues and has the background to intelligently unpick contentious points of a more technical nature. Sometimes the facilitator or recorder is able to satisfy this role. Sometimes several different experts with different domains of expertise need to be present (Louchard *et al.*, 1992). There are many ways of conducting such conferences and these should be customised to reflect methods that are as familiar as possible to the participants.

After an initial presentation and resumee of the purposes of the conference, the first day will typically involve preliminary discussions between participants. The facilitator will encourage the thrust of this discussion which will concern the type of actions that are feasible together with the type and nature of attributes of any utility functions different participants believe should inform the preferences between actions in the group. Each participant is invited to speak for up to ten minutes without interruption. These discussions are electronically recorded in real time by the recorder. The informed expert is used aid the facilitator in clarifying any contentious technical issues.

The participants are then encouraged to freely discuss the various issues that have arisen during the day between themselves over a meal and recreation in the following evening. The facilitator, recorder and domain expert will spend some of the first night drawing together material in the form of a short document of the first day's discussion as a basis for further discussions on the morning of the next day, informed by the reflections of the evening. The domain expert has a dominant role at this point, translating the issues discovered by the recorder so that they are as transparent as possible to the bulk of the participants. By mid-morning of the second day the analysts will hope to have a list of the various pertinent consequences and potential ways of measuring these using attributes, together with documentation of various points of contention.

Provided that the morning is successful, the remainder of the second day will involve eliciting quantitative embellishments of the utility function: for example the elicitation of the conditional utility functions and criterion weights using the methods described earlier in this chapter. Various promising options are then appraised using the elicited scores. Because the DM here is a group there is likely to be disagreement about the relative importance the group should place on different aspects of the consequences. The facilitator must emphasise that these early quantifications are very provisional and can be radically changed and simply give ball-park figures to begin the analysis. The facilitator using customised software then feeds back scores of various candidate decisions. For example if the utility function has via then the effects of different weighting can then be explored through displaying the results on a screen.

Sometimes it is possible for participants to perform their own sensitivity analyses, either by instructing the recorder or, if sufficiently computer literate, on software loaded on to their own laptops. As we have seen in this book even quite dramatic changes to parameter values can often have little impact on the relative efficacy of different decision rules. When this is the case participants will often be drawn into a consensus at least about the type of decisions that are good and those which are not fit for purpose. Although it is unusual to obtain complete consensus about exactly what constitutes the best decision, the number of possible candidate decisions favoured by different groups of participants is often small. Furthermore because of the explicit nature of this analysis the reasons one decision might be preferred to another are straightforward to translate both to the conference and to any outside appraiser. Moreover, if as in our running example there is the expectation that the findings of the conference will be continually refined and reappraised, the group should appreciate that any shared preference of a particular decision rule within the conference in a given hypothetical scenario would be illustrative rather than committing.

I will illustrate how this can be done using the example below. After a final summing up the three analysts must quickly produce a report of the main findings of the conference, the points of agreement and contention, possible ways forward and their supporting logic for future refinement and embellishment. Participants are encouraged to contribute any further relevant to the document, after further reflection. A summary of those issues that the whole set of participants are happy to share is then given to a more general audience or a further conference called to resolve important issues of contention.

6.6.2 A conference evaluating countermeasures after a nuclear accident

Sometimes decision problems are far more sensitive and important even than the ones described above. The possibility of using standard multiattribute utility approaches has now been widely used in a large range of examples. Here I describe one of many elicitation exercises undertaken for European countries to help coordinate emergency response decisions made after an accidental release from a nuclear power plant. The elicitation took the form of a decision conference. The ones described below were facilitated by Simon French soon after the Chernobyl accident in the Ukraine and were designed to help the better coordination of the local real-time response to the event of a repeat of such an accident as well as develop protocol and communication channels between scientists in different countries. The reason I have focused on these decision conferences is that unusually the records of all aspects of their proceeding have been made freely available; see Louchard *et al.* (1992). I have had the good fortune to work with Simon on subsequent analogous studies and helped to design and develop the real-time decision support software informed by these conferences.

To illustrate the type of dynamic development and output of a decision conference, consider the following example. Here participants were drawn together to discuss contingent acts relating to a nuclear accident that might happen at a given plant in a given locality and country. The long-term plan was to provide decision support software to an analogous group of DMs at an emergency control centre when faced with such an eventuality. It was planned that as much pertinent information that could be gathered before an incident took place would be recorded in software and integrated into forecasts of effects of possible countermeasures. As more information became available as the accident progressed, the software could be used to revise forecasts, support any modification of the chosen countermeasure and be the framework round which the developments and possible implications of choices of countermeasures could be explained both to those within the room and to a wider audience.

To emulate such a potential group of users the conference included local and national government representatives representing the political decision makers, military and emergency executives representing the enactors of any countermeasures; representatives of the power station concerned with knowledge of how that plant works; various scientists expert in the diffusion of radioactivity in the atmosphere and its absorption into plants, livestock and human beings and medical experts on the effects on health of the exposure to large quantities of radiation. If an accident were to occur in the near future many of these people would be in the incident room and responsible for informing and taking the countermeasure decisions. Here we will focus on events described based on a conference were participants where representatives of the Republics and All-Union authorities in the USSR, who later kindly agreed that all confidentiality could be lifted including the exact quantifications within the elicited utility.

The possible early countermeasure decisions associated with risks of exposure to radiation after an accidental release could take a number of forms. But broadly they appeared to have the following structure. If the population of any nearby village was likely to be exposed to large amounts of radiation were they to stay at home then these people could be evacuated. If the population could potentially be exposed to a moderately health-threatening amount of radiation a more measured alternative would be to instead issue protective measures like iodine tablets to the inhabitants. Finally if there was only a small risk then there was the possibility of simply telling the population to shelter until the radiation had passed.

There was a challenge about how best to measure exposure. Because much of the current theory was expressed in terms of estimates of the lifetime Y dose of a child born in 1986, it was decided initially to focus on decisions based on the expectation μ of this quantity given the possible countermeasures under the best prediction of the development over space and time of the plume of radiation emitted and released into the atmosphere as it passed over this population. So the conference decided initially to consider only actions denoted by SLa-b which was to tell the population to shelter if $\mu < a$, to deliver protective measures if $a \leq \mu \leq b$ and to evacuate the population if $y > \mu$. Here μ would need to be calculated as a complicated function of the source release profile, the windfield and the geography of the regions near the plant. Clearly in the real-time decision support system the thresholds a, b would need to be a function of the size of the population obtained from demographic information of the threatened population, but for simplicity elicitation of the utility function would consider a scenario where these thresholds were fixed and the plume threatened a single concentrated population. The issue for the participants was then to determine how to set the values of (a, b) for this scenario: i.e. to determine what should constitute a "safe" level a and a danger level b.

After this preliminary commitment the focus of the elicitation exercise could now be directed to eliciting the attributes of the conference's utility function. This was assumed to have via. The most obvious tension perceived by the participants was between the resourcing of the countermeasures and the *radiation health effects* if the population was exposed. French describes differences of perspective between conferences in different Republics of quite how the *resource* attribute should be measured. Some thought that the pure monetary cost was a sufficient whilst others argued that availability of specific resources – like medical treatment – and manpower resources should also be folded into any measure. Call this eventual measure r_1. The analysts must remember that one of the most important outputs of the conference is a level of consensus leading to ownership of aspects of the group judgement by the participants in the room. They therefore need to be sensitive not to impose on the attendees the judgements of others outside the room when these are not agreeable to them. So whilst they might introduce a point made elsewhere they must not try to impose outside preferential judgements but should be as flexible as logic allows to the opinions expressed in the room.

A body of theory outlined predicted health effects on a population predicted to receive an expected life-time dose μ: in terms of both cancerous affects measured by r_5 and genetic effects measured by r_6, and participating health radiologists in the room found themselves able to quantify these broad effects in terms of a number of scenarios. Although these were often necessarily approximations and judgements they nevertheless provided ball-park estimates of the positive effects different countermeasures might provide.

As the conference proceeded it became apparent that another health risk was intrinsic to the event: health risks associated with *stress* – measured by an attribute r_4. The acute stress induced by various acts would have significant effects on both morbidity and mortality within the population. The group with most stress risks was a population that was evacuated with especially high risk to those over 50 years old. Although the issue of protective countermeasures was acknowledged to significantly increase stress, there was some debate as to whether improved communication styles could mitigate against this. This represented a leap of awareness by the group and provided another attribute that needed to be folded into any decision making protocol.

However participants were still uneasy that the full picture of deleterious effect had been achieved. After some lengthy discussion another two attributes were discovered which were labelled under the single label *public acceptability*. It was argued that any strategy involving the relocation of a population had an adverse effect on the quality of life – measured by attribute r_2 – of not only those people evacuated but also the population living in the region of the evacuation. Furthermore there was often a political dimension – measured by r_3 – to relocating a population because to change the demography of a nation could well provoke ethnic tensions throughout that country. This attribute was therefore also added to the utility. The final value tree summarises the discussion until the second morning.

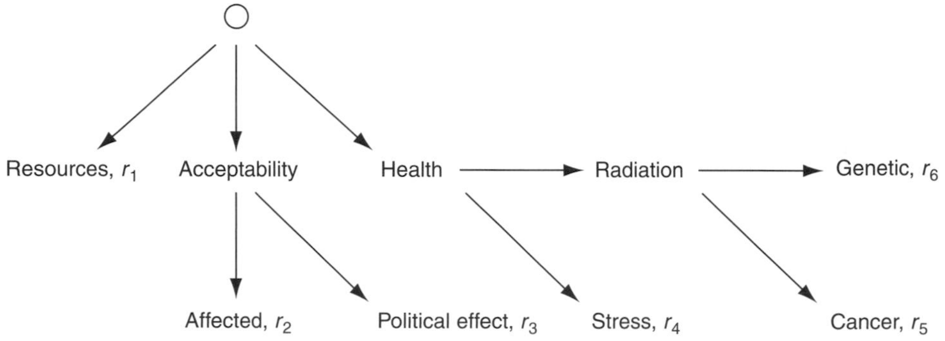

Once the value tree was found, some simple disaster scenarios could be explored: here those where it is known that only one village of a given population is affected by the plume. Various choices d for the parameters using a and b which could be given a vector of preliminary scores $U_i(r_i(d))$ between 0 and 100, $i = 1, 2, \ldots, 6$ were proposed, including various policies that were considered by some to represent wise policy. Such scores were found to be fairly easy to elicit from the group at least ordinally. Preliminary settings let U_i be linear but these setting were changed if they were thought to be inappropriate. Once this was done ball-park figures for criterion weights $(k_1, k_2, \ldots, k_6)$, $\sum_{i=1}^{6} k_i = 1$, could be found. These values provided inputs to software to produce a weighted score $\overline{U}(d) = \sum_{i=1}^{6} k_i U_i(r_i(d))$ between 0 and 100, giving a measure of the benefit of the countermeasure d in the given scenario. Scores for the different plausible countermeasure decision rules could then be calculated as a function of different scenarios and different choices of values of the parameters (a, b) indexing d.

Participants became aware that the score $\overline{U}(d)$ of countermeasures was remarkably robust to changes in the parameters of the utility function. Provided that the sum of the criterion weights k_2 k_3, k_4 was reasonably large and neither k_1 nor $k_5 + k_6$ close to zero or one then moderate values of the thresholds a and b turned out to be optimal. Of course the precise optimal values of (a, b) depended to some extent on the precise values of the utility parameters. However the set of high scoring decisions remained stable over wide changes of parameters and were similar policies. These types of observation gave the group the confidence to agree the broad conclusions of the report. This exercise has now been repeated successfully in many countries across Europe. Each have their own nuances but the types of attribute and weights attributed to them are usually largely similar.

The advantages of these events are both direct and indirect. The indirect effects are a growing appreciation of the points of view of different actors within a decision making framework like this where to be effective all agents have to coordinate their efforts. The direct effect of these events is that they form a framework around which real-time decision making could be supported.

6.7 Real-time support within decision processes

6.7.1 Levels of a decision analysis

As in the problem described above the early elicitation of the form of an appropriate utility function is an intrinsic part of a much wider process. It allows the DM group to identify their needs in the predicted setting. Large-scale problems like these often to need to pass through the stages listed below.

(1) Identify the needs of the DM by the elicitation of her utility function as above.
(2) Identify the uncertain variables whose values define the DM's attributes.
(3) Identify the science or social models that can inform how these attributes link to the values of variables that will be available at the time the decision is made.
(4) Identify the feasible class of decision rules as a function of inputs.
(5) Code up the algorithms that are able to evaluate the efficacy of various different decision rules in terms of their expected scores as a function of the uncertain inputs available just before the decisions need to be made.
(6) Prune out any decision rules which are infeasible or which, for any possible values of the inputs, are always dominated by better decision rules to speed up the code so it can work in real time.
(7) Design a graphical user interface which for the problem faced, presents the DM with the best set of possible decision rules together with their potential consequences as measured by the expected conditional utilities on each attributes.
(8) Develop real-time graphics that summarise the current data linked to the unfolding incident faced.
(9) Develop diagnostics that can indicate whether the statistical models used to predict efficacy are broadly unsurprising given the current unfolding of the events faced in the current incidence, or whether the formal models are producing forecasts that are seriously discrepant with the observed data being collected over real time.

The great of advantage of using the Bayesian paradigm for addressing these types of complex decision problems is that at least formally we know what we are trying to do. We need to identify an expected utility maximising decision rule as a function of the DM's joint probability distribution over all the relevant variables, using as much supporting sampling and experimental evidence as is available and which can respond as a function of information obtained as the DM sees its particular incident unfold. This is of course a very hard thing to do. But at least we know what the practical challenge in front of us is. Furthermore through following the systematic steps given above the DM can prepare well off-line for difficulties she will meet in real time. By systematically working through the first six points many contingencies can be planned beforehand so that the DM is not overwhelmed by a deluge of disparate decisions she needs to coordinate and evaluate on the hoof. This preparation may well need to be customised to a particular genre of potential problem but not in a rush. The DM can then concentrate on the 7th and 8th points when addressing the problem in hand, as more information about the current incident becomes available, only rethinking broad strategy issues if the system signals a failure (the 9th point).

6.7.2 Real-time decision support for nuclear countermeasures

Although the step-by-step procedure described above is quite general it is helpful to flesh it out with a running example. So return to the decision support system designed to inform decisions about the best countermeasures to use in the event of an accidental nuclear release. After a decision conference like the one described above scientific and social models need to be built to predict what will happen if the release occurs. This is extremely complex and depends in part on the countermeasures put in place. First stochastic models need to be built for each given nuclear plant that forecast both in nature, extent and over time the likely profile of the release of radioactive contaminants into the atmosphere. One such model is based on a Bayesian network (Smith *et al.*, 1997) like those described in the next chapter. Other Bayesian probability models (or sometimes deterministic models) describe how the contamination is transported through the atmosphere given the particular windfield and rain events occurring at the plant at a given time. For an example of such a dynamic model based on puff models used in a dynamic setting see Smith and Papamichail (1999). Once the release has stopped and the contamination is largely on the ground or in water the risks to humans becomes related to ingestion, either directly from water or food supplies, or indirectly through contaminated milk or meat. Models are available which predict the probable pathways under the effect of different countermeasures like food bans on the spread of the contamination along the food chain. Finally other medical models are available to help predict the likely adverse effects of these types of exposure on humans.

In fact several models for each of these processes currently exist at least in deterministic forms and many probabilistic analogues are available or currently being researched. However these different models of transport needed to be networked together, ensuring inputs needed for one module are provided by another, to produce a coherent composite picture of the contamination process. This composite then provides the framework for calculating the impact of any given accidental scenario that might be faced, described by its particular covariates $\boldsymbol{x}$, given a promising countermeasure on the attributes of the utility function. We could then – at least in principle – calculate the expected utility of that countermeasure of *all* feasible decision rules in the given scenario faced. In Chapter 8 we will discuss the family of causal Bayesian networks which are particularly useful for this sort of analysis where inference is based on the outputs of a network of modules of probabilistic software.

The challenges for the designer of real-time decision support software, which in this context largely relate to the threat to the population of immediate exposure to an inhaled dose are well defined. The DM simply needs to provide documented predictions of the efficacy of different short-term policies. These will be given in terms of their expected utility scores, where the utilities are provided by the DM group through iterations and refinements of methodologies illustrated above. The expectations will be taken with respect to probability models provided by experts the group trusts to provide good domain knowledge of their particular fields of expertise and, if necessary, probabilities provided by the DM herself.

Of course practical implementation of this programme was challenging and some at the time of writing are only partially solved. Note that the elicitation of their utility function made it possible to translate into natural language why one decision is preferred to another. Thus for example software (Papamichail and French, 2003) can translate the algebraic relationships underpinning the Bayesian preferences of one decision over another as:

"Strategy 11 provides very good decrease of collective dose of radiation in the context of all available strategies."

If Strategy 10 appears a possible good alternative then the system can be asked to explain the preference for Strategy 11.

"Decrease in collective dose of radiation is a significant factor favouring Strategy 10."

If the group require further information, perhaps to override the default adoption of Strategy 11 they can ask for it.

"While decrease of population involved is the main reason for preferring Strategy 10 this is outweighed by considerations of decrease of dose which makes Strategy 11 more preferable."

Note here that the DM could investigate the weight change on the attribute population involved to justify the adoption of Strategy 10 and investigate other candidates with good scores under this reweighting. If it were decided after all to go with Strategy 11 then material to inform a documentation of why the current decision is being taken, perhaps to inform the public through a press release, can also be made available thought the software. Thus

"Strategy 10 provides slightly worse overall benefit than Strategy 11. This judgement takes account of the effects a strategy might have on the decrease of dose to the people involved and the reduction of cost. Whilst the decrease in population involved is the main reason to prefer Strategy 10 this is outweighed by considerations of decrease in collective dose, making Strategy 11 more preferable."

The computer-generated English provided by these types of system is obviously stilted but can be quickly polished up. The formality of the proposed Bayesian method allows the group to move very fast in comparing and evaluating the various courses of action it can direct whilst keeping themselves and others aware of why they prefer one option over another. Of course such systems have to be used flexibly because not all contingencies can be planned for a priori. Nevertheless it can give valuable direct support to the group, keeping them as well informed as they could be about the development of the crisis, and reminding them of decisions they agreed to adopt before the time of the crisis when these are still relevant.

There are many interesting issues associated with setting up this decision support: see Ranyard and Smith (1997); Smith and French (1993); French *et al.* (1995, 1998);

Papamichail and French (2005) and French and Rios Insua (2000) for more details. But I hope in this short summary that I have demonstrated how the careful elicitation of a utility function within a Bayesian analysis can provide the basis of a really powerful tool for real-time decision support.

6.8 Summary

Although sometimes it needs care and in the case of group decisions can be contentious, the elicitation of the attributes of a utility function is often the best way to open up a problem for a decision analysis of a complex problem. The variables of interest to the DM are identified so that any uncertainty management can be focused on those issues impacting on the decision at hand. We have demonstrated above that in large problems this activity is often just the beginning. After the DM has clearly stated the issues she is interested in she then has to somehow structure a narrative that allows her to draw into her model a description of the relevant process and a probabilistic description of how that process links up with data: both data associated with relevant experiments and studies and real-time measurements of the evolution of the process at hand. In the case of large-scale problems this raises important conceptual, inferential and computational issues that will be addressed in the remainder of the book.

The coverage in this book of multiattribute utility elicitation has been necessarily brief. Hopefully I have conveyed some of the general principles of this methodology associated with this fascinating discipline which lies at the intersection of many disciplines: management science, psychology, statistical inference, artificial intelligence and philosophy. The general theory was first expounded by Keeney and Raiffa (1976). Subsequently many excellent and accessible books have been written on this subject, see for example Clemen and Lichtendahl (2002); Goodwin and Wright (2003); Keeney (1992) as well as more technical books in this area, e.g. French and Rios Insua (2000). We shall now leave this area and focus on the use of probabilistic models to address the sort of issues of uncertainty management the are critical inputs to the identification of expected utility maximising decision rules in large decision analyses.

6.9 Exercises

6.1 Prove that when the DM has suia and two attributes her utility function must take the from given in Theorem 6.6.

6.2 You are a member of the company marketing the potentially allergenic product discussed in Chapters 2 and 3. Specify a set of attributes that are:

i) preferentially independent

ii) suia.

6.3 A university needs to choose one of five different brands of photocopying machine $(M_1, M_2, M_3, M_4, M_5)$ for use in the different departments in their institution. A decision analysis has revealed that the university's four mutually utility independent attributes

(x_1, x_2, x_3, x_4) are (reliability, economy, flexibility and size). It is found that this utility function is not linear in its conditional utilities. The values of all attributes other than x_1 can be determined with no uncertainty by simply examining each photocopier and their corresponding conditional utility values are given in the table below.

The client decides that her conditional utility on x_1 is simply the probability θ that a machine will not break down in the first three months of service. This probability θ for each of the five brands is currently uncertain to the university. You have elicited that the distribution of θ that a machine of brand M_j will not break down over the first three months of service is beta $Be(4, 1), j = 1, 2, 3$. The two new models, M_4, M_5, are expected to be more reliable than the other three older models. However their reliability is less certain. Each of these two reliability probabilities θ are therefore given a beta $Be(4.5, 0.5)$ prior distribution.

	Attribute x_1	Attribute x_2	Attribute x_3	Attribute x_4
Utility weights $k_i \times k$	0.5	0.1	0.04	0.02
Conditional utilities	(prior)			
M_1	0.8	0.5	0.5	0.5
M_2	0.8	1.0	0.0	0.0
M_3	0.8	0.0	1.0	1.0
M_4	0.9	0.4	0.25	0.5
M_5	0.9	0.3	0.5	0.5

where k is the unique solution of the equation

$$1 + k = \prod_{i=1}^{4} (1 + kk_i).$$

Calculate the university's Bayes decision under this utility function and briefly explain why this is preferred to the other brands.

6.4 A decision maker has to decide whether or not to exclude a supporter caught on video camera who might have been guilty of attacking fans at a recent football match. Her utility has three value independent attributes. The first is the *probability* that the supporter is excluded or is not excluded but is innocent where q denotes the innocence of that supporter as conveyed by a police expert – elicited using the Brier score. The second attribute measures the *confidence* that the associated supporter will cause future trouble, taking a maximum marginal utility value of 1 if that supporter is excluded and otherwise by the expert's associated expected loss under the Brier score on making her probability statement q. Her third attribute measures the additional financial revenue if the supporter is excluded. Each conditional utility is linear in each attribute, the respective criterion weights of the decision maker are $(k_1, k_2, k_3) = (0.4, 0.3, 0.3)$ and she assumes that the distribution of each attribute is degenerate, her utility associated with each decision taking the value calculated from the above with certainty. Find her Bayes decision rule of whether or not to exclude the supporter as a function of q. Without performing any calculations, can you think of a better way for the decision maker to encode this problem?

6.5 You need to elicit a DM's utility function who is deciding on how to set up a contract with a supplier of oil. You discover that there are three attributes (x_1, x_2, x_3) to her utility function. It will benefit her if the oil is as light and as sulphur-free as possible. So let x_1 denote the amount the supplier is required to spend on lightening the oil, normalised so that $0 \leq x_1 \leq 1$, and x_2 denote the amount the supplier spends on removing sulphur, again normalised so that $0 \leq x_2 \leq 1$. But it is also in her interest to be given preferred customer status $x_3 = 1$ rather than ordinary status $x_3 = 0$. Describe how you would elicit your DM's utility function in this case. She needs to stipulate in her contract the amount her supplier will spend on lightening and desulphuring the oil: i.e. how she sets $d_1 = x_1$ and $d_2 = x_2$. Her marginal utilities for these two attributes are $U_i(x_i) = x_i^2$, $i = 1, 2$. However she believes that the higher she sets (x_1, x_2) the less likely she will be given preferred customer status. In her judgement the probability she will be given this status is

$$P(x_3 = 1) = 1 - 0.5(x_1 + x_2).$$

If the DM has value independent attributes, then find her best course of action as a function of her utility weights.

6.6 a) A DM's utility function has three mutually utility independent attributes (muia) where each attribute X_i, $i = 1, 2, 3$ can take only one of two values. Thus $U_i(X_i) = 1$ denotes the successful outcome of the ith attribute and $U_i(X_i) = 0$ the failed outcome, $i = 1, 2, 3$ where U_i denotes the DM's marginal utilities on $X_i, i = 1, 2, 3$. She tells you that her three criteria weights (k_1, k_2, k_3) are equal and do not sum to one: i.e.

$$k_1 = k_2 = k_3 \neq \frac{1}{3}.$$

Write down the one-parameter family of utility functions consistent with these statements, quoting without proof any result you may need. Briefly describe how you would elicit the parameter k_1.

b) Let d denote any decision with probabilities $\theta(i|d)$ of giving a successful outcome to exactly i of the attributes, $0 \leq i \leq 3$. Prove that the expected utility of this decision is given by

$$\overline{U}(d) = \frac{1}{(s^2 + s + 1)}\theta(1|d) + \frac{[s + 1]}{(s^2 + s + 1)}\theta(2|d) + \theta(3|d)$$

where s, $0 < s = kk_1 + 1 \neq 1 < \infty$, $k = (1 + kk_1)^3 - 1$.

7
Bayesian networks

7.1 Introduction

The last chapter showed how decision problems with many different simultaneous objectives can be addressed using the formal techniques developed earlier in this book. We now turn to a related problem where – as in the last example of that chapter – the processes describing the DM's beliefs is high dimensional. Formally of course this presents no great extension from those described in the early part of this book. The theory leading to expected utility maximising strategies applies just as much to problems where uncertainty is captured through distributions of high-dimensional vectors of random variables as to much simpler ones.

However from the practical point of view a Bayesian decision analysis in this more complicated setting is by no means so straightforward to enact. A joint probability space requires an enormous number of joint prior probabilities to be elicited, often from different domain experts. For the analyst to resource the DM to build a framework that on the one hand faithfully and logically combines the informed descriptions of diverse features of the problem and on the other supports both the calculation of optimal policies and diagnostics to check the continuing veracity of the system presents a significant challenge.

With the increase in electronic data collection and storage many authors have recognised this challenge and developed ways of securely building faithful Bayesian models even when the processes are extremely large. These methods are no panacea. However there is a significant minority of large-scale problems that can be legitimately addressed in this way. The basic principle underpinning these methods recognises a phenomenon already discussed throughout this book and especially in Chapter 4. A DM's beliefs are more faithfully elicited, less ephemeral and more likely to be shared by an auditor when they are structural: expressible in common language rather than by numerical vectors. One such qualitative construct centred on the notion of relevance is particularly useful. Beliefs about the relationship between measurements – or relevance – are more likely to endure over time as a DM learns and to be shared with others.

The idea of relevance was introduced in Chapter 1 where naive Bayesian models were discussed. Recall that in probabilistic models the concept of irrelevance was associated with independence. The identification of irrelevance with conditional independence then permitted the simplification of the description of a problem. That model could then on the one hand be fully specified feasibly and on the other hand had a plausible and explainable rationale behind it. Sadly the class of naive Bayes models has been found to be too restrictive to provide a basis of a faithful representation of the uncertainty between variables in most moderately large problems. But it is possible to extend the ideas behind the naive Bayes models to provide a much more comprehensive technology that is able to faithfully represent many problems.

There is a set of rules – rather pretentiously called the *semi-graphoid axioms* – that define how a rational DM should reason about relevance. In the next section we discuss this logical framework. The beauty of this logic is that it is entirely consistent with families of probabilistic descriptions and can be used as an initial framework (or credence decomposition) around which to elicit probabilities. This means that the elicitation of relevance structure provides the framework of a subjective Bayesian model. Furthermore a Bayesian DM's beliefs about relevance, translated into statements about semi-graphoids, allow the analyst to separate her decision problem into connected collections of subproblems. Elicited beliefs about these much smaller subproblems associated with this decomposition can then be pasted together. In this way a composite picture of the posited non-deterministic relationships between the variables can be presented of a given problem. This provides not only a formal and faithful representation of that problem but also a framework for the fast calculation of optimal policies.

Of course the appropriate framework for performing this sequence of tasks depends heavily on the types of dependence relations the DM believes holds in the context she faces. However one very well-studied structure, which has a wide applicability and is relatively transparent to the DM is the Bayesian network (BN). This chapter will focus on this structure.

7.2 Relevance, informativeness and independence

7.2.1 Rational thinking about relevance

Suppose that, for each decision rule the DM might use, the probability space of a problem can be expressed simply in terms of a product space of a particular set of random variables. In such a context if the DM were to talk about relevance in terms of the relationships between the set of variables then one of the most natural ways she might do this is in terms of the independence or conditional independences between those measurements. This is the motivation for thinking about structures called semi-graphoids and the starting point for many graphical frameworks for uncertainty handling.

We have already noted that from a probabilistic point of view if the client believes that knowing the value of a random vector $\boldsymbol{X}$ is of absolutely no "relevance" to her in guessing the value of another random vector $\boldsymbol{Y}$ then she would simply state that she believed that the measurement vector $\boldsymbol{Y}$ is independent of the measurement vector $\boldsymbol{X}$. Such ideas of relevance are a very important component of a decision analysis. For example, if $\boldsymbol{Y} = (Y_1, Y_2)$ where

Y_d, $d = 1, 2$ were her rewards after taking a decision d and $\boldsymbol{X}$ were some other vector of measurements she could take, then she could conclude that there is no value in observing $\boldsymbol{X}$ since it will not change any of her reward distributions and thus not affect her preferences between the decisions $d = 1$ and $d = 2$. The expected utilities associated with these two decisions will be the same with probability one regardless of the value she observes $\boldsymbol{X}$ to take.

Note in this simple example that to state that she believes that $\boldsymbol{X}$ is irrelevant to predictions about $\boldsymbol{Y}$ does not require the DM to commit to any quantitative statement: it is purely *qualitative*. But this statement is also an extremely useful one. For example, after making this belief statement, the DM justifies a simpler specification of the problem – $\boldsymbol{X}$ need not be evaluated. These judgements are likely to be more stable features of her understanding of a problem. Furthermore, identifying these irrelevant features early on enables her to save time and possible associated financial cost by avoiding the elicitation of useless probabilities.

The idea of eliciting directly whether one set of random variables is irrelevant to another is therefore potentially powerful. However irrelevance on its own is rather too simple an idea to use as a descriptive framework in which complex dependence relationships between many variables can be expressed. Luckily its conditional analogue is much more expressive.

Let $(\boldsymbol{X}, \boldsymbol{Y}, \boldsymbol{Z})$ be arbitrary vectors of measurements in the product space of variables defining the DM's problem.

Definition 7.1. We say that the client believes that the measurement $\boldsymbol{X}$ is *irrelevant* for predicting $\boldsymbol{Y}$ given the measurement $\boldsymbol{Z}$ (written $\boldsymbol{Y} \amalg \boldsymbol{X} | \boldsymbol{Z}$) if she believes now that once she learns the value of $\boldsymbol{Z}$ then the measurement $\boldsymbol{X}$ will provide her with no *extra* useful information with which to predict the value of $\boldsymbol{Y}$.

Note that if the DM is a Bayesian then she will be able to express her beliefs in terms of the structure of her joint probability mass function $p(\boldsymbol{x}, \boldsymbol{y}, \boldsymbol{z})$ on $(\boldsymbol{X}, \boldsymbol{Y}, \boldsymbol{Z})$. Thus if she states that $\boldsymbol{Y} \amalg \boldsymbol{X} | \boldsymbol{Z}$ interpreting this conditional irrelevance statement as a conditional independence statement then it could be concluded that she could write her conditional density $p(\boldsymbol{y}|\boldsymbol{x}, \boldsymbol{z})$ of $\boldsymbol{Y}|\boldsymbol{X}, \boldsymbol{Z}$ so that it did not functionally depend on the value $\boldsymbol{x}$: i.e. for all possible values of $(\boldsymbol{x}, \boldsymbol{y}, \boldsymbol{z})$

$$p(\boldsymbol{y}|\boldsymbol{x}, \boldsymbol{z}) = p(\boldsymbol{y}|\boldsymbol{z}).$$

Another equivalent way of writing this is to stipulate that for all values of $(\boldsymbol{x}, \boldsymbol{y}, \boldsymbol{z})$ her joint mass function would respect the factorisation

$$p(\boldsymbol{x}, \boldsymbol{y}, \boldsymbol{z}) = p(\boldsymbol{y}|\boldsymbol{z})p(\boldsymbol{x}|\boldsymbol{z})p(\boldsymbol{z}). \tag{7.1}$$

A collection of conditional irrelevance statements is important because it is possible to make inferential deductions directly using such a collection. This is not only invaluable as an aid in the construction of faithful models over many variables, by avoiding early spurious quantifications, but also allows the DM and auditor to agree at least about the structure of a model to be analysed. For example whilst two people may often disagree about how to assign the exact value of the probabilities in a joint mass function of a pair of random variables they may well agree using contextual information that those two random variables are independent of each other. Two experts might both believe that a measure of the state of the economy and the measure of aggressiveness of a cancer are independent whilst strongly disagreeing about the probability distribution of each of these measures.

The tertiary irrelevance relationships defined above over sets of measurements provide the foundation of a language that expresses, in a faithful and logical fashion, how dependencies between one collection of sets of measures in a problem induce dependencies in another. Logical demands can be made of this language because it is reasonable for an auditor to expect a DM's statements about irrelevancies to satisfy certain rules.

There is now a considerable body of literature that discusses what those rules should be – see, for example, Dawid (1979); Oliver and Smith (1990); Pearl (1988); Studeny (2005). In this introduction I will discuss only the two most important and universally applicable rules. The first, called the *symmetry* property, demands that for any three vectors of measurements $\boldsymbol{X}, \boldsymbol{Y}, \boldsymbol{Z}$:

$$\boldsymbol{X} \amalg \boldsymbol{Y} | \boldsymbol{Z} \Leftrightarrow \boldsymbol{Y} \amalg \boldsymbol{X} | \boldsymbol{Z}. \tag{7.2}$$

Note, in particular, from the symmetry of the conditional independence equation (7.1) above, the equivalence (7.2) must certainly hold if the DM is a Bayesian. But it is also a property of most other non-probabilistic methods of measuring irrelevance such as ones based on upper and lower probability or belief functions. On the other hand it is not a property of relevance which is obviously vacuous to the uninitiated. For in natural language it allows the DM to conclude that if when forecasting the measurement X the client is confident that once she knows the value of Z there will be no point in keeping the reading Y then whenever she needs to forecast the measurement Y on the basis of X and Z she can confidently discard X as being totally irrelevant. This is not at all transparent. This symmetry property is at the heart of fast-learning algorithms. In a later example we illustrate how this belief and its consequence can be appreciated by the client as operationally different assertions and so can usefully be checked against one another to validate a model.

A second property, called *perfect composition*, although more complicated to write down is, in comparison with symmetry, a very obvious property to demand for statements about irrelevance. It states that for any four disjoint vectors of measurements $\boldsymbol{X}, \boldsymbol{Y}, \boldsymbol{Z}, \boldsymbol{W}$,

$$\boldsymbol{X} \amalg (\boldsymbol{Y}, \boldsymbol{Z}) | \boldsymbol{W} \Leftrightarrow \boldsymbol{X} \amalg \boldsymbol{Y} | (\boldsymbol{W}, \boldsymbol{Z}) \;\&\; \boldsymbol{X} \amalg \boldsymbol{Z} | \boldsymbol{W}. \tag{7.3}$$

A good explanation of why it is necessary to demand this property was given in Pearl (1988). Assume the DM will learn the value of $\boldsymbol{W}$ and is interested in then using the additional information in the measurements $\boldsymbol{Y}$ and $\boldsymbol{Z}$ to help her to predict the value of $\boldsymbol{X}$. Imagine that the information in $\boldsymbol{Z}$ is written in a document called Doc. 1, and the information $\boldsymbol{Y}$ is written in another document called Doc. 2. The assertion now says that the statement: "The DM believes that the two documents give no further useful information for predicting $\boldsymbol{X}$ once $\boldsymbol{W}$ is known" is equivalent to the two statements, taken together, that: "She believes that Doc. 1 gives no further useful information about $\boldsymbol{X}$ once $\boldsymbol{W}$ is known" and "Once $\boldsymbol{W}$ is known and the information in Doc. 1 is fully absorbed, the information in Doc. 2 provides no additional information about $\boldsymbol{X}$ either".

Stated like this it is difficult to see any reason why anyone could regard the statement on the left-hand side of (7.3) and the two statements on the left-hand side of (7.3) as logically different. In an exercise below you are asked to check that in particular any Bayesian DM would automatically follow this rule if she were coherent.

These two properties of irrelevance, defining what is known as a *semi-graphoid*, may seem trite. But in fact they allow the analyst to make surprisingly strong deductions from a collection of statements made by a DM that can then be fed back to her to see if the model really is faithful to her beliefs. Properties of semi-graphoids are now well understood so that exploring logical consequences with the DM in this way is fairly straightforward.

Of course to confront the typical DM directly with the algebra of semi-graphoids would usually be intimidating. So it is not usually a practical option to work with these structures directly. The DM therefore needs to trust the analyst that the features she is asked to confirm really are logical consequences of her original statements about the relationships between variables. The analyst would quickly lose the DM's commitment to the elicitation process if he dwelt too long on these apparent technicalities. However in problems which have a clear hierarchy of causes, like the example below, it is often possible to embed at least a subset of the more important dependence relationships on to a graph called a Bayesian network. Whilst retaining its formal integrity a typical DM appears to interpret appropriately the meaning of the depiction of a Bayesian network and consequently to take ownership of the complex pattern of relationships it embodies.

7.2.2 Deductions about sets of irrelevances*

If the DM is a true Bayesian then it is relatively easy to show – using ideas of relative entropy – that there are an infinite number of conditional independence statements that can be stated – like symmetry and perfect composition – once we know that the DM is thinking probabilistically: see Studeny (1992). So semi-graphoids are quite general rule systems – see Dawid (2001) which simply contains probabilistic models as a special case.

On the other hand, when interpreted probabilistically, certain combinations of irrelevance statements a DM might hold can have strong distributional consequences. Thus for example Feller (1971) p78 proves the following.

Theorem 7.2. *If $X_1 \amalg X_2$ are not both degenerate and $X_1+X_2 \amalg X_1-X_2$ this is equivalent to the statement that X_1 and X_2 are independent normally distributed with the same variance.*

Another interesting result is the following found by Lukacs (1965).

Theorem 7.3. *If X_1, X_2 are non-degenerate positive random variables with both $X_1 \amalg X_2$ and $X_1 + X_2 \amalg Y$ where $Y = \frac{X_1}{X_1+X_2}$ then this is equivalent to saying that X_1 and X_2 are independent and each have a gamma $G(\alpha_1, \beta)$ and $G(\alpha_2, \beta)$ density given in equation*

(5.16). It follows from simple probability arguments that Y then has a beta $Be(\alpha_1,\alpha_2)$ density given in equation (5.12).

Another important characterisation found by Geiger and Heckerman (1997) will be used later. These types of results are important because they demonstrate that commonly used parametric densities can be specified indirectly by eliciting from the DM that a couple of qualitative statements about irrelevance are true.

There are some conditional independence statements which some DMs might like to make like that $X \amalg X + Y$ for some random variable Y with support the whole real line or that $\max_{1\leq i\leq n}\{X_i\} \amalg n^{-1}\sum_{i=1}^{n} X_i$ (Berger (1985); Heath and Sudderth (1989)). There is no standard joint probability distributions over sets of non-degenerate random variables with either of these properties. If the DM really believes these statements then she has to express her beliefs using non-standard semantics: either using finitely additive priors or other methods. We note that finitely additive probability models, or at least their improper prior analogues are widely used as a practical modelling tool – often set as a default in Bayesian software despite the technical difficulties they are known to exhibit; see Gelman *et al.* (1995); O'Hagan and Forster (2004). These are used both to model "uninformative" location priors as in the first example, or to order independence albeit less regularly as in the second; see e.g. Atwell and Smith (1991).

Finally it is interesting to note that there is a Hilbert space (an infinite-dimensional version of a linear space which is completed) that can be used to represent a product probability space like the ones discussed here. Perhaps the most familiar use of this representation is when using Fourier basis representations of probability: for example characteristic functions see e.g. Small and McLeish (1994). In this representation each random variable X can be identified with a closed linear subspace A_X. If $Z \amalg X$ then these two subspaces A_Z and A_X are orthogonal in this Hilbert space: a property often denoted by $A_Z \perp A_X$. Similarly if $Z \amalg X|Y$ then the projection of A_Z into the kernel of Y is orthogonal to the projection of A_X into Y. In this sense use of semi-graphoids can be seen as simply a use of standard ideas about orthogonal projections, ideas commonplace in geometry used on probability judgements.

7.3 Bayesian networks and DAGs

7.3.1 What a Bayesian network does

This section addresses a decision problem whose decision space is simple and whose complexity arises from the fact that the relationships between the variables of the problem are numerous and complex. In such a scenario suppose the analyst wants to help the DM depict her own verbal explanations and those of trusted experts about how these variables might influence one another. Ideally the depiction has the following properties:

- It is *evocative* and understandable to the DM so that it can be owned by her.
- It provides a *faithful* picture of the pattern of relationships the DM believes exists between the salient features of her problem.

- Its topology links to a set of statements about relevance which obey the semi-graphoid properties. This ensures that the graph itself would have a *logical integrity* enabling various logical consequences of a DM's original statements to be fed back to her so that the faithfulness of the graphical representation to her beliefs could be checked by the analyst and auditor and creatively reappraised.
- It is possible to *embellish* the graph directly with further quantitative probability statements provided by the DM, so it provides a consistent depiction of the DM's full probabilistic model. In particular, after this embellishment the graph is still faithful to the originally elicited irrelevance statements.
- In addition this graphically based probability model can be used as a framework to guide Bayesian learning and *fast computation*. These last two topics have attracted a great deal of academic interest, especially over the last two decades, and such techniques are well developed and documented – see, for example, Cowell *et al.* (1999); Jensen and Nielsen (2007); Pearl (1988); Spiegelhalter *et al.* (1993).

Although it appears ambitious to expect that a single construction could have all the properties listed above, there are now several different graphical systems having all these properties for certain types of problem. The most used and developed of these is called a *Bayesian network* (BN). As for any of its graphical competitors, by eliciting a BN first the analyst avoids asking the DM to express, early on, numerical quantifications of her uncertainties, and algebraic specifications of statistical structures. Instead she is encouraged to describe her problem through stating its main features together with the pattern of relationships she believes exists between them. The BN has a directed graph whose vertices represent the features of the problem that the DM considers are important (and possibly as yet unknown). If the DM believes that a first variable labelled by one vertex is informative about a second variable labelled by a second vertex – in a sense which will be formalised later – then the first vertex is connected into the second by a directed edge.

7.3.2 The Bayes net and factorisations

Perhaps the easiest way of thinking of a BN is as a simple and convenient way of representing a factorisation of a joint probability mass function or density function of a vector of random variables $\boldsymbol{X} = (X_1, X_2, \ldots, X_n)$. Henceforth let $\boldsymbol{X}_I$ denote the subvector of $\boldsymbol{X}$ whose indices are $I \subseteq \{1, 2, \ldots, n\}$. From the usual rules of probability the joint mass function or density $p(\boldsymbol{x})$ of $\boldsymbol{X}$ can be written as the product of conditional mass functions. Thus

$$p(\boldsymbol{x}) = p_1(x_1)p_2(x_2|x_1)p_3(x_3|x_1,x_2)\ldots p_n(x_n|x_1,x_2,\ldots,x_{n-1}) \tag{7.4}$$

where $p_1(x_1)$ is the mass function/density of x_1 whilst $p_i(x_i|x_1, x_2, \ldots, x_{i-1})$ represents the density of x_i conditional on the values of the components of $\boldsymbol{x}$ listed before it. The simplest example of such a formula occurs when all the components of $\boldsymbol{X}$ are independent, when we can write

$$p(\boldsymbol{x}) = \prod_{i=1}^{n} p_i(x_i).$$

In most interesting models though not all variables are independent of each other. However many of the functions $p_i(x_i|x_1, x_2, \ldots, x_{i-1})$ will often be an explicit function of components of $\boldsymbol{X}$ whose indices lie in a *proper subset* $Q_i \subset \{1, 2, \ldots, i-1\}$, $2 \leq i \leq n$. Thus suppose

$$p_i(x_i|x_1, \ldots, x_{i-1}) = p_i(x_i|\boldsymbol{x}_{\{1,\ldots,i-1\}}) = p_i(x_i|\boldsymbol{x}_{Q_i}) \tag{7.5}$$

where the *parent set* $Q_i \subseteq \{1, 2, \ldots, i-1\}$ and let the *remainder set* $R_i \subseteq \{\{1, 2, \ldots, i-1\} \setminus Q_i$, where we allow both Q_i and R_i to be empty. Here $\boldsymbol{x}_A$ denotes the subvector of $\boldsymbol{x}$ whose components are indexed by indices in the set A. Under the set of $n-1$ statements (7.5) we obtain from a new simplified factorisation formula of (7.4)

$$p(\boldsymbol{x}) = p_1(x_1) \prod_{i=2}^{n} p_i(x_i|\boldsymbol{x}_{Q_i}). \tag{7.6}$$

Next note that (7.5) can also be written as

$$p_i(x_i|\boldsymbol{x}_{R_i}, \boldsymbol{x}_{Q_i}) = p_i(x_i|\boldsymbol{x}_{Q_i}).$$

The important point to note now is that since the DM is a Bayesian the equation above can also be expressed as an irrelevance statement about the relationship between the measurement X_i the random vector of its parents $\boldsymbol{X}_{Q_i}$ and its remainder vector $\boldsymbol{X}_{R_i}$ viz

$$X_i \amalg \boldsymbol{X}_{R_i}|\boldsymbol{X}_{Q_i}.$$

It follows that the factorisation (7.6) can be seen simply as the set of irrelevance $n-1$ statements

$$X_i \amalg \boldsymbol{X}_{R_i}|\boldsymbol{X}_{Q_i} \quad 2 \leq i \leq n. \tag{7.7}$$

Definition 7.4. A *directed acyclic graph (DAG)* $\mathcal{G} = (V(\mathcal{G}), E(\mathcal{G}))$ *with set of vertices* $V(\mathcal{G})$ and set of directed edges $E(\mathcal{G})$ *is a directed graph having no directed cycles.*

Definition 7.5. A *Bayesian network* (*BN*) on the set of measurements $\{X_1, X_2, \ldots, X_n\}$ is a set of the $n-1$ conditional irrelevance statements (7.7) together with a DAG $\mathcal{G}$. The set of vertices $V(\mathcal{G})$ of $\mathcal{G}$ is $\{X_1, X_2, \ldots, X_n\}$ and a directed edge from X_i into X_j is in $E(\mathcal{G})$ if and only if $i \in Q_j$, $1 \leq i,j \leq n$. The DAG $\mathcal{G}$ is said to be *valid* if the DM believes the conditional irrelevance statements associated with its BN.

Note that the graph constructed above is automatically acyclic because a variable vertex can only lead into another vertex with a higher index. A typical example of such a BN is given below.

Example 7.6. A DM has a problem defined by a set of five measurements (U, V, X, Y, Z). She believes they have a joint mass function such that the conditional densities exhibit the following dependence structure. The mass function $p_2(v|u)$ depends on u, v; $p_3(x|u, v)$ depends on x, v; $p_4(y|u, v, x)$ depends on x, v; and $p_5(z|u, v, x, y)$ depends on x, z.

Taking the variables in the order given above these provide us with four statements of the form (7.7). It follows from the definition of a BN above that its DAG is given below.

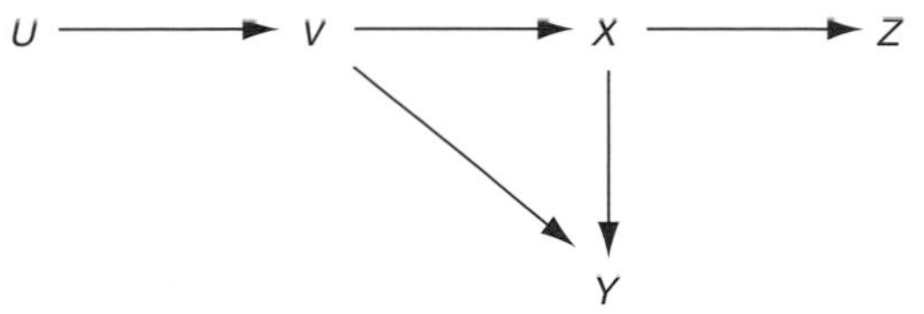

(7.8)

We argued above that the semi-graphoid properties can be used to make logical deductions about such graphs. To illustrate this we show that adding an edge to BN with a valid DAG $\mathcal{G}$ to obtain a new graph $\mathcal{G}^*$ which is also a DAG makes the DAG $\mathcal{G}^*$ valid as well. To prove this suppose the edge (X_j, X_i) is added to $\mathcal{G}$. Then $\{Q_i, R_i : 2 \leq i \leq n\}$ are the parent and remainder sets of $\mathcal{G}$ and $\{Q_i^*, R_i^* : 2 \leq i \leq n\}$ are the parent and remainder sets of $\mathcal{G}$ where $Q_i^* = Q_i \cup \{j\}$ and $R_i^* = R_i \backslash \{j\}$. The only new irrelevance statement introduced into the corresponding list of irrelevance statements is

$$\begin{aligned} X_i \amalg \boldsymbol{X}_{R_i} | \boldsymbol{X}_{Q_i} &\Leftrightarrow X_i \amalg \boldsymbol{X}_{R_i^* \cup \{j\}} | \boldsymbol{X}_{Q_i} \text{ by definition} \\ &\Rightarrow X_i \amalg \boldsymbol{X}_{R_i^*} | \boldsymbol{X}_{Q_i \cup \{j\}} \text{ by perfect composition (7.3)} \\ &\Leftrightarrow X_i \amalg \boldsymbol{X}_{R_i^*} | \boldsymbol{X}_{Q_i^*} \text{ again by definition.} \end{aligned} \tag{7.9}$$

This gives a simple demonstration of how results are proved using the semi-graphoid axioms. Note that this result means that it is the *absence* of edges that is significant in a BN not the existence. The presence of an edge just means that a relationship between the two variables concerned *might* exist. In particular a complete DAG (i.e. one with no missing edges) is totally uninformative and always valid.

There are two good reasons why a BN is useful. First its graph simultaneously depicts many statements about the connections between variables in a formally correct, accessible and evocative way. Second, from a BN alone – and not the order of introduction of the variables used in its construction – it can be proved, using the properties of irrelevance, that it has an unambiguous representation of a set of irrelevance statements.

Example 7.7. Suppose, in an expansion of a mass function we took over the five variables of the example above in a different order $\{U, V, X, Z, Y\}$. Then under this ordering of variables the DAG in that example corresponds to a different factorisation formula

$$\begin{aligned} p(u, v, x, z, y) &= p(u)p(v|u)p(x|u, v)p(z|u, v, x)p(y|u, v, x, z) \\ &= p(u)p(v|u)p(x|v)p(z|x)p(y|v, x). \end{aligned}$$

So it appears that to make a BN unambiguous it is necessary to remember the numbering of the variables defining the problem. However it can be proved (see e.g. Smith (1989b)) that these two different factorisations are actually equivalent in the sense that they code two sets of irrelevance statements each deducible from the other using the semi-graphoid properties. This is true in general: if two BNs support different orderings of variables then they encode exactly equivalent sets of irrelevance statements. So in this sense the BN is a better, more compact, description of a client's irrelevance statements than one of its probability factorisations.

7.3.3 The d-separation theorem

There is a much stronger and more remarkable result proved by Geiger and Pearl (1993); Verma and Pearl (1988) and re-expressed in the form given here by Lauritzen (1996); Laurizen *et al.* (1990). It allows the analyst to find all the irrelevance statements that can be logically deduced from a given BN directly from the topology of its graph. Before we can articulate this result we need a few terms from graph theory. Recall that a vertex X is a *parent* of a vertex Y, and Y is a *child* of X in a directed graph $\mathcal{G}$ if and only if there is a directed edge $X \rightarrow Y$ from X to Y in $\mathcal{G}$. Note that when $\mathcal{G}$ is the DAG of a BN then the set $\boldsymbol{X}_{Q_i}$ is the set of all parents of X_i – henceforth called the parent set of X_i in $\mathcal{G}$.

Similarly we say Z is an *ancestor* of Y in a directed graph $\mathcal{G}$ if $Z = Y$ or if there is a directed path in $\mathcal{G}$ from Z to Y. This term can also be made to apply to all subsets of $V(\mathcal{G})$. Thus let $\boldsymbol{X}$ denote a subset of the vertices $V(\mathcal{G})$ in $\mathcal{G}$ then the *ancestral set* of $\boldsymbol{X}$ – denoted by $A(\boldsymbol{X})$ – is the set of all the vertices in $V(\mathcal{G})$ that are ancestors of a vertex in $\boldsymbol{X}$ together with $\boldsymbol{X}$ itself. The *ancestral graph* $\mathcal{G}(A(\boldsymbol{X})) = (V(\mathcal{G}(A(\boldsymbol{X}))), E(\mathcal{G}(A(\boldsymbol{X}))))$ has vertex set $V(\mathcal{G}(A(\boldsymbol{X})) = A(\boldsymbol{X})$ and edge set

$$E(\mathcal{G}(A(\boldsymbol{X}))) = \{e = X_e \rightarrow Y_e \in E(\mathcal{G}) : X_e, Y_e \in A(\boldsymbol{X})\}.$$

Thus the ancestral graph $\mathcal{G}(A(\boldsymbol{X}))$ is the subgraph of $\mathcal{G}$ generated by the subset of vertices $A(\boldsymbol{X})$.

A graph is said to be *mixed* if some of its edges are directed and some undirected. The *moralised graph* $\mathcal{G}^M$ of a directed graph $\mathcal{G}$ has the same vertex set and set of directed edges as G but has an undirected edge between any two vertices $X_i, X_j \in V(\mathcal{G})$ for which there are no directed edges between them in $\mathcal{G}$ but which are parents of the same child Y in $V(\mathcal{G})$. Thus, continuing the hereditary metaphor, all unjoined parents of each child are "married" together in this operation. If $\mathcal{G}^M = \mathcal{G}$ – so that all two parents of the same child are joined by directed edge for all children in $V(\mathcal{G})$ – then $\mathcal{G}$ is said to be *decomposable*. The *skeleton* $\mathcal{S}(\mathcal{H})$ of a mixed graph $\mathcal{H}$ is one with the same vertex set as $\mathcal{H}$ and an undirected edge between X_i and X_j if and only if there is a directed or undirected edge between X_i and X_j in $\mathcal{H}$. Thus to produce the skeleton $\mathcal{S}(\mathcal{H})$ of a mixed graph $\mathcal{H}$ we simply replace all directed edges in $\mathcal{H}$ by undirected ones.

Finally suppose A, B, C be are any three disjoint subsets of $\{1, 2, \ldots, n\}$ and $\boldsymbol{X}_A, \boldsymbol{X}_B, \boldsymbol{X}_C$ the corresponding sets of the vertices $V(\mathcal{S})$ of an undirected graph $\mathcal{S}$. Then $\boldsymbol{X}_B$ is said to *separate* $\boldsymbol{X}_C$ from $\boldsymbol{X}_A$ in $\mathcal{S}$ if and only if any path from any vertex $X_a \in \boldsymbol{X}_A$ to any vertex $X_c \in \boldsymbol{X}_C$ passes through a vertex $X_b \in \boldsymbol{X}_B$.

We are now ready to state the d-separation theorem. This allows for *all* valid deductions to be read directly from the DAG of a BN. It can be proved using only the semi-graphoid properties of symmetry and perfect composition and so applies even outside a Bayesian framework.

Theorem 7.8. *Let A, B, C be any three disjoint subsets of $\{1, 2, \ldots, n\}$ and G be a valid DAG whose vertices $V(\mathcal{G}) = \{X_1, X_2, \ldots, X_n\}$. Then if $\boldsymbol{X}_B$ separates $\boldsymbol{X}_C$ from $\boldsymbol{X}_A$ in the skeleton of the moralised graph $\mathcal{G}^M(A(\boldsymbol{X}_{A\cup B\cup C}))$ of the ancestral graph $\mathcal{G}(A(\boldsymbol{X}_{A\cup B\cup C}))$ then*

$$\boldsymbol{X}_C \amalg \boldsymbol{X}_A | \boldsymbol{X}_B.$$

The proof of this important result is omitted because it is rather technical and graph theoretic: see the references above. From a practical point of view this result is extremely important because it provides a simple method enabling an analyst to check whether certain interesting statements can be deduced from an elicited BN from the DM. These can then be fed back to the DM for confirmation of their plausibility in a way illustrated below. So the architecture of a BN can be queried to determine whether or not it is requisite (Phillips, 1984) without the analyst having to first elicit further probabilistic embellishments.

Consider the following grossly simplified but nevertheless illuminating example of how eliciting a Bayes net can help the DM express important features of an underlying process that might influence its development.

Example 7.9. A DM needs to predict the costs of programmes of work she will employ contractors to do. An appropriate employee produces a ball-park estimate B of the cost of the work of any potential scheme. On the basis of this the DM may decide to produce a more detailed estimate E from expert civil engineers. She can then decide which programme she puts out to tender and the firm placing the lowest bid T receives the work. The work is then completed and the final out-turn cost O – which may include any unforeseen additional costs – is charged. The evaluations above are used to predict the out-turn costs O of all programmes undertaken and hence her ongoing building costs into the short and medium term.

The DM tells the analyst that she currently uses E to estimate the probability distribution of the winning tender T if this was not yet available, which in turn is used to predict the out-turn price O of tender bid T. It follows that she is implicitly assuming that the BN

$$B \rightarrow E \rightarrow T \rightarrow O$$

is valid. To check this model the analyst can ask two questions respectively related to the two irrelevance statements it contains, namely $T \amalg B|E$ and $O \amalg (B, E)|T$. These can be respectively queried by the analyst by asking the following questions:

- "Can you think of any scenario when estimating the tender price when the ball-park estimate might help refine a detailed estimate?" and
- "Are there any scenarios where the detailed or ball-park estimate might provide additional useful information about the out-turn price over and above that provided by the winning tender price?"

Reflecting on the second query the DM noted that for projects where the tender price was much less than the detailed estimate the out-turn price was often much higher than would normally be estimated from the tender price. She gives the following reason. In periods of economic recession contractors tended to submit artificially low tenders in order to secure the work. To recoup their costs they then would endeavour to find many spurious "add-ons" to boost O.

The BN has provided the DM with a qualitative framework from which a more refined description of her problem can be developed, incorporating a new variable I – an index of the abundance of work available – as a new vertex in the graph. The BN adapted in the light of her comments above is given below.

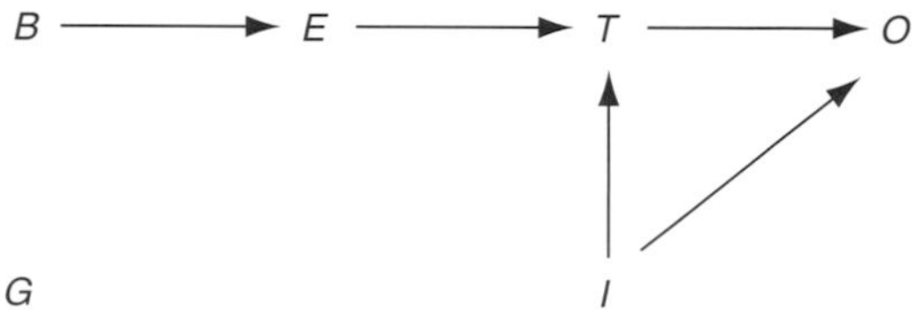

(7.10)

This depicts three new statements. Taken in their natural causal order (I, B, N, T, O), the first $(B, E) \amalg I$ simply says that estimates do not take into account the variation in the market as reflected by the index. The second says that $T \amalg B|\,(I, E)$ – the tender price is independent of the ball-park estimate once the detailed estimate and the index, reflecting any deflation or inflation because of lack or abundance of work, is taken into account. The last says $O \amalg (B, E)|\,(I, T)$ – the two original company estimates are uninformative about the out-turn price once availability of work is factored in.

Let us suppose the DM is initially happy with these statements. Note that if relevant ways of measuring I can be found then this can be folded into her record keeping so that she can make more reliable predictions of the out-turn costs. Further use of the graphoid properties could be used to check whether this new model is requisite. So for example the analyst could ask "If you had to resurrect the rough estimate of a project because this had been lost and you had available both your detailed estimate would the detailed estimate alone be sufficient to resurrect the rough estimate as accurately as possible?" This should be so because the symmetry property demands that the DM has asserted, as above, that $T \amalg B|D$ then it can be deduced that $R \amalg T|E$, i.e. that the answer to the above question is "Yes". But note that this question is not obviously equivalent to the original assertion. So other forgotten dependences might be teased out of the DM just by asking this question. Similarly to check whether our model now implies $O \amalg E|T$ use the transform $\mathcal{G}$ to $\mathcal{S}$ as described above.

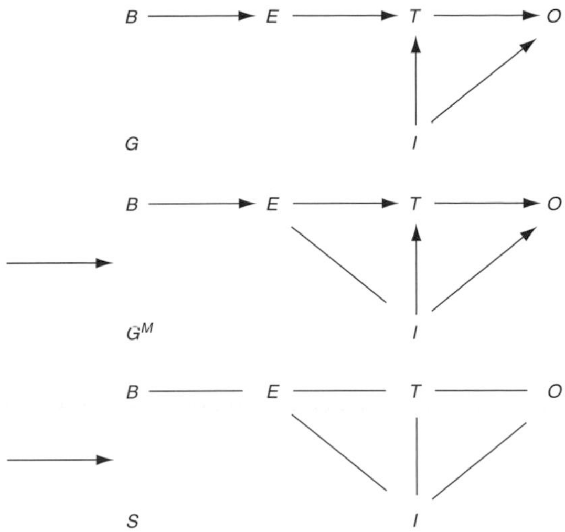

Note that the path (E, I, O) in $\mathcal{S}$ between E and O is not blocked by T. We can therefore conclude that if I is not known for this project, then the detailed estimate E may well be informative about the out-turn price O. This concurs with her earlier uneasiness. However if both T and the index I are recorded then note that (T, I) block all paths between E and O so under the adapted graph it can be deduced that E is assumed unnecessary to record if both (T, I) are recorded.

7.3.4 Completeness and equivalent BNs

In Geiger and Pearl (1990) it was proved that a probability distribution could be constructed that respected all the conditional independence statements deducable by the d-separation but no others. This result was strengthen by Meek (1995) who produced less contrived constructions to prove this point. So the theorem in this sense is necessary as well as sufficient. This is the more remarkable because of the result of Studeny (2005) referred to in the last section. The BN therefore encodes rather special collections of conditional independence statements to enable this "completeness" property to hold. This is further evidence of the compelling descriptive power of a BN from a qualitative point of view and its important position within the class of different graphs especially in the Bayesian paradigm.

Although it is possible to read all the deducible implications from a BN, two BNs can make exactly equivalent sets of conditional independence statements. For example the three DAGs below are topologically the same but all embody just the two conditional independence statements $X \amalg Y|Z$ and $Y \amalg X|Z$.

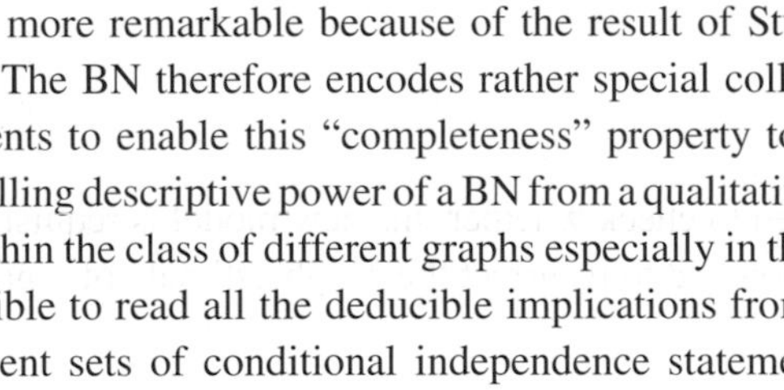

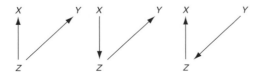

It was proved by Verma and Pearl (1990) that two BNs imply exactly the same set of irrelevance statements if and only if their DAGs share the same "pattern". The pattern $\mathcal{P}$ of a DAG $\mathcal{G} = (V(\mathcal{G}), E(\mathcal{G}))$ is a mixed graph with the same vertex set $V(\mathcal{G})$ and the directed edge $e \in E(\mathcal{G})$ from X_i to Y replaced by an undirected edge between X_i and Y if and only if there exists no other parent X_j of Y which is not connected to or from X_i by an edge. So DAGs say the same thing if they have the same skeleton and the same configurations of unmarried parents. So for example the DAGs $\mathcal{G}_1$ and $\mathcal{G}_2$ have the same pattern $\mathcal{P}$ and so are equivalent in the sense above.

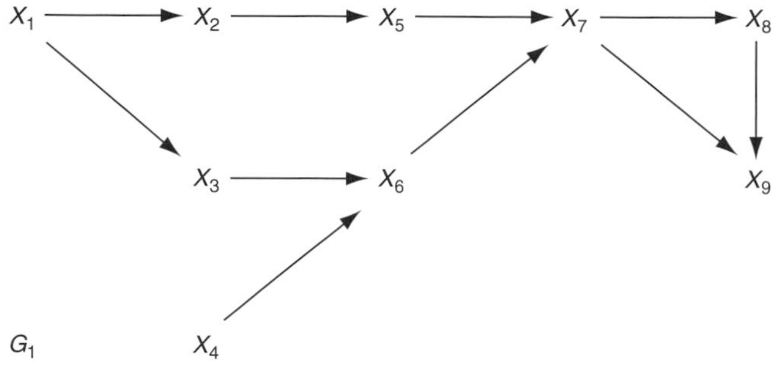

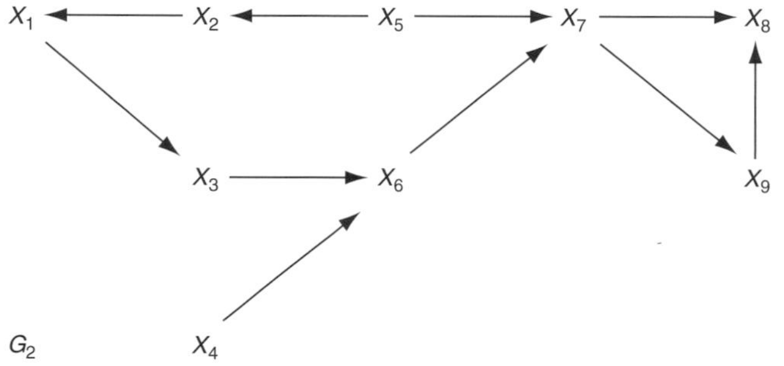

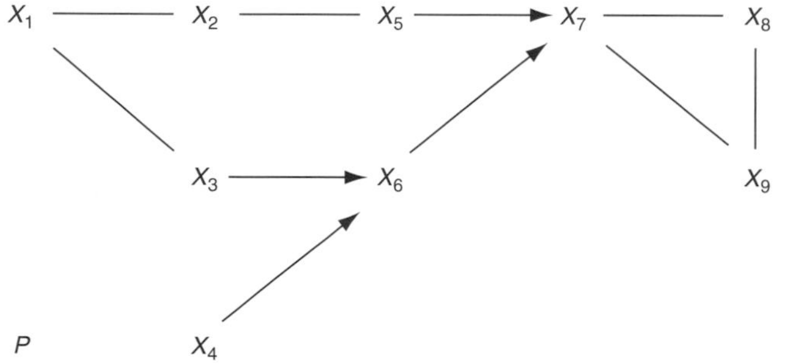

Formally therefore the pattern is a more efficient description of a BN and one that we should use if we are using data to search for a well-fitting BN. Note that a DAG is decomposable if and only if its pattern is equal to its skeleton.

7.3.5 On causal deductions from Bayesian networks

In most aspects the DM will usually interpret the DAG of a BN appropriately. One exception is that she may interpret an edge as a *causal* directionality she can deduce. Clearly from the equivalence relationships above this could only make sense if *all* DAGs with the given pattern had this directionality. The essential graph E can be derived from the pattern and is useful in this regard since any of its undirected edges has two equivalent BNs whose DAGs differ in the direction of this edge. This is not true of the pattern. For example the undirected edge from X_7 to X_8 must be directed $X_7 \rightarrow X_8$. For if a DAG had the arrow pointing in the other direction then the configuration of unmarried parents (and so patterns) would be different. The essential graph $\mathcal{E}$ of $\mathcal{P}$ above is given below.

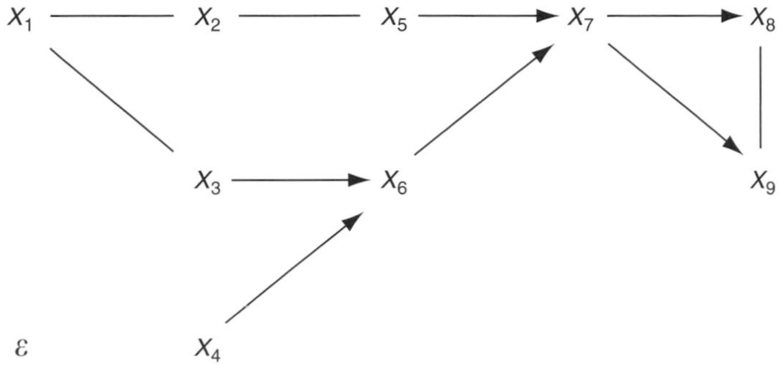

Over the last fifteen years or so different authors have tried to use the BN to describe various sorts of causal structures. From the comments above we see that we should not always interpret all the edges of a BN causally. However if we were to use a huge data set to search over the whole space of BNs – how this can be done is described in Chapter 9 – and we found a BN with an essential graph $\mathcal{E}$ that fitted the data much better than any other, would it be appropriate to deduce that if there was an edge from X_i directed into X_j in $\mathcal{E}$ then, in some sense, the feature measured by X_i "caused" the feature measured by X_j?

Pearl (2000) demonstrated that the existence of a directed edge $X_i \rightarrow X_j$ in a BN, however well supported by data, was not enough to deduce a cause. To appreciate this consider the following example. Consider a BN of the situation in the 1960s where data consists of monthly measures of X_0 a measure of the average affluence of a town, X_1 the sales of washing machines; X_2 the definition of a crime statistic; and X_3 the actual crime figures for that month. In this scenario a plausible BN might have the DAG given by $\mathcal{G}_1$ below.

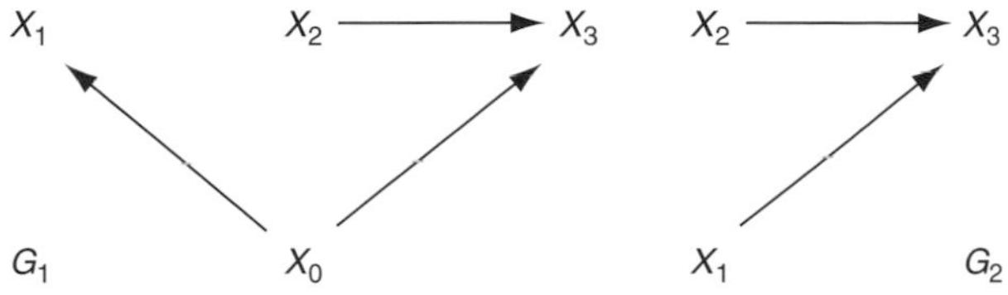

However there are no records of X_0 over this period. So the common cause of both washing machine sales and incidence of crime remains hidden. However from the variables $\{X_1, X_2, X_3\}$ that are observed it is easy to check using the d-separation theorem that the only conditional independence that is valid between set of vertices is that $X_2 \amalg X_1$. So given that $\mathcal{G}_1$ is the DAG of a valid BN we expect when we search over all BNs on the variables $\{X_1, X_2, X_3\}$ that the BN with $\mathcal{G}_2$ as its DAG will be confirmed as the best explanation of the data. But note that $\mathcal{G}_2 = \mathcal{E}$. So in particular we are going to deduce that the increasing sales of washing machines caused increasing crime figures!

It follows that unless we are absolutely certain there are no hidden common causes lurking in the background in our application we cannot deduce that X_i is in any sense a cause of X_j just because there is a $X_i \rightarrow X_j$ edge in the essential graph. Pearl (2000) proves that the only directed edges that might indicate a causal relationship (and not simply be explained by a hidden common cause) are those edges whose direction is implied by the directionality of edges in the pattern. Even then the arguments he needs to prove this depend on the assumption that in the requisite model described by a DM all conditional independences can be *fully* described by a single BN with some vertices hidden.

In fact the DAG can be used to depict causal hypotheses but the semantics of these graphs need to be defined differently to those of the BN: see the next chapter.

7.3.6 A Bayesian network for forensic communication

To illustrate how evidence from different experiments can be integrated using a Bayes net consider the following example. This is based on a case discussed in Puch and Smith (2004) and is currently part of a BN built on a Matlab platform used to help forensic scientists in the UK become more adept at gauging the strength of evidence from matching fibre evidence found in a suspect's hair so that they can faithfully convey this strength to jury members. For many analogous applications see Aitken and Taroni (2004); Dawid and Evett (1997) and Aitken *et al.* (2003) for an example concerning DNA using the software HUGIN.

Example 7.10. A balaclava – a type of whole-head wear – was used and then discarded in a robbery. The balaclava was retrieved and its fibres analysed. A suspect was arrested six hours later and two fibres that matched the fibres on the balaclava were found in his hair. The DM, prosecution or jury need to evaluate the strength of this evidence as it applied to the suspect's guilt given various possible explanations of what might have happened.

In Chapter 2 we argued that the decision maker, prosecution or jury are likely to adopt probability forecasts provided by forensic about the relevant forensic evidence.

The forensic statistician needs to give credible probabilities of the evidence found given the suspect wore the balaclava, and any known facts or hypotheses $\boldsymbol{x}$ – here denoted by $P(Z_3 = 2|G,\boldsymbol{x})$ – and the probability that an innocent person matching the suspect has the two matching fibres on their head given $\boldsymbol{x}$ – here denoted by $P(Z_3 = 2|\overline{G},\boldsymbol{x})$.

Because the forensic scientist's sampling tree is very symmetric in examples like these, the expert's beliefs and all the associated probabilities can be described using the framework of a BN. To illustrate how such a BN can be built we will focus on the elicitation of $P(Z_3 = 2|G)$. There are three components of any story leading to the observation of two matching fibres. The first is a model of the probability of the number of fibres Z_1. A second phase of the story concerns the number of fibres Z_2 remaining on the suspect's head. This depends on the probability $\theta_2(\boldsymbol{x}_2)$ of any one fibre remaining on someone's head as a function $\boldsymbol{x}_2$ of the time t between the incidence and retrieval process, and measurements of the extent of possible physical disturbance (for example running) and head disturbance (such as combing or hair washing). Finally $Z_3|\theta_3(\boldsymbol{x}_3)$ represents the number of fibres of the Z_2 in the suspect's hair actually retrieved as a function of the type of retrieval method used (e.g. combing or taping) and $\theta_3(\boldsymbol{x}_3)$ the probability of the retrieval of any fibre present. A BN of this process is given below.

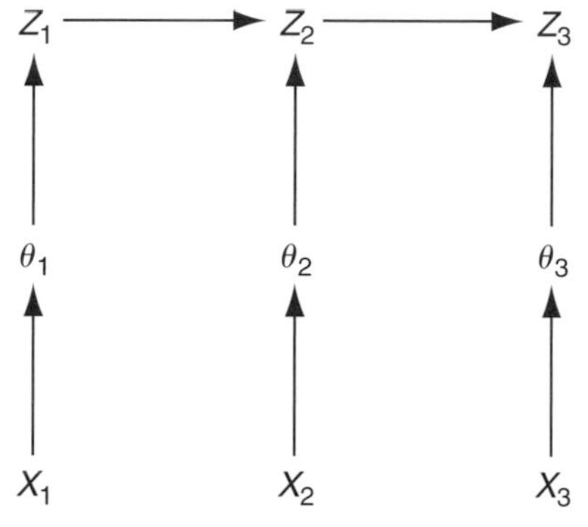

Having decided on this structure of a BN for this part of the story, because it represents a credence decomposition corresponding to a conjecture about what might have happened and follows a natural causal order we are able to embellish this model with probabilities that draw in diverse experimental evidence that actually exists. Thus appeals to randomness make it most natural to assume $Z_1|\theta_1(\boldsymbol{x}_1)$ has a Poisson distribution with rate $\theta_1(\boldsymbol{x}_1)$ where $\boldsymbol{x}_1$ indexes a persons length, style and type (e.g. straight or curly) of hair. An actual designed experiment informs the expert's distribution of $\theta_1(\boldsymbol{x}_1)$ posterior to this experiment. This experiment sampled a vector of random variables $\boldsymbol{Y}_1(\boldsymbol{x}_1) = (Y_{1,i}(\boldsymbol{x}_{1,1}), Y_{1,i}(\boldsymbol{x}_{1,2}), \ldots, Y_{1,i}(\boldsymbol{x}_{1,n_1}))$ where $Y_{1,i}(\boldsymbol{x}_{1i})$ denoted the number of fibres sticking to people's hair, immediately after they took off a balaclava, indexed by different hair types $\boldsymbol{x}_{1i}$, $i = 1, 2, \ldots, n_1$. Since the hair style of the suspect is known the forensic scientist can access the particular rate $\theta_1(\boldsymbol{x}_1)$ of transfer of fibres from the balaclava to a person with a similar hair style to the suspect. In fact the experimental evidence was found $\theta_1(\boldsymbol{x}_1)$ to be well approximated using a posterior $G(\alpha_1(\boldsymbol{x}_1), \beta_1(\boldsymbol{x}_1))$ gamma distribution. Similarly if it is believed that fibres were shaken off or retrieved randomly given their covariates then this leads us to assume that $Z_2|\theta_2(\boldsymbol{x}_2), Z_1 = z_1 \backsim Bi(z_1, \theta_2(\boldsymbol{x}_2))$ and $Z_3|\theta_3(\boldsymbol{x}_3), Z_2 = z_2 \backsim Bi(z_2, \theta_3(\boldsymbol{x}_3))$. The distribution of the parameter $\theta_2(\boldsymbol{x}_2)$ depended on covariates linked to the time between the incident and the arrest, whether the suspect had run during that time and whether he had washed or combed his hair. All these covariates could be matched to different versions of the story corresponding to different paths in the episodic tree of the case at hand. The covariates $\boldsymbol{x}_3$ associated with retrieval in the case at hand are all accepted facts. Available evidence from a different experiment about persistence suggested that it was reasonable to assume that $\theta_2(\boldsymbol{x}_2) \backsim Be(\alpha_2(\boldsymbol{x}_2), \beta_2(\boldsymbol{x}_2))$ – a beta distribution. Finally $\theta_3(\boldsymbol{x}_3) \backsim Be(\alpha_3(\boldsymbol{x}_3), \beta_3(\boldsymbol{x}_3))$ could be estimated using separate conjugate sampling for each pair of retrieval device and hair type. The probability $P(Z_3 = 2|G)$ given each set of combinations of facts and hypotheses $(\boldsymbol{x}_1, \boldsymbol{x}_2, \boldsymbol{x}_3)$ could now be simply calculated by integrating over all the remaining variables. This allowed the expert to provide a single probability for each pertinent vector $(\boldsymbol{x}_1, \boldsymbol{x}_2, \boldsymbol{x}_3)$.

The empirical information behind the assertions above and the technicalities of the associated calculations is given in Puch *et al.* (2004) and the references therein. But the point I am making here is that the BN provides the framework for an extremely useful probabilistic expert system in these sorts of settings. It helps to piece together information from designed experiments to address the pertinent probabilities of possible unfoldings of history in the case at hand in a transparent and flexible way. Moreover the topology of the DAG of the BN also captures some important assumptions that the expert forensic statistician is making that can be held up to scrutiny.

The BN is especially useful for many scenarios where the general structure of dependence is fairly homogeneous. In this domain of application the same dependence structure between variables is likely to apply irrespective of the values of the covariates $(\boldsymbol{x}_1, \boldsymbol{x}_2, \boldsymbol{x}_3)$ and the observed value of the evidence z_3. There is therefore strong motivation for pasting together information from various experiments into software built round the architecture of this BN, because it is likely to be applicable to many cases. Note that a sensitivity analysis can be performed on such a probabilistic expert system to test how sensitive the probability $P(Z_3 = 2|G)$ is to for example whether or not the suspect combed his hair between the incident and being apprehended in the current case. Of course in more complex cases a single BN is unlikely to be able to provide *all* the decision support needed by the DM, just part of it.

These sorts of expert system now inform a wide variety of forensic cases; for example those needed in the sampling subtree of the crime example of Chapter 2. The application to aid the probabilistic forecasts concerning evidence in DNA matches is now particularly well advanced.

7.4 Eliciting a Bayesian network: a protocol

7.4.1 Defining the variables in the problem

Although there are many hierarchical models that could be used to support inferences, unless the DM is fortunate or the information has been designed experimentally so that it fits into the framework of a standard hierarchical structure, usually an off-the-shelf model will not be wholly appropriate as a framework for decision modelling. In the last example we saw how a simple structure following the story of how situations might develop where experimental evidence was available and totally relevant to the inferences that needed to be made. But more usually the analyst will need to elicit prior beliefs about a dependence structure over a much wider domain. It is essential that the DM owns the inferential framework she uses. And this means the analyst needs to elicit a DM's BN so that it is customised to her description.

So how does such an elicitation take place? When modelling for a decision analysis an early objective is to address precisely where the important sources of the DM's uncertainty lie. The measurements of these features will then provide the vertices of the BN and the set of random variables on the product space of the DM's – or possibly a trusted expert's – probability model. In many decision analyses it is usually most efficient to encourage the DM to work backwards from the point when she receives her vector of attributes of her reward space, see Chapter 6, imagining herself at a point where the die has been cast and she has observed the outcome of her chosen policy decision. This will encourage her to focus on those sources of uncertainty that impinge on her decision problem and not describe her problem too generally. Henceforth in this section assume the DM is exploring the possible consequences of following a particular fixed policy or decision rule. The probability model constructed as illustrated below – loosely based on work some of which is reported in Freeman *et al.* (1996) – can be used on each decision rule she considers employing and the best policy identified as one that maximises her expected utility function. If there is a degree of homogeneity in the problem then different decision rules will have similarly associated BNs and it is possible to use the framework of influence diagrams discussed later in this book.

The initial random variables considered – the attributes of the DM's utility function – will be called Level 1 quantities. Thus suppose a water company wants to renovate a portion of its pipework. It could simply assess the cost implication of a particular scheme of work. However, more realistically it would also be interested in the effect that scheme of work might have on the quality measured by purity, the security of its water supply measured by the number of reported leaks after the work had taken place, and also the public acceptability of the level of disruption of supply whilst the work was taking place measured by the number and nature of complaints and breaking of EC directives.

Having elicited the attributes of the decision problem the analyst encourages the DM to consider other features of the problem that might have a direct influence on at least one of these attributes. This next tranche of features we will call Level 2 features. In the example above, costs will depend on as yet uncertain economic conditions. One of the important

Level 2 variables here would be the local availability of subcontractors to employ. Notice that a particularly large and geographically concentrated programme of work could drive up costs because of this feature of the problem. A second Level 2 feature of this problem would be the degree of degradation of metal pipework which would not only have financial implications but also impinge on water purity and the annoyance caused to the public through interruptions of supply caused by major leaks. A third set of features influencing the public acceptability of the work might be the quantity of properties affected by the proposed work and measures of traffic congestion, and disruption of retail outlets whilst the repairs are being made.

The elicitation process continues to trace back these dependencies to their sources. The next layer of features, called Level 3 features, will have a direct influence on the Level 2 items you have identified and so have an indirect impact on the attributes of the problem. Thus in the above example both the age of the metal pipes and the nature of the soil in which they lie will give an indication of the likely degree of degradation and should be listed as Level 3 features. The decision analyst will continue to work down the levels of uncertainty with the client until she is content that all sources of significant uncertainty have been traced back. For example, the Level 3 feature, "soil type", mentioned above could well be known to the company. Alternatively the Level 3 feature, "age", may be known only to the nearest year, but it is realised that no other sources giving better information on this feature are easily available. In either of these cases it would probably be expedient to stop at Level 3 on this path in the explanatory regress, at least in the first instance. Determining at which level to stop, however, often has subtle consequences and needs to be sensitively handled by the analyst. As a general rule it is wise to allow the DM to explore unnecessary depths of sources of uncertainty than to stop too early in the process. Even if the elicited deeper level features are not used directly they can form useful reference points for more systematic judgements elicited later. Later interrogation of an initial structure in ways illustrated below will enable the DM to review her model-building process in ways to be illustrated below. A partially drawn trace-back graph of the features discussed in the water company example is given in the figure below.

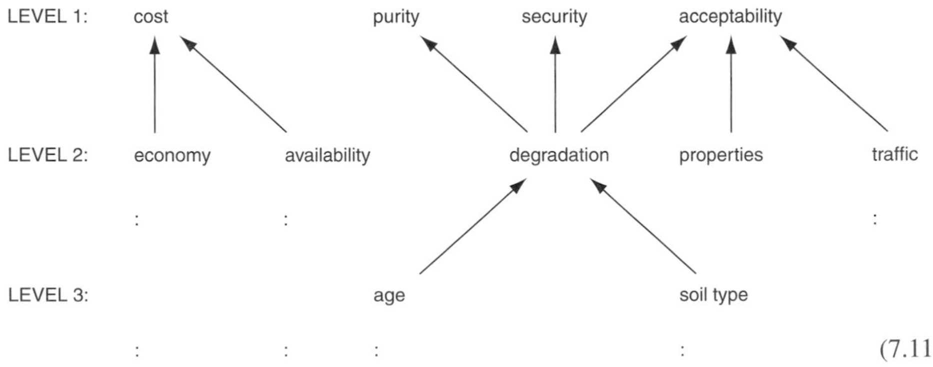

(7.11)

What we now have elicited is a provisional list of features, including the attributes of the DM's utility function, that might be relevant to assessing the efficacy of a proposed policy or decision rule together with a partial order of how these features influence one another.

7.4.2 Clarifying the explanatory variables

The next stage of the elicitation process is to divide the list of features elicited in the way described above into those which have a clear and unambiguous meaning – called explicit features – and those which defy crystallisation – called implicit features. The ideas behind this division were first discussed by De Finetti (1974) and developed into practical methodology by Howard (e.g. Howard (1988, 1990)). We have already discussed this issue in Chapter 3 when we addressed the need to define attributes in a measurable way. We now simply extend this to the whole description of the problem.

We demand that it is possible to ensure that the value of any explicit features could be measured unambiguously at some future time and seen to take a certain real value. Thus, in the example above, the total cost of a particular scheme of work, whilst being currently uncertain, will in the future be known. So this is an explicit feature of the problem. Similarly the age of a particular length of pipe, while being uncertain, could in principle be determined if we had access to the appropriate records, so this too is an explicit feature. However, "public acceptability" is not as it stands an explicit feature because we have yet to say precisely how we intend to measure the extent to which a policy may or may not be acceptable.

Explicit features can be treated as if they were an uncertain measurement and their lack of ambiguity of meaning allows them to be the objects of logical manipulation. The next task is to look at the set of implicit features and to attempt to redefine these features in a precise and verifiable way; i.e. to transform these into explicit features. Thus consider the feature public acceptability for the water company. In its mission statement, through various regulatory bodies or in its explicit undertakings to customers it is likely that the company has committed itself to various undertakings that can regulate quality of service and which can and will be monitored. Many of these can be used as a measure of public acceptability, including, for example, the number and extent of interruptions to service. Each implicit feature should be examined and a set of explicit features substituted whenever possible. In practice this process tends to unearth previously unconsidered features of the problem which may unearth other lower level quantities which are now seen as pertinent to the problem in hand. New elements are then included in the trace-back graph.

On the other hand, it may not be possible to substitute all implicit features by explicit features in this way. When this happens these residual implicit features are recorded. However, rather than explicitly including these features as vertices in the graphical framework described below in the subsequent analysis they are instead referred to indirectly when any new judgements are required for input. Such judgements will include whether relationships exist between the explicit features identified in the problem, used to document why certain significant relationships between features might exist and, much later, the estimation of the values of the probabilities needed for numerical evaluations of the consequences of the policy under examination.

So, to summarise, at the end of Stage 2 the analyst has obtained from the DM a prototype set of clearly defined measures linked together through a trace-back graph and which relates explicitly to the objectives of the analysis. This set is still provisional: it may well be adjusted when the relationships in the problem are examined more closely. Notice that at no point has the analyst requested any numerical quantifications from the DM but only a verbal description of her problem. Eventually numbers may need to be elicited, but not at such an early stage of the process.

It is critical to realise that the trace-back graph is *not* a BN. It does however give us a list of variables – the explicit features – which the client thinks constitute the important variables together with a partial order induced by the levels which is, in a loose sense, causal and therefore easier for the DM to think about.

7.4.3 Drawing your Bayesian network

The third stage of the process is to draw a provisional BN. The vertices of the BN will be the measurements – the list of explicit features. It is useful to order these variables consistent with these levels – variables at levels with a higher index appearing before variables in levels with lower indices. A BN is now drawn in the way described below, taking the explicit features that appeared in the trace-back graph as nodes and drawing a new graph which encodes irrelevance relationships: a missing edge between two nodes expressing the type of irrelevance relationship satisfying properties of symmetry and perfect composition discussed in the last section. This stage of the elicitation is much better documented than the stages described above and can be found in example Cowell *et al.* (1999); Jensen and Nielsen (2007); Pearl (1988) and Kjaerulff and Madsen (2008).

To demonstrate this process consider our running example of the water company. To simplify this illustration we will pretend that the only explicit features of the problem in the client's description are the following, taken in an order consistent with the trace-back graph above.

X_1 = *Soil type*.
X_2 = *Age* of pipework.
X_3 = *Availability* of contractors to do the work.
X_4 = State of *degradation* to the pipework.
X_5 = Disruption of supply to *properties* caused by the chosen programme of work.
X_6 = Disruption of *traffic* by the chosen programme of work.
X_7 = Total *cost* of the work.
X_8 = *Purity* level of the water after work completed.
X_9 = *Security* of supply after work completed.
X_{10} = *Acceptability* (measured by number and seriousness of complaints by customers).

Note that any listing of features consistent with the reverse ordering of their levels can be chosen. So, for example, we could equally have chosen to list the features with X_1

permuted with X_2. Each feature will be a node of the BN. Note that we have omitted the feature *economy* in the trace-back diagram because the DM could not be precise enough to convert this into a measurable quantity.

Starting with the second listed feature the analyst asks the DM which subset, Q_i, of indices of earlier listed features are relevant for predicting the measurement of X_i, $2 \leq i \leq n$ for each of the remaining $n - 1$ explicit features. The set Q_i may be empty, in which case knowing the values of the measurements earlier in the list is totally unhelpful for predicting the measurement X_i. Note that this will always be the case if the value of X_i will be known to the DM before she commits to her policy because then no other variables will be useful in predicting it (it will be known anyway!). At the other extreme, all of the previously listed features might provide its own additional helpful information about X_i. In the simplified water company example above, the analyst first asks the DM whether the soil type classification and the age of the pipe are related in any way. The DM can think of no reason for there to be any relationship here, so no edge is placed between X_1 and X_2 in the BN. The analyst then asks whether X_1 or X_2 could be used to predict X_3. Again the answer is no, so no edge is drawn leading into X_3. Moving on, she states the DM believes X_1 and X_2 could be useful to help predict X_4, but could see no reason why information concerning the availability of contractors X_4 would help her to improve her prediction of the current degradation of the pipework. So the BN has a directed edge from X_1 to X_4 and from X_2 to X_4 but not from X_3 to X_4.

When asked to consider the impact of X_5 on X_6, the disruption variables, she realised that the unexpectedly high levels of degeneracy of the pipework might affect disruption because of the possible delays it would cause, as would the non-availability of contractors. Also X_5 would be related to weather-related delays which could also affect traffic. So X_5 and X_6 are related and need to be joined by an edge in the BN. However, given these, knowing X_1 and X_2 would provide no additional information for predicting X_5, X_6. The analyst's queries of the types above finally result in a BN whose DAG is given below.

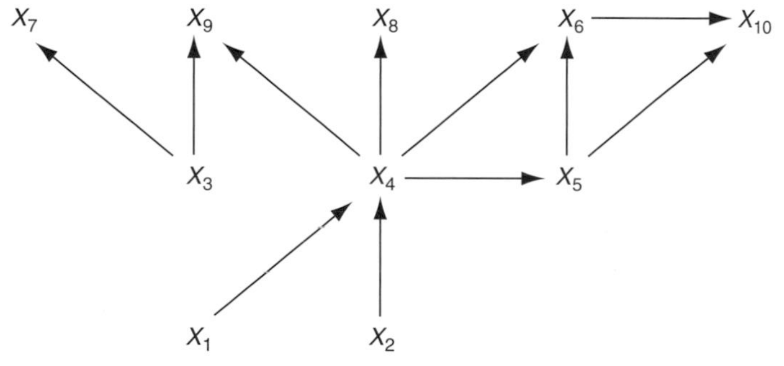

(7.12)

There are two issues of note in this process. The first is one that helps us to simplify this elicitation. If the value of an explicit feature will be known with certainty at the time a decision is taken then obviously no other information is needed to help forecast it. It follows that it therefore needs to have no parents, i.e. no vertex connected into it: it is a *root vertex*. Identifying such explicit features is operationally very helpful because the analyst then need not enquire what deeper features might influence this. For example a company will have its own databases often recording the values of many known or knowable features in the DM's description of the process. So the early elicitation of this information from the DM can speed an analysis up considerably.

Second it is very important that the analyst actively engages in all the three stages of this process not just the last, otherwise the probabilistic inputs will be difficult for the DM to specify faithfully. Note that tracking back influences on general features is not the same as asking relevance statements that can be used as a framework for a probability model. In particular, as we have demonstrated the trace-back graph is not necessarily a BN. For example, we have added an edge (X_5, X_6) not originally appearing in the trace-back graph. Here, the earlier elicitation encouraged us to miss these dependencies – particularly when they lay on the same level, because the trace-back graph describes positive relationships, i.e. that features *should* be related, through *including* an edge in the diagram. In contrast the BN *denies* the possibility of a relationship between two features through the *absence* of an edge. Implicit features will often induce new dependences. For example the dependence associated with the edge (X_5, X_6) arises from a newly elicited implicit feature – delay. If the DM were able to code this feature as a measurable quantity then it could be included as another explicit feature in a new and more refined BN. In the terminology of Pearl (2000) weather delay X_0 would then be a "common cause" which would connect into X_5 and X_6 but may allow us to remove the (X_5, X_6) edge in this elaborated BN. The new BN would then be

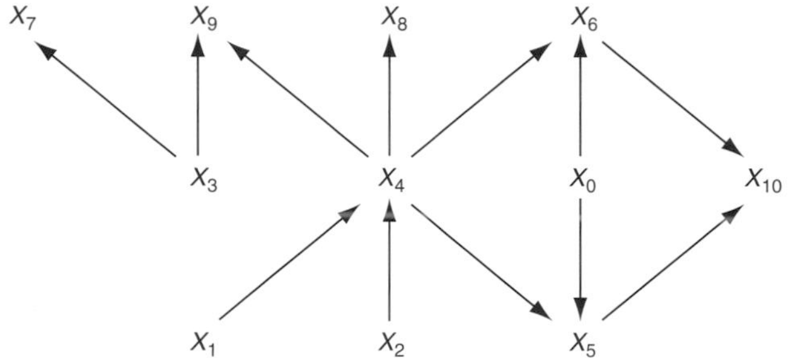

(7.13)

But otherwise the edge must stay.

7.4.4 Towards a requisite Bayesian network

We have already seen how the d-separation theorem can be used on the DAG of a simple BN to verify whether the DM believes the statements it implies are valid or whether they need to be adapted. This is also so for larger BNs such as the water company BN elicited above. Suppose we have been told the security X_9 of a piece of work and we are interested in reconstructing its cost X_7. Will learning about X_2 provide us with any useful information with which to help refine our forecast? Following the construction of the d-separation theorem we have the ancestral graph $\mathcal{G}(X_7, X_2, X_9)$, the moralised graph $\mathcal{G}^M(X_7, X_2, X_9)$

and the skeleton of this graph $\mathcal{S}(X_7, X_2, X_9)$ respectively given in the figure below.

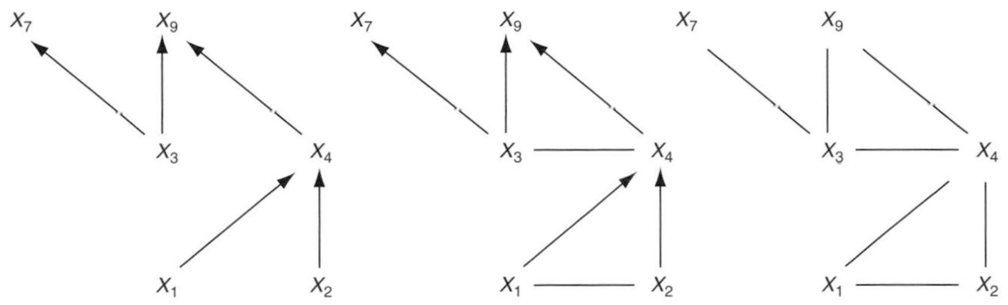

It is simple now to check whether it is valid to deduce from what we have been given: i.e. from what is coded into the BN, the assertion that $X_7 \amalg X_2 | X_9$. If a path in $\mathcal{S}(X_7, X_2, X_9)$ from X_7 to X_2 does not pass through X_9 then the DM cannot deduce that X_2 is uninformative about X_7 once she learns the value of X_9. Otherwise she can. So $X_7 \amalg X_2 | X_9$ is not a valid deduction. There is a path (X_2, X_4, X_3, X_7) from X_2 to X_7 which does not pass through X_9. This is actually a logical consequences of what the DM has already stated, although the reasoning is quite subtle. From the original BN it can be deduced that X_3 might be informative about X_7 and also that different combinations of values of X_3, X_4 might give rise to a different distribution for X_9. Furthermore from the original BN X_2 might be useful for the prediction of X_4. From the above if we know the value of X_9 then X_4 in turn might provide additional information about X_3. Hence X_2 may give new indirect information about X_7.

Note first that this argument can be constructed simply by following the information path causing the violation: in this case (X_2, X_4, X_3, X_7). But also note how difficult it would be to try to construct this argument without guidance. As an exercise you might like to check that if the DM was not told the value of X_9 then X_2 would no longer be any use in helping the client forecast X_7.

The d-separation theorem can be used to check the validity of someone's initially elicited assertions by using the BN to derive implied, but not transparent statements and double checking by asking about these. In the example above, it is easy to check that one of the client's direct statements is that once she has been told the state of degradation of the pipework its age would provide no extra information about the purity of that water. But by symmetry an equivalent question to ask is whether, to guess the age of some pipework, it would be useful, after being told of its level of degradation, to know what the purity levels of the water were. The second question may evoke in the mind of the client considerations like the materials used in making the pipework, for example lead, that the purity measures would detect. If the degradation index was not refined enough to include the material of the pipe then this may lead the client to question her first assertion that age was not useful to predict purity given degradation. This would encourage her to adjust her BN either by including the material of the pipework explicitly as another variable or to add an edge (X_2, X_8) to the original BN.

After a number of iterations a BN can be drawn that is requisite to the DM. She is happy that this BN represents her beliefs about how all the relevant features of a problem are connected to one another. She is also at a point where she is happy to share her reasoning for why she has chosen the BN she has. Of course in the light of further quantitative elicitation or any sampling she might undertake before she commits to a decision she might still want to refine her model. But for now she is content that it represents her honest beliefs.

Obviously the BN cannot express *everything* a DM might like to communicate and is no panacea for modelling. We have already seen how the decision tree can also be a useful tool and encode quite different information. To be an efficient representation of significant portions of a DM's belief structure needs some level of homogeneity of dependences independent of the values of influencing factors associated with combinations of values of parental random variables. For example if for some values of explicit features there is a great deal of dependence between variables but for others there are none, or if the existence of one feature logically precludes others then the BN begins to lose its predictive power.

However many decision problems can be expressed elegantly within the framework of a BN. Because of its transparency and underlying logic it is therefore an extremely valuable tool to help the client explore the relationships between the features of her problem. As illustrated above, it not only efficiently codes what she currently believes about how the uncertain factors in her problem are related, but it can also be used to help her adapt her current beliefs to ones which more precisely fit the circumstances she is trying to model. Most importantly, when the process is finished the model she has built will be her own description of the problem and not one imposed by the analyst.

7.5 Efficient storage on Bayesian networks

7.5.1 Storing conditional probabilities

Once a valid BN has been faithfully elicited, it can be used as a framework on which to store a probability distribution and to calculate various marginal distributions of interest. There is now a vast literature on this important topic much of it driven by researchers in the machine learning community. In this short volume it is only possible to skim some of the more basic material in this area. It is however very important for the decision analyst to be aware of how this is done and what the capabilities of these methods are. We discuss below some of the more important ideas from this research as it impinges on decision modelling.

Because there are far fewer technical issues to address, although most of the points we demonstrate here have a much more general validity, most ideas are easiest to explain when all variables in the net are discrete and have a finite state space. We therefore will concentrate on this case. So suppose the DM believes that a particular discrete BN with variables $\{X_1, X_2, \ldots, X_n\}$ determined by the $n-1$ conditional independence statements $X_i \amalg \boldsymbol{X}_{R_i} | \boldsymbol{X}_{Q_i}$ $2 \leq i \leq n$ is valid. As a Bayesian it follows that her full joint mass function

$p(\boldsymbol{x})$ over her measurement will respect the factorisation.

$$p(\boldsymbol{x}) = p_1(x_1) \prod_{i=2}^{n} p_i(x_i|\boldsymbol{x}_{Q_i}). \tag{7.14}$$

So to fully specify her model we need to elicit $p_1(x_1)$ – the marginal probability mass function of X_i together with the conditional mass function $p_i(x_i|\boldsymbol{x}_{Q_i})$ of each of the variables conditioned on each possible configuration of values of its parents that might occur. Note that if we follow the construction of a BN as described in the last section then we will be eliciting the probability of each variable conditional on possible values of other variables that might influence its value. So usually these variables are quite simple for a DM to think about. A discussion of how this might proceed is given in Renooij (2001). The practical difficulty is of course that the number of different configurations of parents, and hence the number of probability vectors that need to be elicited can be extremely large. BNs for which this is not so are therefore the most viable to fill out into full probability models.

Example 7.11. In Chapter 1 we saw the naive Bayes model. When there is one indicator of the disease X_1 and symptoms $\{X_2.X_3, \ldots, X_{n+1}\}$ then this model has the "star" BN model whose DAG when $n = 8$ is given in the figure below.

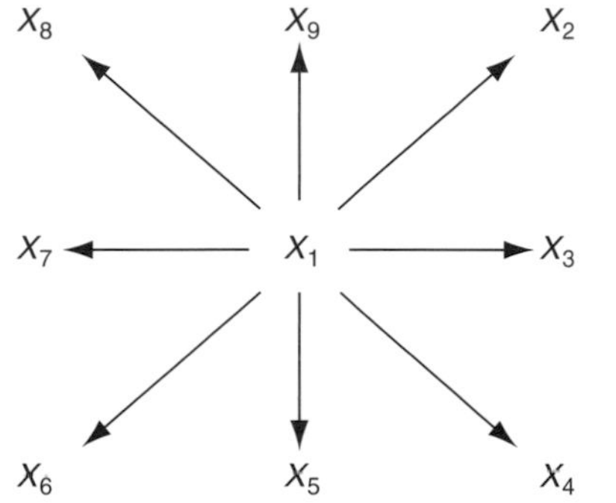

If the number of possible diseases/levels of X_1 is m_1 and the number of possible levels the symptom X_i could take is m_i, $i = 2, 3, \ldots, n+1$ then the number of probabilities we need to elicit $p_1(x_1)$ is $m_1 - 1$ (subtract one because the last probability must be chosen so that the probabilities add to one) and to elicit all the conditional tables $p_i(x_i|\boldsymbol{x}_{Q_i})$ we need $(m_i - 1)\, m_1$ – i.e. $m_i - 1$ for each possible disease. Summing these gives us that the total number of probabilities that need to be elicited is

$$m_1 \left\{ \sum_{i=2}^{n+1} (m_i - 1) + 1 \right\} - 1$$

compared with the number of probabilities $\prod_{i=1}^{n+1} m_i - 1$ – orders of magnitude larger – we would need to elicit if the star BN was not assumed. Thus with 8 symptoms taking one of three levels and 10 possible diseases we have 169 probabilities to elicit in the idiot Bayes model – which is just about practically feasible, but 65,609 probabilities in the saturated model.

Very different BNs share these practical advantages. Consider the following BN with a very different topology to the star model above.

Example 7.12. This is a rather gross simplification of a BN associated with the adverse affect on an individual exposed to radiation. Here the random variable X_1 measures ten categories of exposure to radiation of workers at a particular nuclear plant that has experienced malfunction. Random variable X_2 has three levels, – no adverse affects, moderate biological disruption, severe biological disruption. The variables $\{X_3, X_4, \ldots, X_7\}$ are all binary. Variable X_3 indicates whether or not exposure has disrupted cell division, X_4 whether the person develops a detected tumour, X_5 whether surgery is given, X_6 indicates whether or not the exposed person is over fifty and X_7 whether she survives ten years after the exposure event. Finally the variable X_8 measures whether or not there will be adverse hereditary effects as a function of severity. The suggested BN has DAG $\mathcal{G}$ given below.

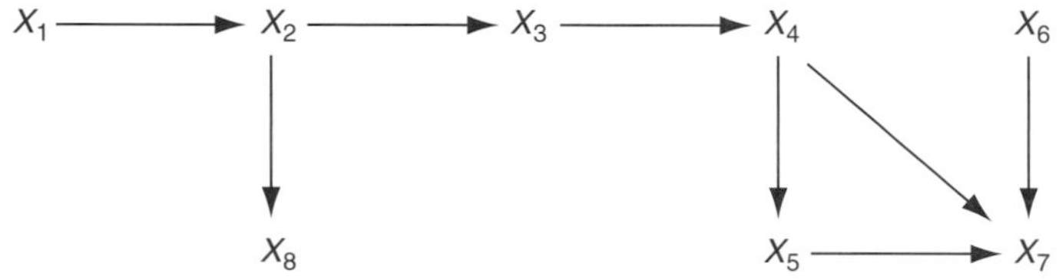

If the DM believed this BN was valid then the number of probabilities that need to be elicited to embellish it to a full probability model – taking account that probabilities must sum to one – is

$$9 + 20 + 3 + 2 + 2 + 1 + 8 + 27 = 72.$$

These are elicited to be as follows.

x_1	1	2	3	4	5	6	7	8	9	10
p_1	.4	.15	.1	.09	.08	.07	.05	.03	.02	.01

x_1	1	2	3	4	5	6	7	8	9	10
$p_2(x_2 = 1\|x_1)$	1	.9	.8	.7	.5	.4	.3	.2	.1	0
$p_2(x_2 = 2\|x_1)$	0	.1	.2	.3	.4	.5	.5	.4	.2	.1
$p_2(x_2 = 3\|x_1)$	0	0	0	0	.1	.1	.2	.4	.7	.9

x_2	1	2	3	x_3	0	1	x_4	0	1
$p_3(x_3 = 0\|x_2)$	1	.8	.5	$p_4(x_4 = 0\|x_3)$	.9	.6	$p_5(x_5 = 0\|x_4)$	1	.6
$p_3(x_3 = 1\|x_2)$	0	.2	.5	$p_4(x_4 = 1\|x_3)$	.1	.4	$p_5(x_5 = 1\|x_4)$	0	.4

$$p_6(x_6 = 0) = 0.8, p_6(x_6 = 1) = 0.2$$

$\boldsymbol{x}' = (x_4, x_5, x_6)$	000	001	010	011	100	101	110	111
$p_7(x_7 = 0 \mid \boldsymbol{x}')$	.95	.8	N/R	N/R	.2	.2	.5	.1
$p_7(x_7 = 1 \mid \boldsymbol{x}')$	.05	.2	N/R	N/R	.8	.8	.5	.9

x_2	1	2	3
$p_7(x_7 = 0 \mid x_2)$	1	.7	.6
$p_7(x_7 = 1 \mid x_2)$	0	.3	.4

Note that actually two of these probabilities are unnecessary to elicit since they could never happen. Their cells are therefore marked N/R = "not relevant". Eliciting joint probabilities from scratch not using the BN would need the elicitation of 38,399 probabilities. This would be totally infeasible, even in this grossly simplified model of this process and even disregarding the many small difficult to elicit probabilities in this set.

The sort of efficiency gains illustrated above obtained by using the elicited conditional probabilities needed to fully embellish a BN into a full distribution against trying to elicit these probabilities directly are often huge. All that is needed is for the number of possible configurations of values of parents of any variable not to be too large. These gains are most dramatic when the underlying connected graph is a tree so that each of its n vertices has no more than one parent. In an exercise you are asked to prove that if each variable takes r levels than the tree needs $(r-1)\{r(n-1)+1\}$ probabilities to be elicited to embellish it rather than $r^n - 1$.

So if a BN is a valid description of a problem then elicitation of information to describe that problem is usually orders of magnitude easier. This saving is essential if we are to address even moderately large decision analysis problems.

7.5.2 Storing probabilities on cliques

Although it is often sensible to elicit the joint probabilities associated with a BN in terms of the conditional probability tables $p_i(x_i|\boldsymbol{x}_{Qi}))$ the d-separation theorem shows that learning the values of some variables can destroy the originally specified conditional independences. For this reason it is often more convenient to store the joint probability distribution as a function of marginal mass functions over certain subsets of variables. This is always possible for a BN and can usually be achieved with little loss of efficiency. We construct such a storage methodology in this section. (Details of such algorithms can be found in: Cowell *et al.* (1999) and Jensen and Nielsen (2007).) Henceforth it will be convenient to label the joint probability mass function/density of the subvector $\boldsymbol{X}_A$ whose components lie in the index set $A = \{1, 2, \ldots, n\}$, by $p_A(\boldsymbol{x}_A)$.

Definition 7.13. A *clique* of an undirected graph H is a maximally complete subset of vertices in H.

Example 7.14. The undirected graph

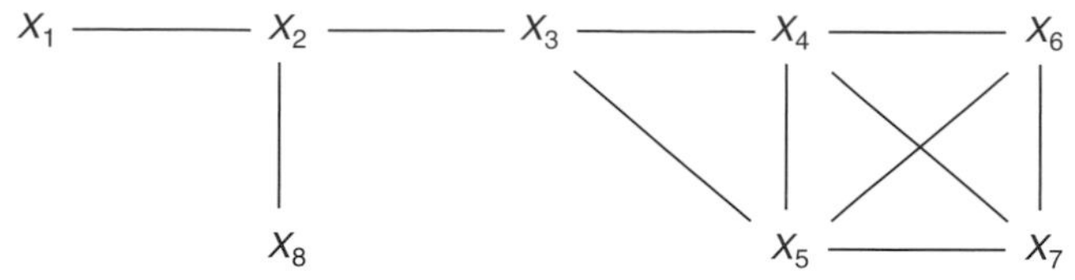

has cliques $\{\boldsymbol{X}_{\{1,2\}}, \boldsymbol{X}_{\{2,3\}}, \boldsymbol{X}_{\{3,4,5\}}, \boldsymbol{X}_{\{4,5,6,7\}}, \boldsymbol{X}_{\{2,8\}}\}$. Note for example that $\boldsymbol{X}_{\{5,6,7\}}$ is not a clique for although all its components are connected by an edge in G and so it is complete, it is not maximal because there is another complete vector of vertices $\boldsymbol{X}_{\{4,5,6,7\}}$ that strictly contains it.

Let $\mathcal{S}(\mathcal{G})$ denote the skeleton of the moralised graph of a DAG $\mathcal{G}$ of a valid BN. The cliques of $\mathcal{S}(\mathcal{G})$ are important because if we store the marginal distributions over these then we can use the BN to reconstruct the full joint distribution. To see this note that we need to construct the marginal distribution of X_1 together with the joint distribution of the subvectors $\boldsymbol{X}_{\{i,Q_i\}}$, $i = 2, 3, \ldots, n$. The marginal distribution of X_1 can be obtained by summing probabilities over the values of all other variables in a clique containing it: and there must be at least one such clique by definition. Furthermore for each $i = 2, 3, \ldots, n$ all variables in $\boldsymbol{X}_{\{i,Q_i\}}$ must lie in a single clique. This is because they must all be connected to each other by an edge in $\mathcal{S}(\mathcal{G})$. To see that this must be so first note that Q_i is the parent set of X_i so certainly connected to X_i by an edge in $\mathcal{S}(\mathcal{G})$. However in the moralisation step we have joined all the previously unconnected members with indices in Q_i together as well. So the set of vertices of $\boldsymbol{X}_{\{i,Q_i\}}$ of G^M, $i = 2, 3, \ldots, n$ all form complete subsets of this graph.

It follows that we can calculated the marginal distribution of each $\boldsymbol{X}_{\{i,Q_i\}}$, $i = 2, 3, \ldots, n$ simply by identifying any clique containing it and summing over the value of other variables in that clique. The conditional mass function $p_i(x_i|\boldsymbol{x}_{Q_i})$ of $X_i|\boldsymbol{X}_{Q_i}$ can be obtained using the usual rules of probability. Thus for example if the margin on $\overline{p}_{Q_i}(\boldsymbol{x}_{Q_i}) > 0.$ for all configurations $\boldsymbol{x}_{Q_i}$ of the parent then

$$p_i(x_i|\boldsymbol{x}_{Q_i}) = \frac{p_{\{i \cup Q_i\}}(\boldsymbol{x}_{\{i \cup Q_i\}})}{\overline{p}_{Q_i}(\boldsymbol{x}_{Q_i})}$$

Example 7.15. The skeleton $\mathcal{S}(\mathcal{G})$ of the moralised DAG $\mathcal{G}$ of the BN given in our introductory example (7.8) is given by

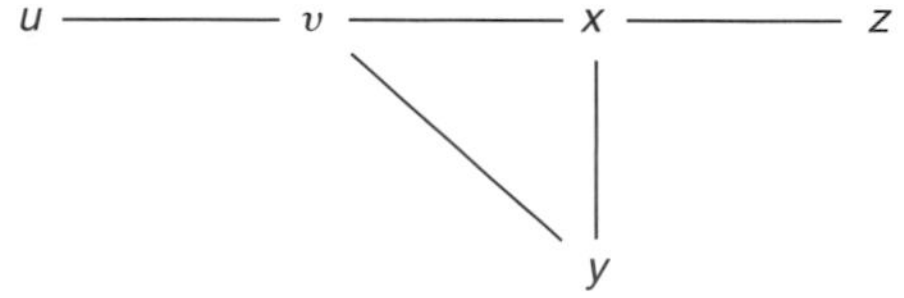

Here $\mathcal{S}(\mathcal{G})$ is just the skeleton of $\mathcal{G}$ because all the parents of its vertices are already married. The associated factorisation can be written

$$\begin{aligned}p(u,v,x,y,z) &= p_1(u)p_2(v|u)p_3(x|v)p_4(y|v,x)p_5(z|x)\\ &= p_1(u)\frac{p_{\{u,v\}}(u,v)}{p_1(u)}\frac{p_{\{v,x\}}(v,x)}{\overline{p}_{\{v\}}(v)}\frac{p_{\{v,x,y\}}(v,x,y)}{\overline{p}_{\{v,x\}}(v,x)}\frac{p_{\{x,z\}}(x,z)}{\overline{p}_{\{x\}}(x)}\\ &= \frac{p_{\{u,v\}}(u,v)p_{\{v,x,y\}}(v,x,y)p_{\{x,z\}}(x,z)}{\overline{p}_{\{v\}}(v)\overline{p}_{\{x\}}(x)}\end{aligned}$$

for all configurations for which $p_1(u)\overline{p}_{\{v\}}(v)p_{\{v,x\}}(v,x)\overline{p}_{\{x\}}(x) > 0$ and is zero otherwise. It follows that $p(u,v,x,y,z)$ is fully specified from probability tables on the clique margins $p_{\{u,v\}}(u,v), p_{\{u,x,y\}}(v,x,y)$ and $p_{\{x,z\}}(x,z)$ of $\mathcal{S}(\mathcal{G})$ The quotient probabilities $\overline{p}_{\{v\}}(v), \overline{p}_{\{x\}}(x)$ can be obtained from $p_{\{u,v\}}(u,v), p_{\{v,x\}}(v,x)$ by summing over u and v respectively.

Example 7.16. Note that the BN of the adverse effects of radiation exposure has $\mathcal{S}(\mathcal{G})$ given by the undirected graph in Example 7.14 above and so has cliques

$$\{(X_1,X_2),(X_2,X_3),(X_3,X_4,X_5),(X_4,X_5,X_6,X_7),(X_2,X_8)\}.$$

With the sample spaces defined in this example, if we chose to store the elicited probabilities in terms of these margins then this requires storage of

$$29+5+7+15+29=85$$

probability values. So for a small loss of efficiency it is possible to re-express the elicited conditional probabilities in terms of probabilities over clique margins over $\mathcal{S}(\mathcal{G})$.

In an exercise below you are asked to prove that for a directed tree with n variables where each variable takes r levels than the $n-1$ cliques of that tree require $(n-1)(r^2-1)$ probabilities to store, slightly $(n-2)(r-1)$ more than the number $(r-1)\{r(n-1)+1\}$ probabilities of elicited conditional probabilities $(r-1)\{r(n-1)+1\}$ when $n>2$. More generally if $\mathcal{S}(\mathcal{G})$ has k cliques of no more than r binary variables then the clique probability tables will need $\leq k.(2^r-1)$ storage points rather than 2^n-1. The clique tables have a storage saving usually of orders of magnitude over direct storage of the joint mass function.

7.6 Junction trees and probability propagation

7.6.1 Triangulation for propagation

In the last section we saw that it was possible to store enormous joint mass functions as a function of a set of much more manageable clique tables. The joint table could be constructed as rational functions of the stored clique tables and so are – at least in principle – recoverable. In particular margins of vectors of variables all lying in a single clique could then be extracted quickly by summing over some of the components of the containing clique. But what if we subsequently learn of the value of certain functions of the random variables in the BN? Is it possible to use the BN as a framework with which to update these clique probability tables directly without resurrecting the whole joint mass function? If this were not possible then the coding of a multivariate problem in the way we have described above would be of limited value.

The answer to this question is however usually affirmative. Furthermore in the case when the underlying BN is decomposable and what we learn is a vector of values whose components are random variables conditionally independent of all variables given the values of variables in a clique, propagating information from them is extremely straightforward and usually very fast. We will focus on this case here. Note that although most BNs we might elicit are not decomposable, by forgetting some conditional independences when we code it up we can always make it so. This process was called *triangulation* by Flores and Gamez (2006); Lauritzen and Speigelhalter (1988). The construction of a valid decomposable DAG from an elicited one is illustrated below.

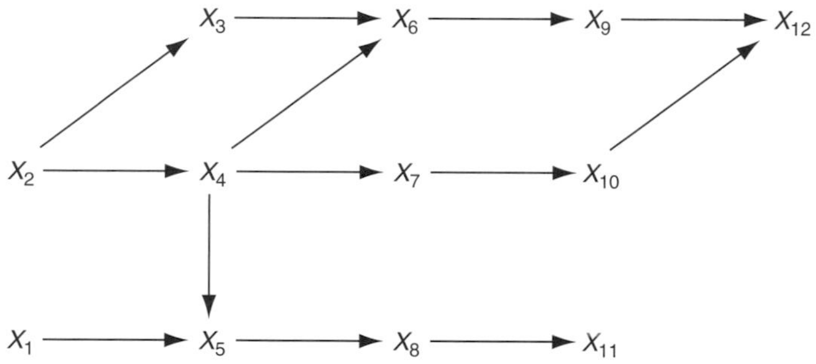

We now add directed edges to this graph in a way that keeps it acyclic. We have seen in (7.9) that if the DAG $\mathcal{G}$ of the original BN were valid then one with an added edge will be. By iterating this argument a new DAG is obtained from $\mathcal{G}$ where several edges are added. An economical way of adding edges that usually ensures that none of the cliques are too large is as follows. We first moralise the graph. We replace any undirected edges by a directed one. We are free to choose any direction here provided we do not introduce a cycle and this is always possible by directing edges so that lower indexed variables are attached to higher indexed variables (see an exercise below): although this may not be the best choice. Note however that a triangulation is not usually unique. We then have a new DAG which, by the above argument, is valid because the original one was. We then repeat this step on the new graph and keep on doing this until we end up with a decomposable BN. This must happen at some stage because the complete DAG is decomposable. Thus the moralised graph of

the last example is given by

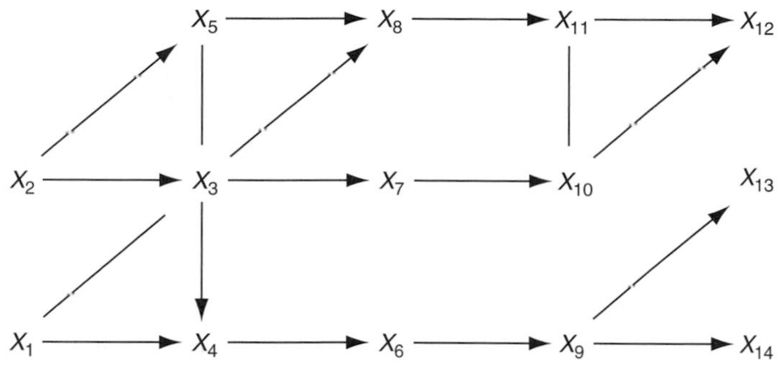

So one choice of implied DAG is

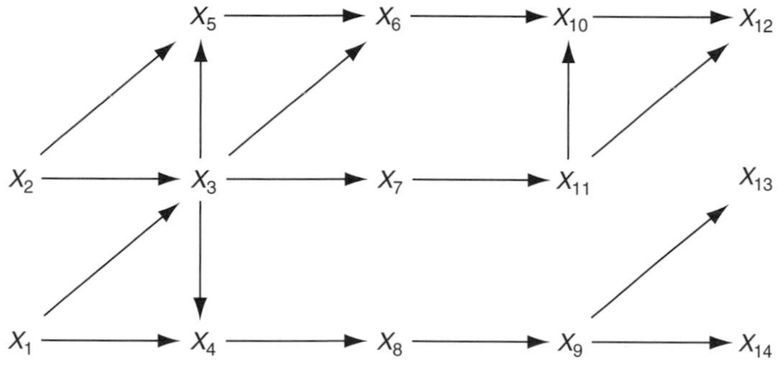

First moralise this graph. As drawn this DAG is not decomposable so we repeat this process to obtain

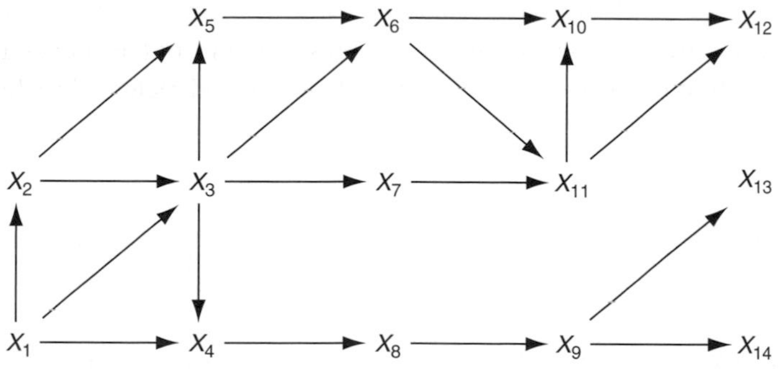

Repeating the process once again gives

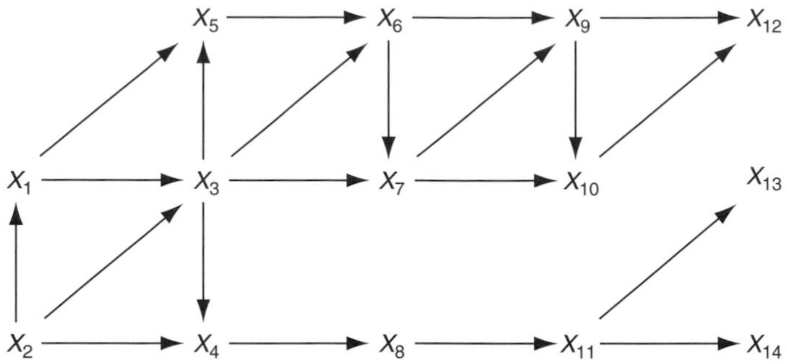

A quick check now confirms that the parents of all children in this graph are married so this graph is decomposable and the triangulation process is complete. Note that the skeleton of this graph has eleven cliques $\{C(j) : j = 1, 2, \ldots, 11\}$ of no more than three variables

$$C(1) = \{1, 2, 3\}, \quad C(2) = \{1, 3, 5\}, \quad C(3) = \{2, 3, 4\}, \quad C(4) = \{3, 5, 6\},$$
$$C(5) = \{3, 6, 7\}, \quad C(6) = \{4, 8\}, \quad C(7) = \{6, 7, 9\}, \quad C(8) = \{7, 9, 10\},$$
$$C(9) = \{8, 11\}, \quad C(10) = \{9, 10, 12\}, \quad C(11) = \{11, 13\}, \quad C(12) = \{11, 14\}.$$

There are various ways of performing this task as efficiently possible – outside the scope of this book – to obtain probability tables with a least number of aggregated cells. However, as illustrated above, a straightforward application of the algorithm above, whilst not optimal, can be performed with a few iterations usually with only a moderate loss of efficiency

Decomposable DAGs support simple propagation algorithms because their cliques can be totally ordered to have the running intersection property. Let $\{C(j) : j = 1, 2, \ldots, m\}$ be the cliques of a decomposable DAG and let the *separators* $\{B(j) : j = 2, 3, \ldots, m\}$ be defined by

$$B(j) = C(j) \cap C^{j-1}$$

where $C^{j-1} = \bigcup_{i=1}^{j-1} C(i)$ is the set of indices of all components of $\boldsymbol{X}$ appearing in a clique listed before $C(j)$.

Theorem 7.17. *A decomposable graph has cliques that can be indexed so that they exhibit the* running intersection property *that $B(j) \subset C(j_s)$ for some index j_s such that $1 \leq j_s < j$.*

The proof of this theorem is given in Lauritzen (1996) p18. In the example above the indexing of the cliques actually used satisfies the running intersection property with separators

$$B(2) = \{1, 3\}, \quad B(3) = \{2, 3\}, \quad B(4) = \{3, 5\}, \quad B(5) = \{3, 6\}, \quad B(6) = \{4\},$$
$$B(7) = \{6, 7\}, \quad B(8) = \{7, 9\}, \quad B(9) = \{8\}, \quad B(10) = \{9, 10\}, \quad B(11) = B(12) = \{11\}$$

with

j	2	3	4	5	6	7	8	9	10	11	12
j_s	1	1	2	4	3	5	7	6	8	10	10 or 11

The cliques of decomposable DAGs always exhibit more than one ordering of cliques with the running intersection property. It is usually quite simple to discover one for small DAGs following a compatible order of its vertices. For larger DAGs it is better to use one of the algorithms for finding such an index from the topology of a DAG – for example maximal cardinality search (Tarjan and Yannakakis, 1984). Note that even when the order is fixed there is often a choice of mother for a given clique. For example in the problem above, $C(12)$ has either $C(10)$ or $C(11)$ as its mother. Henceforth assume that all indexing of the cliques of a DAG is compatible – i.e. they follow an index that satisfies the running intersection property.

Such an indexing allows us to draw a simple undirected tree and devise algorithms which propagate information around it.

Definition 7.18. A *prejunction tree* $\mathcal{J}$ of a decomposable DAG $\mathcal{G}$ is a directed tree with vertices the cliques of $\mathcal{G}$ and a directed edge from $C(i)$ to $C(j)$ if $i = j_s$ i.e. i is the mother of j. A *junction tree* $\mathcal{J}$ of a decomposable DAG $\mathcal{G}$ is an undirected tree with vertices the cliques of $\mathcal{G}$ and an edge from $C(i)$ to $C(j)$ if $i = j_s$ i.e. i is the mother of j.

One junction tree of the DAG G above is given below.

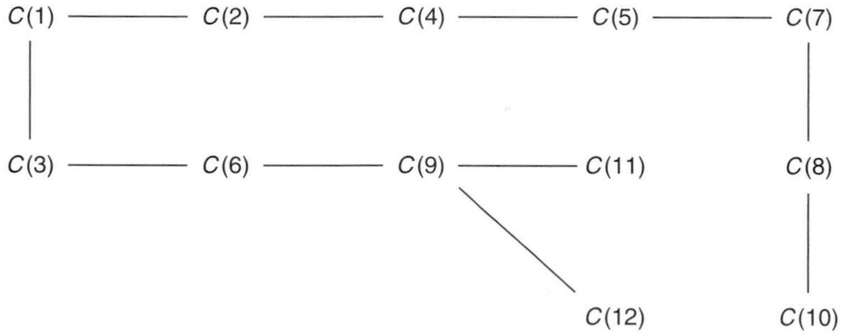

(7.15)

Notice that the edges can be thought of as labels of the separators linking each clique to its mother. Then it is easy to check from applying the d-separation theorem to $\mathcal{G}$ that for $j = 2, 3, \ldots, m$

$$\boldsymbol{X}_{C(j)\backslash B(j)} \amalg \boldsymbol{X}_{C^{j-1}\backslash B(j)} | \boldsymbol{X}_{B(j)}.$$

It follows by the property of extended conditioning that

$$\boldsymbol{X}_{C(j)} \amalg \boldsymbol{X}_{C^{j-1}\backslash B(j)} | \boldsymbol{X}_{B(j)} \tag{7.16}$$

and so by perfect composition that in particular

$$\boldsymbol{X}_{C(j)} \amalg \boldsymbol{X}_{C^{j-1}\backslash C(j_s)} | \boldsymbol{X}_{C(j_s)}.$$

It follows that a prejunction tree is a valid BN whose vertices are the random vectors $\{\boldsymbol{X}_{C(j)} : j = 1, 2, \ldots, m\}$ and the pattern of this prejunction tree is the junction tree. But we can actually say more than this. For note that from (7.16) the joint mass function of $\boldsymbol{X}$ can be written on the form

$$p(\boldsymbol{x}) = p_{C(1)}(\boldsymbol{x}_{C(1)}) \prod_{j=2}^{m} p_j(\boldsymbol{x}_{C(j)} | \boldsymbol{x}_{B(j)}). \tag{7.17}$$

By the usual rules of probability, provided that $\boldsymbol{x}$ is a vector whose probabilities satisfies $p(\boldsymbol{x}) > 0$ – so that in particular $p(\boldsymbol{x}_{B(j)}) > 0, j = 2, 3, \ldots, m$ – is equivalent to saying $p(\boldsymbol{x})$ respects the algebraic form

$$p(\boldsymbol{x}) = \frac{\prod_{j=1}^{m} p_{C(j)}(\boldsymbol{x}_{C(j)})}{\prod_{j=2}^{m} p_{B(j)}(\boldsymbol{x}_{B(j)})}. \tag{7.18}$$

7.6.2 Using a junction tree to propagate information

An important point to note here is that if we had another DAG $\mathcal{G}^*$ where the skeleton of its prejunction tree was also $\mathcal{J}$ but whose cliques were indexed differently then the same argument would tell us that its associated density would also satisfy the equation (7.18). In particular, whenever $p(\boldsymbol{x}) > 0$ for all $\boldsymbol{x}$, all decomposable graphs with junction tree $\mathcal{J}$ are equivalent and simply assert that $p(\boldsymbol{x})$ satisfies (7.18). Notice that there are exactly m different such prejunction trees defined by the clique vertex we choose as a root. For by definition, no two edges can direct into the same child so once the root is set all other edges in the prejunction tree must be directed to form a path away from the root.

Now suppose we learn the value of some random vector $\boldsymbol{Y}$ which is independent of $\boldsymbol{X}$ given the values of $\boldsymbol{X}_C$ for some clique C of a BN with decomposable DAG $\mathcal{G}$ whose junction tree is $\mathcal{J}$. For example $\boldsymbol{Y}$ could be simply some subvector $\boldsymbol{X}$. From the comments above without loss we can choose to index C as $C(1)$ the root in the prejunction tree. So for values of $(\boldsymbol{x}, \boldsymbol{y})$ for which $p(\boldsymbol{x}, \boldsymbol{y}) > 0$

$$p(\boldsymbol{x}, \boldsymbol{y}) = p_Y(\boldsymbol{y} | \boldsymbol{x}) p(\boldsymbol{x})$$

which by the hypotheses above can be written

$$\begin{aligned} p(\boldsymbol{x}, \boldsymbol{y}) &= p_Y(\boldsymbol{y} | \boldsymbol{x}_{C(1)}) p_{C(1)}(\boldsymbol{x}_{C(1)}) \prod_{j=2}^{m} p_j(\boldsymbol{x}_{C(j)} | \boldsymbol{x}_{B(j)}) \\ &= p_Y(\boldsymbol{y}) p^*_{C(1)}(\boldsymbol{x}_{C(1)}) \prod_{j=2}^{m} p_j(\boldsymbol{x}_{C(j)} | \boldsymbol{x}_{B(j)}) \end{aligned}$$

where $p^*_{C(1)}(\boldsymbol{x}_{C(1)}) = p_{C(1)}(\boldsymbol{x}_{C(1)}|\boldsymbol{y})$. Thus

$$p(\boldsymbol{x}|\boldsymbol{y}) = p^*_{C(1)}(\boldsymbol{x}_{C(1)}) \prod_{j=2}^{m} p_j(\boldsymbol{x}_{C(j)}|\boldsymbol{x}_{B(j)}).$$

So after this information has been accommodated into the joint distribution respects the same factorisation (7.17) as it did before. It follows that equation (7.18) is still valid, its just that the clique probability tables – and hence the probability tables of the separators – have changed. Thus

$$p(\boldsymbol{x}|\boldsymbol{y}) = \frac{\prod_{j=1}^{m} p^*_{C(j)}(\boldsymbol{x}_{C(j)})}{\prod_{j=2}^{m} p^*_{B(j)}(\boldsymbol{x}_{B(j)})}.$$

Here $p^*_{C(1)}(\boldsymbol{x}_{C(1)})$ is calculated simply by using Bayes rule. By definition the separator $B(2)$ of the clique $C(2)$ must be such that $B(2) \subset C(1)$ so that the new mass function of this separator can be obtained from the probability table $p^*_{C(1)}(\boldsymbol{x}_{C(1)})$ by summing those components with indices outside $B(2)$. Explicitly

$$p^*_{B(2)}(\boldsymbol{x}_{B(2)}) \triangleq p_{B(2)}(\boldsymbol{x}_{B(2)}|\boldsymbol{y}) = \sum_{\boldsymbol{x}_{C(1)\backslash B(2)}} p^*_{C(1)}(\boldsymbol{x}_{C(1)}).$$

It follows that

$$\begin{aligned} p^*_{C(2)}(\boldsymbol{x}_{C(2)}) &\triangleq p_{C(2)}(\boldsymbol{x}_{C(2)}|\boldsymbol{y}) = p_2(\boldsymbol{x}_{C(2)}|\boldsymbol{x}_{B(2)})p^*_{B(2)}(\boldsymbol{x}_{B(2)}) \\ &= p_{C(2)}(\boldsymbol{x}_{C(2)})\frac{p^*_{B(2)}(\boldsymbol{x}_{B(2)})}{p_{B(2)}(\boldsymbol{x}_{B(2)})} \end{aligned}$$

by the definition of conditioning where division in the last ratio is performed term by term as illustrated below.

Similarly, calculating the new cliques consistently with their indexed order, so that thc associated new separator tables can be calculated from an earlier clique we have that for $j = 2, 3, \ldots, m$

$$p^*_{C(j)}(\boldsymbol{x}_{C(j)}) \triangleq p_{C(j)}(\boldsymbol{x}_{C(j)}|\boldsymbol{y}) = p_{C(j)}(\boldsymbol{x}_{C(j)})\frac{p^*_{B(j)}(\boldsymbol{x}_{B(j)})}{p_{B(j)}(\boldsymbol{x}_{B(j)})}. \tag{7.19}$$

So the information we learnt about the first clique is passed along the edges of the junction tree sequentially using the formula above until all the clique tables are revised. Note that even with problems where there are thousands of cliques, provided the number of elements in each clique table is not large this simple update will be almost instantaneous to enact on a laptop.

7.6.3 An example of propagation

We now return to the example of the BN associated with the adverse affect on an individual exposed to radiation. Note that this is not decomposable but if triangulated can be contained in a decomposable graph whose pattern is the one given in (7.15) whose cliques are

$$\{\boldsymbol{X}_{\{1,2\}}, \boldsymbol{X}_{\{2,3\}}, \boldsymbol{X}_{\{3,4,5\}}, \boldsymbol{X}_{\{4,5,6,7\}}, \boldsymbol{X}_{\{2,8\}}\}$$

whose separators are $\{X_2, X_3, \boldsymbol{X}_{\{4,5\}}, X_2\}$ and has junction tree

$$\begin{array}{ccccc} \mathcal{J} & & \boldsymbol{X}_{\{2,8\}} & & \boldsymbol{X}_{\{4,5,6,7\}} \\ & & | & & | \\ \boldsymbol{X}_{\{1,2\}} & - & \boldsymbol{X}_{\{2,3\}} & - & \boldsymbol{X}_{\{3,4,5\}} \end{array}$$

Using the usual rules of probability we can quickly calculate the joint clique and separator probability tables which are given below.

x_1	1	2	3	4	5	6	7	8	9	10	p_{x_2}
$p_{\{1,2\}}(1, x_1)$	.4	.135	.08	.063	.04	.028	.015	.006	.002	0	.769
$p_{\{1,2\}}(2, x_1)$	0	.015	.02	.027	.032	.035	.025	.012	.004	.001	.171
$p_{\{1,\}2}(3, x_1)$	0	0	0	0	.008	.007	.01	.012	.014	.009	.060

x_2	1	2	3	p_{x_3}
$p_{\{2,3\}}(x_3 = 0, x_2)$	.769	.137	.03	.936
$p_{\{2,3\}}(x_3 = 1, x_2)$	0	.034	.03	.064

(x_3, x_4, x_5)	000	001	010	011	100	101	110	111
$p_{\{3,4,5\}}$	.8424	0	.0842	.0094	.0384	0	.0154	.0102

x_4, x_5	00	01	10	11
$p_{\{4,5\}}$	.8808	0	.0996	.0196

(x_4, x_5, x_6)	000	001	010	011	100	101	110	111
$p_7(x_7 = 0, \boldsymbol{x}')$	.6694	.1409	0	0	.0159	.0040	.0078	.0004
$p_7(x_7 = 1, \boldsymbol{x}')$	.0352	.0352	0	0	.0637	.0159	.0078	.0035

(x_2, x_8)	$1, 0$	$2, 0$	$3, 0$	$1, 1$	$2, 1$	$3, 1$
$p_{\{2,8\}}$	.7690	.1197	.0360	0	.0513	.0240

Suppose that we learn that a worker has not has died after ten years and you are concerned about the hereditary effects of his possible exposure and the probabilities of exposure he is likely to have suffered. The only thing the DM is told is that he survives, i.e. that $X_7 = 0$. So we label a clique containing this variable as a root vertex of the new directed tree and propagate information around the cliques in an order consistent with the edges in the prejunction tree given below, with all arrows pointing away from this containing clique.

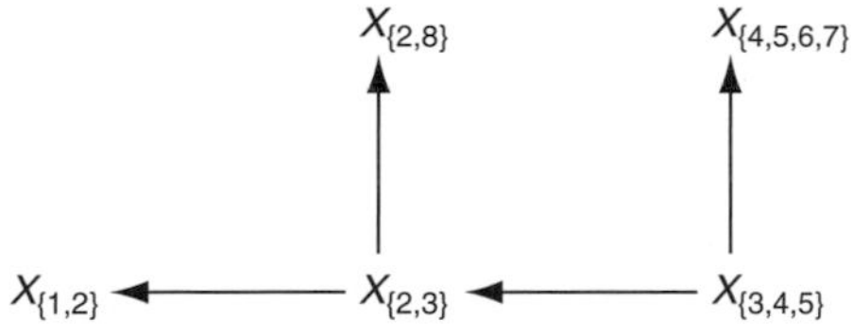

The new clique table for $\boldsymbol{X}_{\{4,5,6,7\}}$ can be calculated by Bayes theorem to be

(x_4, x_5, x_6)	000	001	010	011	100	101	110	111
$p_7(x_7 = 0, \boldsymbol{x}')$	.7984	.1681	0	0	.0190	.0048	.0093	.0005
$p_7(x_7 = 1, \boldsymbol{x}')$	0	0	0	0	0	0	0	0

The distribution of $\boldsymbol{X}_{\{4,5\}}$ on the separator can be calculated by summing over x_6 in this new table and compares with the old one as

x_4, x_5	00	01	10	11
$p^*_{\{4,5\}}$	.9665	0	.0237	.0098
$p_{\{4,5\}}$	.8808	0	.0996	.0196
$\frac{p^*_{\{4,5\}}}{p_{\{4,5\}}}$	1.097	N/R	.238	.500

We can now calculate the clique margin of the adjacent clique $\boldsymbol{X}_{\{3,4,5\}}$ using the formula (7.19) to be

(x_3, x_4, x_5)	000	001	010	011	100	101	110	111
$p_{\{3,4,5\}}$	.8424	0	.0842	.0094	.0384	0	.0154	.0102
$\frac{p^*_{\{4,5\}}}{p_{\{4,5\}}}$	1.097	N/R	.238	.500	1.097	N/R	.238	.500
$p^*_{\{3,4,5\}}$	.924	0	.020	.005	.042	0	.004	.005

We can now calculate how the distribution on the separator on X_3 has changed viz

x_3	0	1
$p^*_{\{3\}}$	.949	.051
$p_{\{3\}}$	.936	.064
$\frac{p^*_{\{3\}}}{p_{\{3\}}}$	1.014	0.797

(x_2, x_3)	1, 0	2, 0	3, 0	1, 1	2, 1	3, 1
$p_{\{2,3\}}$	.769	.137	.030	0	.034	030
$\frac{p^*_{\{3\}}}{p_{\{4,5\}}}$	1.014	1.014	1.014	0.797	0.797	0.797
$p^*_{\{2,3\}}$	.780	.139	.030	0	.027	.024

x_3	1	2	3
$p^*_{\{2\}}$	.780	.166	.054
$p_{\{2\}}$	.769	.171	.060
$\frac{p^*_{\{2\}}}{p_{\{2\}}}$	1.0143	0.9708	0.9000

x_1	1	2	3	4	5	6	7	8	9	10	$\frac{p^*_{\{2\}}}{p_{\{2\}}}$
$p_{\{1,2\}}(1, x_1)$	.4	.135	.08	.063	.04	.028	.015	.006	.002	0	1.0143
$p_{\{1,2\}}(2, x_1)$	0	.015	.02	.027	.032	.035	.025	.012	.004	.001	0.9708
$p_{\{1,2\}}(3, x_1)$	0	0	0	0	.008	.007	.01	.012	.014	.009	0.9000
$p_{\{1\}}$	.4	.15	.1	.09	.08	.07	.05	.03	.02	.01	

x_1	1	2	3	4	5	6	7	8	9	10	$\frac{p^*_{\{2\}}}{p_{\{2\}}}$
$p^*_{\{1,2\}}(1, x_1)$	.4057	.1369	.0811	.0639	.0406	.0284	.0152	.0061	.0020	0	1.0143
$p^*_{\{1,2\}}(2, x_1)$	0	.0146	.0194	.0262	.0311	.0340	.0242	.0116	.0038	.0010	0.9708
$p^*_{\{1,2\}}(3, x_1)$	0	0	0	0	.0072	.0063	.0090	.0108	.0126	.0081	0.9000
$p^*_{\{1\}}$	.4057	.1515	.1005	.0901	.0789	.0687	.0484	.0285	.0184	.0091	
$p_{\{1\}}$	.4000	.1500	.1000	.0900	.0800	.0700	.0500	.0300	.0200	.0100	

We can therefore conclude that the probabilities of being in the lower four categories has increased (although not by much) whilst the probabilities of being exposed to higher amounts of radiation have slightly gone down. Similar effects can be seen on the probabilities associated with adverse hereditary effects that are also slightly lower on learning he has survived ten years – reducing from about 0.075 to 0.072: see below.

(x_2, x_8)	1, 0	2, 0	3, 0	1, 1	2, 1	3, 1
$p_{\{2,8\}}$	.7690	.1197	.0360	0	.0513	.0240

(x_2, x_8)	1, 0	2, 0	3, 0	1, 1	2, 1	3, 1
$p_{\{2,8\}}$	.7690	.1197	.0360	0	.0513	.0240
$\frac{p^*_{\{2\}}}{p_{\{2\}}}$	1.014	0.971	0.900	1.014	0.971	0.900
$p^*_{\{2,8\}}$	.778	.116	.032	0	.050	.022

Typically – but not always – the further two cliques are apart from each other in a junction tree the less learning about something in one affects the other. In non-pathological cases this communication tends to die out exponentially fast the further the information travels. When we learn information concerning many different individual cliques then a brute-force method of propagating this information would be component by component using the algorithm above. However this would be very inefficient and there are now myriads of clever message-passing algorithms for junction trees where such data can be input simultaneously into the junction tree. Although we have described only BN algorithms on discrete systems here there are simple analogues that use analogous algorithms on many parametric families of joint distributions such as on Gaussian vectors of random variables. There is now a vast selection of software supporting these faster methods; see Cowell *et al.* (1999); Jensen and

Nielsen (2007), and approximate methods especially designed for mixtures of continuous and discrete variables; see e.g. Neil *et al.* (2008).

The algorithms discussed above do not work however if the data we have cannot be expressed as a vector of functions, each function being conditionally independent of all other clique variables given a variable in a single one. This is because if you observe the system in this way the conditional independence in your original BN may no longer be valid after sampling. Although the joint posterior distributions of components of variables all lying in a single clique are trivial to calculate using these methods more work needs to be done to discover the joint distributions of variables lying in different cliques.

None of the algorithms can work fast if the number of cells in the cliques become unmanageably large: although even current software seems to cope with problems where the number of cells is of the order of hundreds of thousands. Perhaps the most pertinent word of caution is that if evidence seems to have a large effect on the inference then this is often because it was a priori very improbable in the light of any distribution consistent with the given BN. We saw how this can distort even a very simple Bayesian inference in the introduction. The shear scale of a BN model can greatly magnify these effects. So a BN should be used with a diagnostic to check whether this is happening. Happily such a diagnostic now exists (see Cowell *et al.* (1999); Spiegelhalter *et al.* (1993)). Despite all these caveats, BNs have revolutionised modelling of large highly structured systems and allowed countless analyses of high-dimensional systems.

7.7 Bayesian networks and other graphs

7.7.1 Discrete Bayesian networks and staged trees

In Chapter 2 we defined the event tree $\mathcal{T}$ each of whose non-leaf vertices – called a situation – represented a distinct point in the unfolding of a unit's history. We noted that there were many classes of problem where the set $S(\mathcal{T})$ could be partitioned into parallel situations. Here the DM would believe that the immediate development of a unit finding itself in one or other of the situations in the same set in the partition – with the appropriate identification of edges – would respect the same probability distribution

Definition 7.19. A *staged tree* $\mathcal{T}$ is an event tree together with a partition $\mathbb{U}(\mathcal{T})$ of its situations $S(\mathcal{T})$ such that two situations v_1 and v_2 are in the same set $u \in \mathbb{U}(\mathcal{T})$ if and only if they are parallel situations.

Now somewhat surprisingly it can be shown that any finite discrete BN is a staged tree. In fact for a staged tree to admit a BN representation not only does the event tree $\mathcal{T}$ need to have a very special topology but also the partition $\mathbb{U}(\mathcal{T})$ needs to take a very special form. To see this let the vertices of the DAG G of the BN be $\{X_1, X_2, \ldots, X_n\}$ where the variables are listed in a compatible order, i.e. where each parent is listed before each of its children.

Example 7.20. Let $n = 3$ and X_1, X_2, X_3 take values on $2, 2, 3$ levels respectively where x_1 and x_2 take the values 0 or 1 and x_3 the values $1, 2, 3$. Then their sample space can be represented by the event tree $\mathcal{T}$ given below.

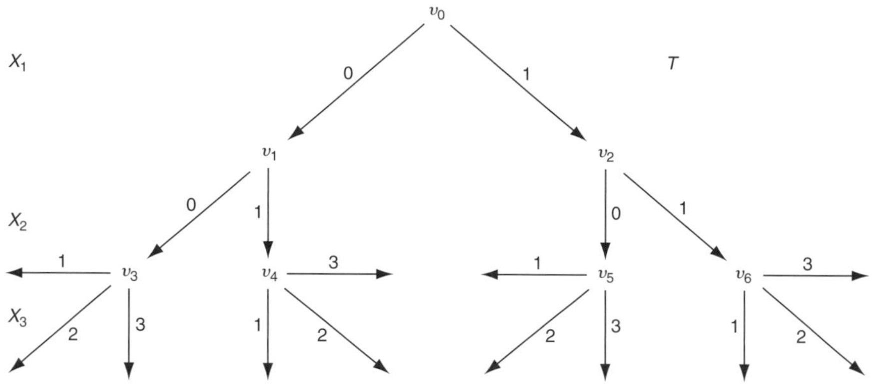

Suppose the valid DAG $\mathcal{G}$ of the BN is $X_1 \rightarrow X_2 \rightarrow X_3$ expressing the conditional independence statement $X_3 \amalg X_1 | X_2$. This simply means that

$$p(x_3|x_1 = 0, x_2 = 0) = p(x_3|x_1 = 1, x_2 = 0)$$

and

$$p(x_3|x_1 = 0, x_2 = 1) = p(x_3|x_1 = 1, x_2 = 1).$$

In terms of the staged tree this simply means that with the obvious association of edges v_3 and v_5 are parallel situations as are v_4 and v_6. It follows that the pair of the DAG $\mathcal{G}$ and its sample space can be identified with the tree $\mathcal{T}$ above and the partition of $\mathbb{U}(\mathcal{T})$ into its stages where

$$\mathbb{U}(\mathcal{T}) = \{\{v_0\}, \{v_1\}, \{v_2\}, \{v_3, v_5\}\{v_4, v_6\}\}.$$

In general it is easily checked – see Smith and Anderson (2008) – that any BN with valid DAG $\mathcal{G}$ can be represented by its event tree $\mathcal{T}(\mathcal{G})$ constructed as above and stage partition $\mathbb{U}(\mathcal{T}(\mathcal{G}))$ where $v, v' \in u \in \mathbb{U}(\mathcal{T}(\mathcal{G}))$ if and only if v, v' are associated to the same configuration $\{\boldsymbol{X}_{Q_i} = \boldsymbol{x}_{Q_i}\}$ of parents of a vertex $X_i \in V(\mathcal{G}), i = 2, 3, \ldots, n$.

Whilst the staged tree is a more cumbersome graphical description of a problem it is expressive enough to form the platform of many different generalisations. For example the *context specific* BN (McAllister *et al.*, 2004; Poole, 2003; Bonet, 2001a, 2001b; Salmaron *et al.*, 2000) has a DAG $\mathcal{G}$ which has additional symmetries expressible in terms of a stage partition which is not simply defined by the set of parent coefficients of each variable.

Example 7.21. Consider the BN whose DAG $\mathcal{G}'$ is $X_1 \rightarrow X_3 \leftarrow X_2$, where the levels of $\{X_1, X_2, X_3\}$ are as in the last example so that the event tree $\mathcal{T}(\mathcal{G}') = \mathcal{T}(\mathcal{G})$ depicted above but whose partition is

$$\mathbb{U}(\mathcal{T}(\mathcal{G}')) = \{\{v_0\}, \{v_1, v_2\}, \{v_3\}, \{v_5\}, \{v_4\}, \{v_6\}\}.$$

Suppose we have additional contextual information that $p(x_3|x_1, x_2)$ is always the same unless $x_1 = x_2 = 0$. Then this defines a context-specific BN whose stage partition is coarser and equal to

$$\mathbb{U}(\mathcal{T}(\mathcal{G}')) = \{\{v_0\}, \{v_1, v_2\}, \{v_3\}, \{v_5, v_4, v_6\}\}.$$

We will see later that because fewer conditional probability distributions are needed to define the system, this structure is not only easier to elicit but easier to estimate.

In fact we have seen that event trees often do not have the symmetry of structure associated with that of a BN because their associated sample space is not in fact a product space: see French *et al.* (1997). Obviously we can still define a staged tree. In a *chain event graph* (*CEG*) the stage partition can be generally defined on its situations. Conditional independences can be read from its topology (Smith and Anderson, 2008) and its framework used for propagation just as a BN (Thwaites *et al.*, 2008). The graph of a CEG depicts as much as possible of the stage structure of a staged tree. Two vertices of the tree combined into one of the subtrees rooted at these two situations are isomorphic. So situations are identified if the distribution of a unit's development having reached either situation is the same.

Example 7.22. Let a development be expressed in terms of the asymmetric event tree $\mathcal{T}''$ given below whose stage partition is

$$\mathbb{U}(\mathcal{T}) = \{u_0 = \{v_0\}, u_1 = \{v_1, v_2\}, u_2 = \{v_3, v_5\}, u_3 = \{v_4, v_6\}\}.$$

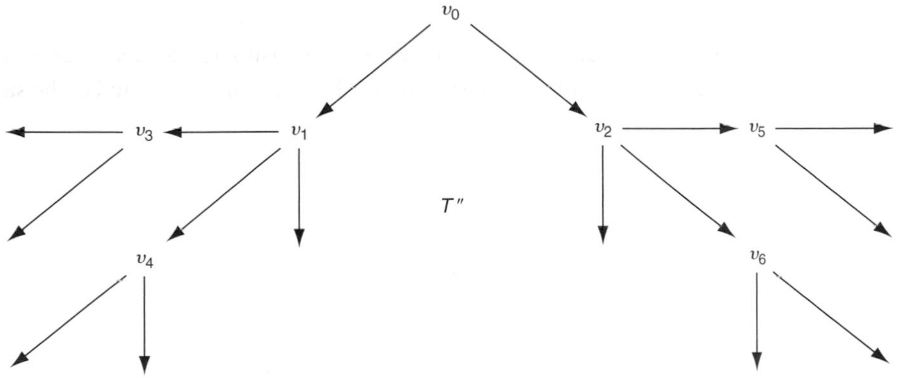

The graph $\mathcal{C}(\mathcal{T}'')$ of its CEG is then

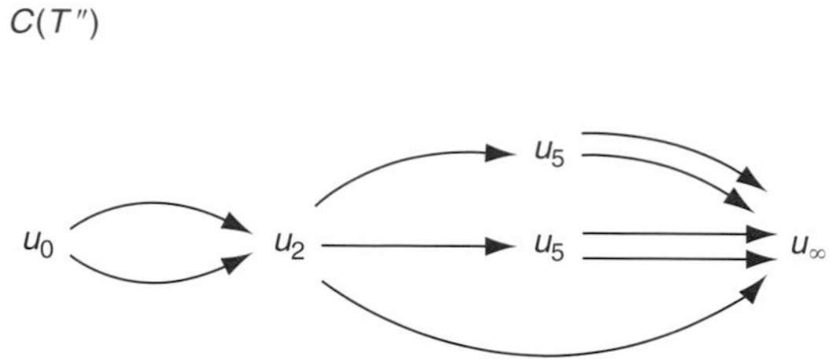

Some problems have many zeros because of logical constraints in the system. In the illustration above for example surgery was only performed on individuals with detected tumours. If the conditional tables are very sparse or there are symmetries in the probability distributions not close to a decomposable BN then there are much faster and more transparent methods based on CEGs that can do the job much better. It is important to note that there are also many other competing graphical representations of problems like this, for example, Covaliu and Oliver (1995); Jaegar (2004).

7.7.2 *Other graphical models*

There are many other graph-based methods for representing a variety of different probabilistic symmetries over a set of random variables $\boldsymbol{X} = \{X_1, X_2, \ldots, X_n\}$ which make up the vertices of these graphs. One important class is the object-orientated BNs (e.g. Koller and Friedman (2004) and Korb and Nicholson (2010)). These generalise the BN taxonomy of an expert system so that elements of the graphical system can be matched to the case at hand.

The most common alternative to the BN for one-off applications is the undirected Markov graphical (UG) models usually defined on a set of strictly positive distributions where there is an undirected edge between X_i and X_j if and only if $X_i \amalg X_j | \boldsymbol{X} \backslash \{X_i, X_j\}$. These are particularly natural for expressing certain families of multivariate logistic models (Lauritzen, 1996; Whittaker, 1990) and spatial models where the indexing of the variables is often arbitrary and meaningless. Two other very useful classes are chain graphs (CG) (Lauritzen, 1996; Lauritzen and Wermuth, 1989) and the AMP model class (Andersson *et al.*, 1997) which are each both a hybrid and a generalisation of the BN and the UG model classes. All these classes have their own associated separation theorem which allows the structure to be interrogated before it is fleshed out with probability models. The CG class has since been further generalised to joint response chain graphs (Cox and Wermuth, 1996). Reciprocal graphs (Koster, 1996) again generalise the class of BNs and have close links with important explanatory classes of models used in econometrics. There are many others all suited to different genres of applications (see e.g. Richardson and Spirtes (2002); Drton and Richardson (2008) and Studeny (2005) for a good review of some of these).

Many important large-scale problems – like the application mentioned at the end of the last chapter – have variables whose development over time effects the efficacy of any proposed action. There has been a great deal of effort to develop stochastic analogues of the BN for use in a Bayesian analysis. Some of these will be briefly discussed later in the book.

But despite all these and many other variant graphical models, at the time of writing the BN is still the most used framework for encoding expert judgements required for a decision analysis.

7.8 Summary

Bayesian networks are one of a number of very useful frameworks that guide the DM towards building a structured probability distribution which is sympathetic to a customised credence decomposition. They have a long history Canning *et al.* (1978); Cowell *et al.* (1999); Howard and Matheson (1984); Jensen and Nielsen (2007); Lauritzen and Speigelhalter (1988); Oldmsted (1983); Pearl (1988) and their properties are now well understood and documented and they are now implemented in many pieces of software. They have the fortunate property that they on the one hand seem to give an accessible and evocative representation of a problem that many DMs enthusiastically take ownership of and on the other are compatible with the widely used Bayesian hierarchical model which is increasingly used to build probabilistic models of processes and expert systems. They are of course limited in their application. Nevertheless it has been repeatedly demonstrated that the basic rationale behind them can be generalised to many more complicated domains.

However for a decision analysis it is not only necessary to capture what a DM or domain expert believes will happen in an observed system but also what she believes will happen when the system is controlled by her acts. In the next chapter we will address how this gap might be filled.

7.9 Exercises

7.1 Check that if a DM's beliefs can be fully expressed through her distribution of n random variables $\{X_1, X_2, \ldots, X_n\}$ and she interprets conditional irrelevance as conditional independence then she satisfies the semi-graphoid axioms.

7.2 One possible way to define irrelevance might be to say that $\boldsymbol{Y}$ is irrelevant to forecasting $\boldsymbol{Z}$ given $\boldsymbol{X}$ if the expectation of $\boldsymbol{Z}$ depends only on $(\boldsymbol{X}, \boldsymbol{Y})$ only through its dependence on $\boldsymbol{X}$. Demonstrate that this definition of irrelevance can violate the symmetry property of the semi-graphoid axioms.

7.3 (Simpson's paradox). Because the technical meaning of the term "independence" is not fully equivalent to the statistical one an elementary but plausible inferential error is to assume that if $\boldsymbol{Z} \amalg \boldsymbol{Y} | \boldsymbol{X}$ then $\boldsymbol{Z} \amalg \boldsymbol{Y}$. Construct a simple example demonstrating this is not a deduction that can legitimately be made in general. How does the terminology of irrelevance help in explaining why this deduction might not be sound?

7.4 (Graphoids) A graphoid is a semi-graphoid with the additional demand on our subsets of variables that

$$\boldsymbol{Z} \amalg \boldsymbol{Y} | \boldsymbol{X}, \boldsymbol{W} \quad \text{and} \quad \boldsymbol{Z} \amalg \boldsymbol{X} | \boldsymbol{Y}, \boldsymbol{W} \Leftrightarrow \boldsymbol{Z} \amalg (\boldsymbol{Y}, \boldsymbol{X}) | \boldsymbol{W}.$$

Demonstrate that this axiom will not necessarily hold when deterministic relationships might exist between the variables by considering the case when $\boldsymbol{X} = \boldsymbol{Y}$. By using the factorisations of the joint densities these relationships simply demonstrate that if all joint probabilities of a collection of discrete variables are strictly positive then the collection forms a graphoid structure.

7.5 Draw the DAG of a non-homogeneous Markov chain $\{X_i : i = 1, 2, \ldots, n\}$ so that for $i = 3, 4, \ldots, n$

$$X_i \amalg \{X_1, X_2, \ldots, X_{i-2}\} | X_{i-1}.$$

Use the d-separation theorem to prove the global Markov property that

a) for $i = 2, 3, \ldots, n-1$

$$\left\{X_1, X_2, \ldots, X_{i-1}\right\} \amalg \left\{X_{i+1}, X_{i+2}, \ldots, X_n\right\} | X_i$$

and

b) for $i = 3, 4, \ldots, n-2$

$$X_i \amalg \left\{X_1, X_2, \ldots, X_{i-2}\right\} \cup \left\{X_{i+2}, X_{i+2}, \ldots, X_n\right\} | \left\{X_{i-1}, X_{i+1}\right\}.$$

7.6 Draw the DAG of the conditional independence statements listed below.

$$X_1 \amalg X_2,$$
$$X_4 \amalg (X_1, X_2) | X_3,$$
$$X_5 \amalg (X_1, X_2, X_4) | X_3$$
$$X_6 \amalg (X_1, X_2, X_3) | (X_4, X_5)$$
$$X_7 \amalg (X_1, X_2, X_3, X_4, X_5) | X_6$$
$$X_8 \amalg (X_1, X_2, X_3, X_4, X_6, X_7 | X_5$$

Use the d-separation theorem to determine whether or not the following two statements are valid deductions from a BN with the DAG you have drawn.

$$X_4 \amalg X_5 | (X_3, X_8)$$
$$X_4 \amalg X_5 | (X_3, X_7)$$

7.7 Write down a minimal collection of conditional independence statements that are represented by the DAG $\mathcal{G}$ valid where $\mathcal{G}$ is given below.

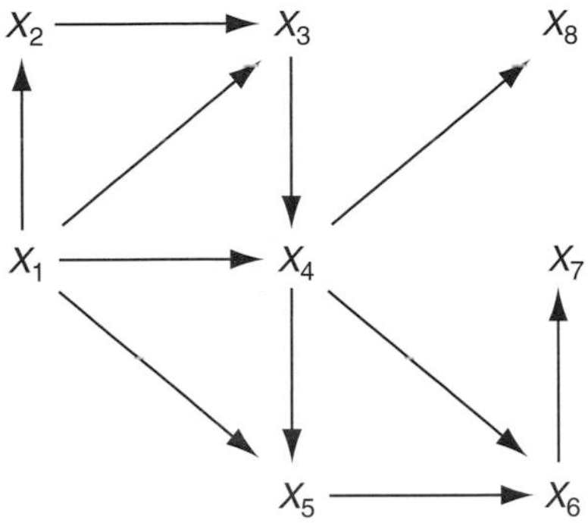

Use the d-separation theorem to check whether $X_2 \amalg X_8 | X_1, X_3$. Draw the pattern of $\mathcal{G}$. A researcher has "deduced" that $\mathcal{G}$ is valid on the basis of a very large random sample of units respecting $\mathcal{G}$. Because there is a directed arrow from X_1 to X_3 he now wants to deduce that X_1 "causes" X_3. Use pattern equivalence to demonstrate that is a not a valid deduction that can be made even if $\mathcal{G}$ is valid.

7.8 Consider the BN whose DAG is given below.

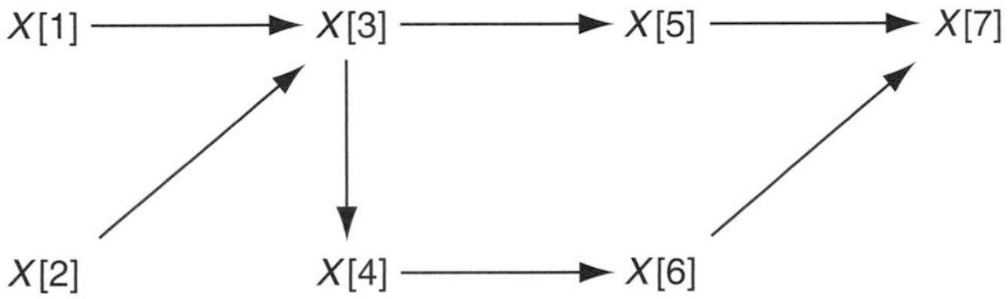

State the conditional independence relation between $X[6]$ and $\{X[1], X[2], X[3], X[4], X[5]\}$ implied by this graph. Use the d-separation theorem to prove which, if either, of the following two conditional independence statements are true.

$$X[3] \amalg X[6] | X[1], X[2], X[4]$$

$$X[3] \amalg X[6] | X[4], X[5], X[7]$$

7.9 The DAG of a BN on two diseases, D_1 and D_2, and five other related conditions and symptoms (G = exposure to great stress, H = chronic headache, V = vomiting, Z =

dizziness, C = doctor called) for any particular patient is given below.

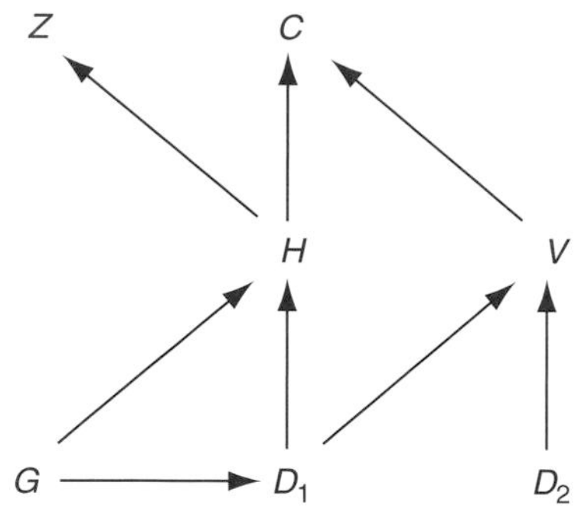

Write down a set of five conditional independence statements from which any other conditional independence statements implied by this DAG can be derived. Use the d-separation theorem to determine which (if either) of the following conditional independence statements can be derived from this DAG.

$$D_2 \amalg G | V, Z$$

$$D_2 \amalg G | Z$$

Use the d-separation theorem to prove that $D_2 \amalg G | D_1, C$ cannot be deduced. From your construction, produce a verbal argument to your client to demonstrate why the value of G might be relevant for the prediction of D_2 in the given circumstances. Triangulate the DAG to obtain a decomposable DAG Γ and identify its cliques and separators. Draw a junction tree J of the decomposable DAG Γ and briefly describe how J is used to update the probability tables of Γ after observing C – that a doctor is called.

7.10 The management team of a nuclear power plant are concerned to build an inferential system to guide countermeasures taken after a possible terrorist bombing attack of a given kind. After long discussion they describe the critical features of the problem using seven variables. The variable X_1 measures the size of the ensuing explosion within a system, X_2 is an indicator of whether or not a cooling fan is working, X_3 is an indicator of whether or not a water coolant system is working, X_4 measures the extent of damage to the external casing of the core, X_5 is the extent of nuclear particle release from a breach in the casing, X_6 is the extent that the core becomes overheated and so liable to release and X_7 is the total release of nuclear contaminant into the atmosphere outside the plant, due to both the breach in the casing and the heating of the core. The DAG of a BN representing the relationship between the variables $\{X_1, X_2, \ldots, X_7\}$ is given below.

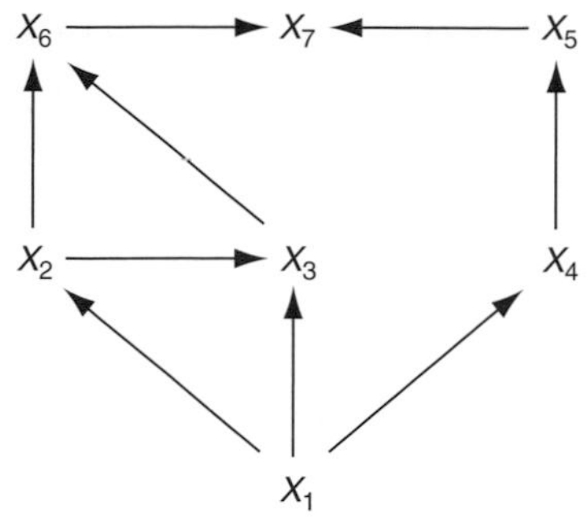

Write down a set of four conditional independence statements from which any other conditional independence statements implied by this DAG can be derived. Use the d-separation theorem to demonstrate that

$$X_4 \amalg X_6 | X_1$$

but that it is not true that

$$X_6 \amalg X_1 | (X_2, X_3, X_7) .$$

Using the context give a verbal argument which follows the d-separation construction that explains how it might be possible that X_1 gives useful information additional to (X_2, X_3, X_7) for predicting X_6.

8

Graphs, decisions and causality

8.1 Influence diagrams

8.1.1 Introduction

In Chapter 2 we discussed the decision tree which extended the semantics of an event tree so that the full decision problem could be expressed. In this section we discuss how a similar extension can be made to the semantics of a BN. This diagram $\mathcal{I}$ is called an *influence diagram* (ID) and is very useful for representing a decision problem and for providing a framework to discover optimal decision rules.

An influence diagram cannot be effectively used to represent all decision problems since they depend to a significant degree on a certain type of symmetry being present. However the conditions in which they are a useful tool are met quite often in practice and Gomez (2004) catalogues over 250 documented practical applications of the framework before 2003.

Unlike the decision tree whose topology represents relationships between events and particular decisions taken, the influence diagram represents the existence of relationships between *random variables* – represented by $\circ$ vertices, *decision spaces* – denoted by $\Box$ vertices and a utility variable – denoted by a $\diamond$ vertex. When appropriate they have many advantages over the decision tree. First they are usually much simpler to draw. Second like the BN they represent qualitative dependences exhibited by a problem and so the structure they express can be quite general, transparent and easy to elicit early in an analysis. Third we have seen how useful and intrinsic the conditional independence relationships between variables can be and the influence diagram expresses these directly through its topology. Fourth they are equally good at representing decision problems on mixtures of continuous and discrete variables whereas the decision tree can only depict a discrete decision problem. Finally the influence diagram, like the tree, can be used as a transparent framework for calculation of an optimal decision rule.

In this chapter, rather than describing the technical features of fast algorithms for easy calculation of Bayes decision rules which is well documented elsewhere: see Cowell *et al.* (1999); Jensen and Nielsen (2007); Marshall and Oliver (1995); Shachter (1986) and Kjaerulff and Madsen (2008), I will focus my discussion on the use of influence diagrams for *problem representation*. Of course, once a problem has been faithfully represented as an influence diagram the algorithms mentioned above can be implemented and calculations made, see Smith (1988b); Smith (1994a) and Smith and Thwaites (2008b).

Problems can be represented by an ID if the vertices in its graph can be indexed so that the following conditions apply:

Product decision space Decisions can be represented as an ordered sequences of choices $\boldsymbol{d} = (d_1, d_2, \ldots, d_k)$, taking values in a product space $D_1 \times D_2 \times \cdots \times D_K$ where decision $d_j \in D_j$ is taken before decision $d_{j+1} \in D_{j+1}$. The spaces D_i, $i = 1, 2, \ldots, K$ can be continuous or discrete. Henceforth let $\boldsymbol{d}^{(j)} \triangleq (d_1, d_2, \ldots, d_j) \in D_1 \times D_2 \times \cdots \times D_j \triangleq D^{(j)}, j = 1, 2, \ldots, K$.

No forgetting The vector of decisions $\boldsymbol{d}^{(j-1)} \in D^{(j-1)}$ is remembered when the decision $d_j \in D_j$ is chosen $j = 2, \ldots, K$.

Compatibly ordered Suppose a random variable Y is represented as a vertex in the graph of the ID and the value of X will be known before the decision associated with the space D_j is taken. Then Y is listed before D_j. On the other hand if X is listed before D_j then the joint distribution vector of Y depends only on $\boldsymbol{d}$ through $\boldsymbol{d}^{(j-1)}$, $j = 2, \ldots, K$. The random variables represented in the ID can be either discrete or continuous.

Note that these conditions demand quite fierce homogeneity in the coding of the structure of a problem, so are quite often not met, at least in the variables as they are originally defined. They are nevertheless satisfied in a wide range of problems. If there exists an ordering such that these three conditions are met then the problem is called a *uniform decision problem*.

Suppose the DM faces a uniform decision problem and let the set of vertices of the ID $\mathcal{I}$ be denoted by $V(\mathcal{I}) = V'(\mathcal{I}) \cup V''(\mathcal{I}) \cup \{U\}$ – where $V'(\mathcal{I})$ denotes the set of all random variable vertices Y and $V''(\mathcal{I})$ denotes all the decision space vertices D – indexed as above in their compatible order. This means that a compatible order can be chosen so that the vertices $V(\mathcal{I})$ can be listed thus:

$$\begin{aligned} &Y_{01}, Y_{02}, \ldots, Y_{0k(0)}, D_1, Y_{11}(d_1), Y_{12}(d_1), \ldots, Y_{1k(1)}(d_1), D_2, \ldots \\ &D_K, Y_{K1}(\boldsymbol{d}), Y_{K2}(\boldsymbol{d}), \ldots, Y_{Kk(K)}(\boldsymbol{d}), U(\boldsymbol{d}, \boldsymbol{X}) \end{aligned} \tag{8.1}$$

where the utility vertex $U(\boldsymbol{d}, \boldsymbol{X})$ appears last on the list. Note here that $\{Y_{0i} : i = 1, 2 \ldots, k(0)\}$ denote the random variables listed before D_1 and

$$\{Y_{ji}(\boldsymbol{d}^{(j)}) : i = 1, 2, \ldots, k(K)\},$$

$j = 1, \ldots, K$ denote the random variables in the list that appear after D_j – but before D_{j+1}, $j = 1, 2, \ldots, K - 1$, where, from the compatibility condition, the distribution of $Y_{ji}(\boldsymbol{d}^{(j)})$ can depend only on $\boldsymbol{d} \in D_1 \times D_2 \times \cdots \times D_K$ through the value $(d_1, d_2, \ldots, d_j) \in D^{(j)}$. Let $\boldsymbol{X}_{Q(X)}, \boldsymbol{Y}_{Q(X)}, \boldsymbol{D}_{Q(X)}$ denote the subset of vertices in $V(\mathcal{I})$, $V'(\mathcal{I})$, $V''(\mathcal{I})$ connected to a vertex $X \in V(\mathcal{I})$ respectively. Call $\boldsymbol{X}_{Q(X)}$ the parent set of X. Similarly let $\boldsymbol{X}_{R(X)}, \boldsymbol{Y}_{R(X)}, \boldsymbol{D}_{R(X)}$ denote the subset of vertices in $V(\mathcal{I})$, $V'(\mathcal{I})$, $V''(\mathcal{I})$ listed before X but not connected to the vertex $X \in V(\mathcal{I})$ respectively.

We are now ready to draw the ID of the DM's problem starting from the compatible order (8.1). To complete the definition of the graph of an ID $\mathcal{I}$ we need to specify its set $E(\mathcal{I})$ of edges. To do this it is sufficient to define the parent set of $\boldsymbol{X}_{Q(X_j)}$ each of its vertices $X \in V(\mathcal{I})$: i.e. the set of edges directed into each of its vertices. Like the BN the parent set of each vertex is defined to be a subset of the vertices listed before it. Note that, as for the BN, this ensures that the graph of an ID is also a DAG. The graph of a valid influence diagram whose ordered vertices are given by (8.1) has parent sets that satisfy the following properties.

(1) The parent set $\boldsymbol{X}_{Q(U)} = \boldsymbol{Y}_{Q(U)} \cup \boldsymbol{D}_{Q(U)}$ of U is the set of random variables and arguments of decision spaces that appear explicitly as arguments of the utility function $U(\boldsymbol{d}, \boldsymbol{X})$.
(2) The parent set $\boldsymbol{X}_{Q(X_{ji}(\boldsymbol{d}^{(j)}))} = \boldsymbol{Y}_{Q(X_{ji}(\boldsymbol{d}^{(j)}))} \cup \boldsymbol{D}_{Q(X_{ji}(\boldsymbol{d}^{(j)}))}$ of $X_{ji}(\boldsymbol{d}^{(j)})$, $i = 1, 2, \ldots, k(K), j = 0, 1, \ldots, K$ must be such that

$$X_{ji}(\boldsymbol{d}^{(j)}) \amalg \boldsymbol{Y}_{R(X_{ji}(\boldsymbol{d}^{(j)}))} | \boldsymbol{Y}_{Q(X_{ji}(\boldsymbol{d}^{(j)}))}$$

for all $(d_1, d_2, \ldots, d_j) \in D_1 \times D_2 \times \cdots \times D_j$ where the distribution of $X_{ji}(\boldsymbol{d}^{(j)}) | \boldsymbol{Y}_{R(X_{ji}(\boldsymbol{d}^{(j)}))}$ is an explicit function of $\boldsymbol{d}^{(j)}$ only through the arguments in $\boldsymbol{D}_{Q(X_{ji}(\boldsymbol{d}^{(j)}))}$.
(3) For any decision vertex D_j, $j = 1, 2, \ldots, K$ the parent set $\boldsymbol{X}_{Q(D_j)} = \boldsymbol{Y}_{Q(D_j)} \cup \boldsymbol{D}_{Q(D_j)}$ where $\boldsymbol{Y}_{Q(D_j)}$ consists of all those random variables whose values are known when D_j is taken and $\boldsymbol{D}_{Q(D)}$ decision spaces $\{D_i : 1 \leq i < j\}$ whose decisions have already been committed to and remembered at the time a decision $d_j \in D_j$ is taken.

Definition 8.1. An influence diagram with DAG $\mathcal{I}$ is *valid* if applied to a uniform decision problem where the three conditions above hold.

Remark 8.2. Partly because of the cross-disciplinary nature of their development, the semantics of IDs is still not fully agreed, especially about how to define the parents of decision vertices. A useful way Shachter (1986) proposed to simplify the graph of an ID to the *reduced graph* $\mathcal{I}^R$ of the ID. In the reduced graph certain edges into decision vertices are removed from the DAG $\mathcal{I}$. Explicitly no edge from a parent of D_j is included if it has appeared as a parent of D_i where $i < j$. Since the no-forgetting condition demands all earlier decision spaces and random variables seen when the earlier decision was taken being connected to D_j in $\mathcal{I}$ this convention allows us to reconstruct $\mathcal{I}$ from $\mathcal{I}^R$. However, I will use this convention only when $\mathcal{I}$ would otherwise have many edges. Some authors (e.g. Dawid (2002a)) add no parents to decision vertices, which simplifies the diagram further but can faithfully represent only a much smaller set of decision problems. I have chosen the convention above because then the topology of the DM's ID relates directly to the BN of the analyst in a sense explained below.

Like an event tree it is usually best from a descriptive point of view to draw an ID using a listing of variables which is consistent with when the events associated with different random variables actually happen. The ID is then said to be *historic*. On the other hand to calculate an optimal policy it is often then best to transform this structure, either so that the variables are re-ordered so that they are listed in the order they are seen or alternatively to transform the structure into a generalised junction tree containing tables of utilities as well as tables of probabilities over its cliques.

Example 8.3. In the dispatch problem of Chapter 2 the product is first manufactured and is faulty ($Z = 0$) or not ($Z = 1$) the first decision $D_1 = \{d_0, d_1, d_2\}$ is whether to dispatch d_0, to perform one scan d_1, or to perform two scans d_2. She will then not observe, observe $+$ or observe $-$ as a result of the first or second scan. Let Y_i take the value 0 if the scan is not observed, 1 if it is positive and 2 if it is negative, $i = 1, 2$. The DM must then take the decision $D_2 = \{a_1, a_2\}$ where a_1 is to then dispatch or overhaul and dispatch a_2. The ID of this problem has vertices introduced in their causal order $(Z, D_1, Y_1, Y_2, D_2, U)$ where U represents the DM's utility function. As in a BN the parents of any vertex must be chosen from those listed before that vertex. There is no edge from Z into D_1 because whether or not the product is faulty is not known when a decision in D_1 is taken. What scans we see and the results Y_1 is a function of D_1 which determines whether or not we do a scan and Z – whether or not the product is faulty. Y_2 is a function that has an edge from D_1 and Z for the same reason as for Y_1. However there is no edge from Y_1 to Y_2 because whatever decision in $d \in D_1$ is taken $Y_2 \amalg Y_1 \mid Z, D_1$. Thus if the decision d is either d_0 or d_1 then no scan is taken so we know Y_2 takes the value 0 whatever the value of Z and Y_1. Furthermore if the decision is d_2 then the scans will be independent conditional on Z. Finally the utility function $U \amalg Y_1, Y_2 | Z, D_1, D_2$ can depend only on D_1 – through the cost of any scans – D_2 – through the cost of any overhaul – and Z – through whether or not the product is faulty.

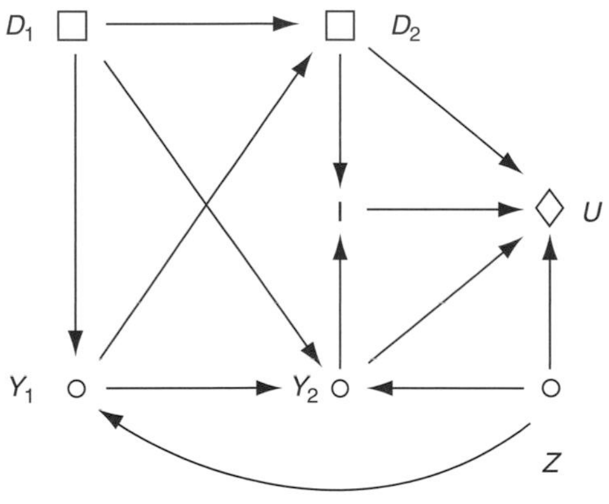

Here is another problem that can be represented by an ID.

Example 8.4. A company plans to market a new shampoo. It contains a new chemical and the higher the concentration the better the shine on the hair. However it is suspected that if its concentration is above a currently unknown acceptable threshold Y_{01} it might cause an allergenic reaction in some users. If the company launches the product when it is possibly allergenic then they will be forced to withdraw the product at a given time Y_{21} with a loss of profit Y_{22} and reputation for responsible sales Y_{23}. They therefore plan to first chose to trial the product on a subpopulation over a certain period of time D_1. The longer the trial is allowed to go on the better the quality of information Y_{11} about the potential allergenicity as a function of concentration but the less profit from the product, since competing products marketed by others can be expected to appear in the medium term. The second subdecision is the concentration D_2 to use in the final marketed product that can be chosen after seeing the results of the trial. It is easily checked that this is a uniform decision problem admitting the compatible order $(Y_{01}, D_1, Y_{11}, D_2, Y_{21}, Y_{22}, Y_{23}, U)$ and the DM may well choose to represent his problem with DAG $\mathcal{I}$ of the ID of this problem as given below.

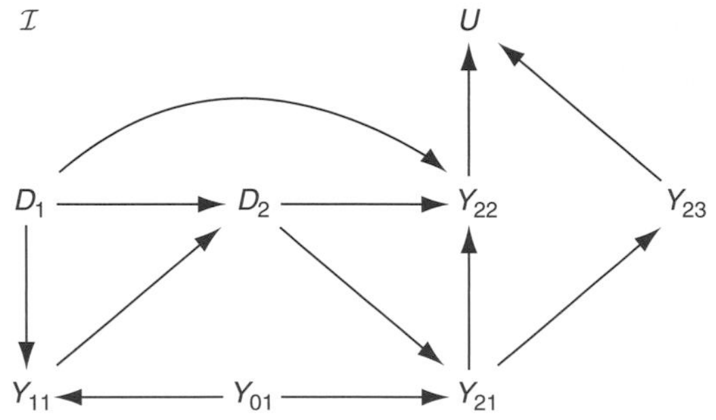

Notice first that all the random variables in this problem have a continuous rather than a discrete distribution but that this causes no problem. Second the attributes (Y_{22}, Y_{23}) of the utility function have been represented as the only parents of U: a property that can always be shown on a utility function with more than one attribute. Third as for the BN, the random variables can be substituted by random vectors and the integrity of the ID is maintained. For example instead of Y_{21} representing the time of withdrawal of the product it could include a second component measuring how effectively the withdrawal was enacted and the ID would still be valid.

Once an ID has been elicited, refined and simplified as above it is immediately apparent that like the BN its graph provides an elegant framework round which to store the elicited probabilities and (expected) utilities that are relevant to identifying the DM's Bayes decision rule.

8.1.2 The uses and limitations of influence diagrams

As the DAG of a BN provides a succinct and transparent representation of a probabilistic structure so the ID provides the same for a uniform decision problem. Even for simple decision problem like the one illustrated above the topology of the DAG is significantly simpler than the associated decision tree. Furthermore if the DM wanted to embellish the problem, for example assuming the results of the scan were graded – perhaps on a continuous numerical scale – instead of by binary division $(-, +)$ then the DAG of the ID would remain unchanged. This is also so if the DM wanted to introduce a gradation of the degree of faultiness of the product resulting in different costs into her description. So the DAG of this ID gives a very general picture of this type of problem. By suppressing local features associated with the sample and decision spaces it can be used as a vehicle through which the DM is able to represent the broad dependence structure over all the variables and decision spaces in the model. In particular the qualitative implications of certain types of embellishment can be examined with little qualitative cost. For example the possibility of using a third scanner in the last example would just involve adding a single new vertex and some connecting edges. Notice that representing these embellishments in the decision tree would either make the tree become much more bushy, or – when the embellishments involve the introduction of continuous variables – make the tree impossible to draw.

A second advantage the ID enjoys over the decision tree is that certain types of information can be expressed through the topology of an ID which cannot be expressed in the tree. For example in the illustration above the fact that conditional on the type of sampling we have that the scans are independent of each other conditional on the state of the machine is expressed by the fact that certain edges are missing. This property only appears implicitly in the topology of the tree once its edges are adorned with probabilities.

However these advantages have been gained at a cost. First the way in which the sampling decisions impinge on the information gathering associated with the relationship between D_1, Y_1, Y_2 and Z, neatly described by the decision tree, no longer appears in this representation. Also the conditional independence represented in this structure depends on the careful definition of Y_1 and Y_2. Note that we had to introduce a dummy level corresponding to "no reading taken" before the conditional independence could be exploited. It is something of an art for an analyst to transform a raw problem as a uniform decision problem in a way that expresses important dependence information believed by the DM. Furthermore such a transformation can tend to make the diagram more opaque and so less easy for the DM to own. Finally not all decision problems, even simple ones, have a useful ID representation.

So the ID often provides an additional representational resource rather than a single all-encompassing framework for a problem description. In particular for discrete problems it provides a complementary description to the decision tree, rather than subsuming it. On the other hand for large-scale uniform decision problems with or without continuous variables it can provide an excellent qualitative representation, with its own internal logic that can be queried. Moreover we will discuss later how, like the BN it can be used as a framework for embellishing a qualitative structure into full Bayesian description and its structure used to guide fast calculations of optimal policies.

8.1.3 The BN and ID and the movement to a requisite structure

One reason the ID is very useful is that its close link to the BN makes it amenable to formal qualitative deductions. Of course all BNs are special cases of an ID when we omit the decision variables and draw the BN of $\{Y_{01}, Y_{02}, \ldots, Y_{0k(0)}, U\}$. But more subtly an

ID of a uniform decision problem can always be thought of as simply a collection of BNs. To see this simply let D_1 denote the (enormous) set of all possible decision rules d_1 the DM might enact. A *normal form* ID can now be drawn which uses the variable ordering $\{D_1, Y_{11}(D_1), Y_{12}(D_1), \ldots, Y_{1k(1)}(D_1), U(D_1)\}$. Note that because there is a single component of the decision space and we can always specify a decision rule before we observe any information then the conditions for an ID are automatically satisfied. So in this sense all decision problems are uniform decision problems. It is just that its topology may not reflect any structure. For example if the decision rules concerned different orders of sampling then the BN associated with different rules may have completely different graphs, leading to all the chance nodes being connected to each other and the sorts of ways variables need to be defined can be very contrived: see the comments on the last section. So from a practical perspective this representation is often unhelpful.

However at least from a formal perspective, the use of an ID can be identified with using a set of BNs to encode the expectation of the last random variable in each BN. To discover of an optimal policy we then simply evaluate the expected utility of all the decision rules and thus to identify a Bayes decision rule $d_1^* \in D_1$ which will have the highest such value. Of course, like the normal form decision trees, the description of a problem using a normal form ID is often not a practical possibility. This representation is useful only if the number of feasible decision rules considered by the DM is fairly limited, or some clever screening preprocessing is performed to sieve away poor candidate rules and the underlying relationships between most of the uncertain quantities are invariant to the choice of decision rule. However in problems like the asset management example in Chapter 7 the company was content to consider only a relatively small number of candidate rules. Again for the evaluation of countermeasures to a nuclear accident decision rules are first sieved and most discarded as being either infeasible, illegal or clearly poor (Papamichail and French, 2005). So again in this context the number of remaining decision rules was relatively small although the normal form ID was not ideal to use in this context because different countermeasure strategies could have profoundly different impacts on exposure and hence led to very different dependences (French *et al.*, 1997).

In settings like the asset management problem of Chapter 7, by simply drawing a set of BNs corresponding to different potential decision rules – and often the qualitative structure of the BNs indexed by different decision rules are very similar – and embellishing each of these with their probabilities – many of which are shared by the different BNs – enables an optimal policy to be identified using evaluation methods discussed in Chapter 7. So in such cases much of the computation technology originally designed for inference problems can be used almost directly to calculate optimal policies.

There is a third link between the ID of a uniform decision problem and the BN. This particular relationship can be used to check the model to determine whether or not it is requisite, and preprocess the structure to determine whether there are any redundant qualitative features that need not be included. These in turn can greatly simplify the structure used to describe and calculate a Bayes decision rule, and in deriving algebraic forms that an optimal decision rule should take when the decision space D is finite. Thus suppose that an analyst wants to build his own BN of the DM's problem. He adopts all the DM's subjective conditional distributions over the random variables $\boldsymbol{Y}(\boldsymbol{d})$ as his own. He does not know the decisions the DM might choose $\boldsymbol{d} = (d_1, d_2, \ldots, d_k)$, taking values in a product space $D_1 \times D_2 \times \cdots \times D_K$ but only on what each component decision depends. He therefore places a mass function over all $d_j \in D_j$ given any configuration of parents $\boldsymbol{X}_{Q(D_j)}$. It is easy to check now that if the DM's problem has a valid ID with DAG $\mathcal{I}$ then $\mathcal{I}$ must also be the DAG of a valid BN for the analyst: see Smith (1989b) for further discussion. This can often help the DM simplify her problem, identifying which features might be redundant for the purposes of finding an optimal decision rule: see the next section.

8.1.4 Some ways to simplify an influence diagram

The last link between the BN and the ID can sometimes help to simplify and hence clarify the structural information embedded in the topology of that elicited ID. Recall that by the no-forgetting condition in an ID with DAG $\mathcal{I}$ the set of ancestors $\boldsymbol{X}_{Q(D_j)}$ are equal to the parents $\boldsymbol{X}_{Q(D_j)}$ of a decision space D_j represented by a vertex in that DAG.

Theorem 8.5 (Sufficiency theorem). *Let the subset $\boldsymbol{X}_{C(D)}$ – called the* core *– of the parents' $X_{Q(D)}$ of a decision space D be such that U is d-separated from $\boldsymbol{X}_{B(D)} \triangleq \boldsymbol{X}_{Q(D)} \backslash \boldsymbol{X}_{C(D)}$ given $\boldsymbol{X}_{C(D)} \cup \{D\}$. Then the DM has a Bayes decision $d^* = (d_1^*(\boldsymbol{X}_{Q(D_1)}), d_2^*(\boldsymbol{X}_{Q(D_2)}), \ldots, d_K^*(\boldsymbol{X}_{Q(D_K)}))$ d_j^* that can be written as $\boldsymbol{d}^* = (d_1^*(\boldsymbol{X}_{C(D_1)}), d_2^*(\boldsymbol{X}_{C(D_2)}), \ldots, d_K^*(\boldsymbol{X}_{C(D_K)}))$ where each component of this rule is a function of its arguments only.*

Proof Since the DAG $\mathcal{I}$ is a valid BN to the analyst he can deduce that for each D_j

$$U \amalg \boldsymbol{X}_{B(D)} | \boldsymbol{X}_{C(D_j)} \cup \{D_j\}.$$

It follows that he can deduce that the distribution of the utility score U – and so in particular its expectation – may depend on how the DM chooses the decision D_j as a function of $\boldsymbol{X}_{C(D_j)}$ but that how it is otherwise chosen as a function of $\boldsymbol{X}_{B(D)}$ will not impact on this expectation. It follows in particular that there must be a Bayes decision $\boldsymbol{d}^*$for the DM where the component d_j^* is chosen to be a function only of $\boldsymbol{X}_{C(D_j)}$, this being true for all $j = 1, 2, \ldots, K$. The result follows. □

One role of an analyst is to help the DM be as *parsimonious* as possible: i.e. to help discover a smaller class of decision rules than the ones she has identified as feasible which always contain an optimal expected utility maximising decision rule. The ID is then simply used to deduce another uniform description of the problem which has the DAG of its ID. Of these we find the equivalent ID whose DAG has the smallest subset of the original vertices and of these one with the smallest number of edges which will still contain a Bayes decision rule in the original problem as one of its Bayes decision rules.

Using the original ID to simplify the original statement of the problem will be useful for the DM in three ways. First it gives a simpler framework to describe how good decisions need to relate to available information. Second suppose the DM is uneasy about the deductions the analyst makes about how certain decision rules can be safely ignored. The deductions the analyst makes will be logically valid. So by explaining why she is uneasy the DM is provoked into discussing new potential measurement variables or dependences she accidentally omitted to mention in her earlier description of the problem in hand. This creative dialogue using the qualitative logic behind the Bayesian model's description should now be very familiar to the reader. The dialogue continues to modify the ID until it is requisite. Third it can help to simplify the full decision analysis when this involves the elicitation of probabilities and calculation of explicit Bayes decision rules.

Thus if certain measurement variables are available but do not need to be known in order for the DM to be able to act optimally then their probabilities do not need to be elicited and relevant data does not need to be collected to support these judgements. Furthermore by reducing the dimension of the space describing the problem we will usually simplify the algorithm used to calculate the optimal policy. Consider the following example.

Example 8.6. Suppose an ID is elicited from a DM whose reduced DAG $\mathcal{I}^R$ is given below.

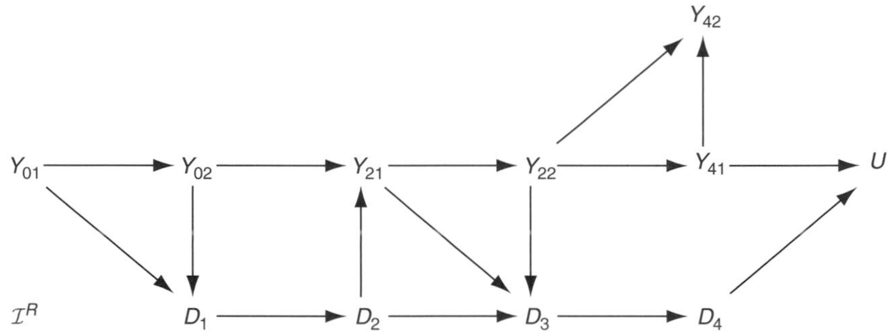

The no-forgetting condition implies that the parent sets $\{X_{Q(D_i)} : i = 1, 2, 3, 4\}$ of the four decision spaces in the full DAG $\mathcal{I}$ of this ID are

$$X_{Q(D_1)} = \{Y_{01}, Y_{02}\},$$
$$X_{Q(D_2)} = \{Y_{01}, Y_{02}, D_1\},$$
$$X_{Q(D_3)} = \{Y_{01}, Y_{02}, D_1, D_2, Y_{21}, Y_{22}\},$$
$$X_{Q(D_4)} = \{Y_{01}, Y_{02}, D_1, D_2, Y_{21}, Y_{22}, D_3\}.$$

Now suppose the analyst helps the DM to be parsimonious. It follows from the sufficiency theorem that her decisions will only depend on parents, i.e. features of the past, that form the core of the parent set. Here these cores are

$$\begin{aligned} \boldsymbol{X}_{Q(D_1)} &= \{Y_{02}\}, \\ \boldsymbol{X}_{Q(D_2)} &= \{Y_{02}\}, \\ \boldsymbol{X}_{Q(D_3)} &= \{Y_{22}\}, \\ \boldsymbol{X}_{Q(D_4)} &= \{Y_{22}\}. \end{aligned}$$

We can therefore conclude that decisions d_1 and d_2 of a parsimonious DM will depend only on Y_{02} and d_3 and d_4 only on Y_{22}. Now note that the expected utility $\overline{U}(\boldsymbol{d})$ given the parsimonious DM chooses a decision $\boldsymbol{d} \in D_1 \times D_2 \times D_3 \times D_4$ is given by

$$\overline{U}(\boldsymbol{d}) = \sum_{y_{22}} \overline{U}(d_4(y_{22}), y_{22}) p_{22}(y_{22}|d_1(y_{02}), d_2(y_{02}))$$

where

$$\overline{U}(d_4(y_{22}), y_{22}) = \sum U(d_4(y_{22}), y_{41}) p_{41}(y_{41}|y_{22}).$$

Now if the analyst uses the d-separation theorem she will see that

$$Y_{22} \amalg D_1 | Y_{02}, D_2$$

whence we can write

$$p_{22}(y_{22}|d_1(y_{02}), d_2(y_{02})) = \sum_{y_{22}} p_{22}(y_{22}|y_{02}, d_2(y_{02})) p_{02}(y_{02})$$

and where, because the analyst inherits the DM's elicited probabilities

$$p_{02}(y_{02}) = \sum_{y_{01}} p_{02}(y_{02}|y_{01}) p_{01}(y_{01}).$$

Thus a parsimonious DM will choose a $\boldsymbol{d} \in D_1 \times D_2 \times D_3 \times D_4$ maximising the function

$$\overline{U}(\boldsymbol{d}) = \sum_{y_{22}} \overline{U}(d_4(y_{22}), y_{22}) \sum_{y_{22}} p_{22}(y_{22}|y_{02}, d_2(y_{02})) p_{02}(y_{02}).$$

This expectation is an explicit function only of $U, D_4, Y_{22}, D_2, Y_{02}$ where the analyst's DAG $\mathcal{I}$ of a valid BN (not in reduced form) is given by

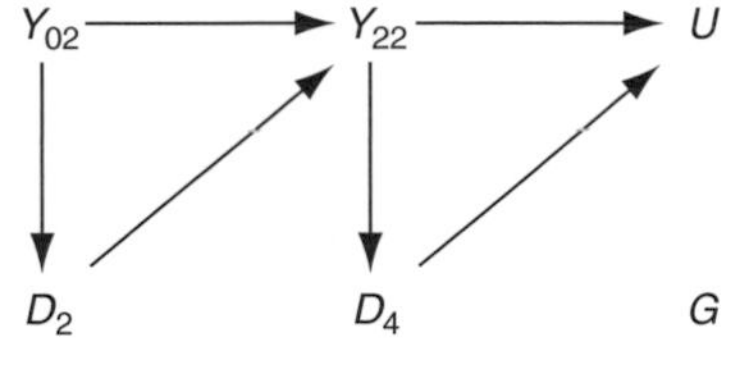

The analyst can therefore conclude that the DM needs to specify only $p_{02}(y_{02})$, her $p_{22}(y_{22}|y_{02}, d_2)$ for each $d_2 \in D_2$ and her utility $\overline{U}(d_4, y_{22})$ in order to calculate a Bayes rule optimal for her. If she has correctly represented her problem then the analyst can assure her that the decisions she chooses in $D_1 \times D_3$ will be quite irrelevant to her performance in terms of her expected utility score. In fact there is an unexpected bonus. The missing edge between D_2 and D_4 in the analyst's BN tells him that in fact the DM can safely forget the decision D_2 once she has taken it. This is because its only impact on the efficacy of the final consequence will be on its beneficial effect on Y_{22} whose value will be known when the DM chooses her decision from D_4. This deduction can be computationally helpful because it reduces by an order of magnitude the number of decision rules that need to be evaluated before an optimal policy can be identified. Note in this example that $\mathcal{I}$ is a valid reduced ID of the DM's original problem but with all irrelevant detail pared away.

It is easily checked that following an algorithm like the one above to find the analyst's BN and then adding some additional edges into decision vertices if necessary so that the no-forgetting condition is satisfied gives a valid graph of an ID of the DM's problem. This valid description omits any unnecessary detail (Smith, 1989a). In our example to obtain such an ID $\mathcal{I}$ we need simply to add the edges $Y_{02} \to D_4$ and $D_2 \to D_4$ to $\mathcal{G}$. Note here that only the arguments of U need to be known – as reflected through its attributes – not its functional form. Similarly only the dependence structure is needed as inputs not explicit values of probabilities. We have noted earlier in the book how carefully these quantitative inputs need to be elicited in a given decision analysis. It is therefore extremely irritating to discover only half way through the quantitative part of the elicitation process that the structure of the problem is not requisite and some new variables are actually relevant or some of the originally stated ones not relevant, because then much time can be wasted. Personally I have found the sort of logical simplifying processes extremely useful for making sure that the DM finds the qualitative implications of her original specification requisite early in a decision analysis. In the example above she can be asked to reflect on whether she thinks it plausible that the decisions d_1 and d_3 are irrelevant to the success of her chosen decision rule and that the only variables that need to be taken account of directly in any decision rule are Y_{02} and Y_{22}. If this is not so then the analyst can conclude that she has omitted some potential dependences between the variables she has presented or omitted other variables pertinent to her analysis. In the latter case the analyst can then help her search for these other variables.

Of course it may not be possible to make a useful simplification of this type. Furthermore it can be helpful to retain technically redundant random variables in the model description in order to improve the elicitation because these present a credence decomposition that naturally reflects the DM's thinking – see Chapter 4 – or to make the supporting narrative more compelling – see Chapters 2 and 7. But even when this is the case the analyst should reflect on whether the redundant variables have their distributions coded into the evaluation software. Certainly in the example above it would not be appropriate to keep the variable Y_{42}.

We note that several other techniques are now available that help this preprocesses. In particular powerful methods that exploit assumptions of utility separation and value independent attributes can both simplify the representation of an ID and vastly speed up the calculation of the identification of an expected utility maximising policy: see e.g. Jensen and Nielsen (2007); Shachter (1986); Tatman and Shachter (1990); Vomelelova and Jensen (2002).

8.1.5 Using an influence diagram as a computational framework

A Bayes decision rule can now be calculated using an analogous algorithm to the rollback algorithm described above for decision trees: see Chapter 2. This algorithm again uses the fact that a Bayes decision rule exists which assumes that from all situations arrived at in a decision sequence, future decisions can be assumed to be expected utility maximising.

Like the decision tree an influence diagram can also be used as computational aid as well as a representational tool. The first application of these techniques were to use it as a framework for a rollback or backwards induction algorithm exactly analogous to that used in the decision tree. However the ID introducing variables in the order they happened, as recommended for representational purposes is sometimes not appropriate for this purpose unless the ID happens to be in extensive form. So sometimes the original ID needs to be transformed to one in extensive form just like the tree. This is always possible but some conditional independence structure may be lost in the course of the transformation.

Definition 8.7. An *extensive form influence diagram* of a uniform decision problem has its random variable vertices listed so that Y_{ik} is always a parent of D_j, $k = 1, 2, \ldots, k(i)$, $0 \leq i \leq j \leq K$ so that it is more like the dynamic programming tree (see above).

Sometimes an ID is already in extensive form as in the ID of the simplified graph in the last example with $\mathcal{G}$ as its reduced graph. However in the dispatch example this is not so because the compatible listing in that problem was $(Z, D_1, Y_1, Y_2, D_2, U)$ and Z – labelling whether or not the product is faulty is not known before the scanning or dispatch and so is not a parent of either D_1 or D_2. It can be put into extensive form however by introducing variables into the problem in the order $(D_1, Y_1, Y_2, D_2, Z, U)$. It is set as an exercise to check that the d-separation theorem now allows the analyst to conclude that the figure below is the DAG of a valid ID. Note that an additional edge has appeared because the conditional independence $Y_2 \amalg Y_1 | D_1, Z$ cannot be represented when introducing the variables of the problem in this order.

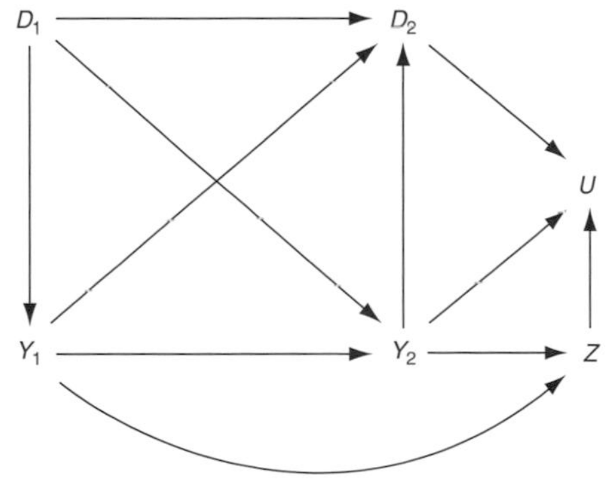

(8.2)

The evaluation of decisions using an extensive form ID follows the same rollback procedure as for a tree. This is illustrated on the example of the last section below.

Example 8.8. Suppose d_2 and d_4 can each take three values, and both y_{02} and y_{22} two and three values respectively with $P(Y_{02} = 1) = 0.7$ and $p_{22}(y_{22}|y_{02}, d_2)$ and expected utility $\overline{U}(d_4, y_{22})$ given below

$\overline{U}(d_4, y_{22})$		y_{22}		
		1	2	3
d_4	1	**0.6**	0.1	0
	2	0.2	**0.8**	0
	3	0.2	0.1	**1**

$p_{22}(y_{22}|y_{02}, d_2)$		(y_{02}, d_2)					
		$(0, 1)$	$(0, 2)$	$(0, 3)$	$(1, 1)$	$(1, 2)$	$(1, 3)$
y_{22}	1	0.3	0.4	0.5	0.8	0.1	0
	2	0.3	0.4	0.5	0.1	0.8	0
	3	0.4	0.2	0	0.1	0.1	1

Clearly here the Bayes decision for d_4 is to set this equal to y_{22} with vector of expected utilities 0.6, 0.8 and 1 associated with $\{Y_{22} = 1\}$, $\{Y_{22} = 2\}$, $\{Y_{22} = 3\}$ respectively. The expected utilities $\overline{U}^*(d_2|y_{02})$ associated with the two possible values of y_{02} are therefore given by

$\overline{U}^*(d_2|y_{02})$		y_{02}	
		0	1
d_2	1	**0.82**	0.66
	2	0.76	0.8
	3	0.70	**1**

where the expected utilities $\overline{U}^*(d_2|y_{02})$ are obtain by averaging over the expected utilities associated with the DM's optimal choice for d_2 using the appropriate condition probability for y_{22} given in the table above. For example

$$\overline{U}^*(1|0) = 0.3 \times 0.6 + 0.3 \times 0.8 + 0.4 \times 1 = 0.82.$$

We can read from this matrix that the DM maximises her expected utility by choosing $d_2 = 1$ – and subsequently $d_4 = y_{22}$ – if $Y_{02} = 0$ with an expected utility value of 0.82 and $d_2 = 3$ – and subsequently $d_4 = y_{22}$ – if $Y_{02} = 1$ with an associated utility value of 1. The expected utility for following this decision rule is

$$\overline{U}(\boldsymbol{d}^*) = 0.82 \times 0.3 + 1 \times 0.7 = 0.946.$$

Notice from the previous analysis of the larger description of this problem that any choice for the pair (d_1, d_3) with $\left(d_2^*, d_4^*\right)$ chosen as above will give this expected utility.

Many methods are now available which use the topology of the graph of an ID to guide the fast calculation of optimal policies. Because these are a more fitting topic for an AI or Optimality text they lie outside the scope of this book. Some good texts with reviews of various techniques can be found in Cowell *et al.* (1999); Jensen and Nielsen (2007); Kjaerulff and Madsen (2008). It has subsequently been discovered that sometimes it is possible to evaluate a Bayes decision rule more efficiently using methods not based on rollback. However for the practitioner it suffices to know that software algorithms are available to automatically calculate a Bayes decision rule for any given historic influence diagram.

8.2 Controlled causation

8.2.1 Introduction

Intrinsic to most successful decision analyses is the elicitation of a framework which captures how the DM believes one feature of a problem influences or impacts upon another. We have argued in earlier chapters that once such a framework is in place it is possible, with care, to help the DM embellish her model into a full probabilistic model, enabling her to contemplate and then evaluate how to act. Such a process could be seen as determining a causal framework to help her structure her reasoning. The structure of Western languages and of English in particular – with its framework of subject–verb–object encourages us to think of someone or something (subject) acting (verb) so that something (object) changes. It is therefore inevitable – at least within Western culture – that many supporting explanations of a probabilistic model use causal reasoning.

Now until recently statisticians in particular have been extremely wary of using causal reasoning in their explanations. The reason for this has already been eluded to in this book. We demonstrated in Chapter 7 that it is often simply not logically possible to discriminate between various causal hypotheses simply on the basis of statistical properties – like apparent conditional independences – exhibited in data sets no matter how large and well collected those data sets are. So in this sense, whilst a statistical analysis can be informative about different causal models it cannot support it on its own. Of course this has not stopped the uninformed using statistical analysis erroneously in this way. Indeed many spurious statistical analyses are used specifically to come to policy decisions. But this has made principled statisticians all the more concerned to treat causal analyses with suspicion. Causal modelling needs a narrative of what is happening.

However, in contrast to the discipline of statistical inference, a Bayesian decision analysis *is* often based on a narrative of what has happened and will happen, why this happened and a description of how this has come about. For example the event tree discussed in Chapter 2 does exactly this. When statistical information is drawn into the study of probabilities in event trees or decision trees this information is thus being accommodated into a causal narrative.

In this book we have argued that a Bayesian decision analyst needs to build a *subjective* probabilistic model for each decision rule or policy she might follow. Any available data she can collect helps to *refine* her structural hypotheses and the embellishment of her original probabilistic judgements. It also enables her to perform *diagnostic checks* as to whether what information she collects appears to contradict one or more of her hypotheses and to *explain* to an auditor why she believes what she does. But an intelligent and well-informed DM will appreciate that some of the preconceptions she brings into her analysis will not be refutable or supported by the data she collects. Therefore, for the Bayesian analyst, different causal conjectures are just more concepts the DM brings to the table to explain to others and herself what is happening and might happen in the world of her problem. They may or may not be refutable by data already collected just as other parts of the DM's model. Obviously the part of causal reasoning most important to a DM – and the one we focus on here – will concern conjectures about the effect any decision she makes will have on her expected utility. Her chosen actions will "cause" certain chains of events to be excited which will either be to her benefit or to her loss.

8.2.2 Representing causality using trees

In this book I will draw mainly on the work of Shafer (1996). This uses the event tree as the framework for expressing causal hypotheses. In Chapter 2 we have already encountered such trees, where the "historic tree" describes the history of a unit as it passes from one situation to the next. In one example the historic tree described what might happen when an individual is exposed to a particular allergenic compound. Another example described the possible unfolding of events in a crime and yet another described what might happen as a product passed from manufacture to quality control to dispatch.

The focus of interest in these *controlled causal* models is in the links that can be legitimately made between the probabilistic description of the history of a unit as it develops in an uncontrolled way and the probabilistic description of the history of a unit when it is subject to control or manipulation. This phenomenon was briefly introduced in Chapter 2, where in the allergenic compound example the tree was extended to include the possibility that a unit of skin was subject to a manipulation in which there was a wounding which resulted in the breaching of the epidermis. In the crime case another interesting manipulation to investigate might be the effect on the outcome of the case of the police planting evidence that the brother could not have gone to the house. In the allergenic compound example, given that the historic event tree was acceptable to all parties as a description it would be plausible to assume that under a manipulation – here the wounding – the only change to the tree would be local and affect only the immediate development of the unit. This led to the conjecture that the two situations v_4 and v_6 were parallel: i.e. after the dermis has been exposed to the compound, either naturally or through a wound, whether or not this caused inflammation had the same probability. Furthermore we argued that the two subsequent situations v_5 and v_7 of whether the unit would be sensitised would also be parallel.

Obviously it is often necessary in a decision analysis to forecast the probable effects of a control you might impose. Sometimes a domain allows a historic tree to be the framework from which these forecasts can plausibly be made. Here a tree is drawn of the system when it is not subjected to control – i.e. in *idle*. If the DM asserts this to be a *causal tree*, $\mathcal{T}$, on the set of situations $C(\mathcal{T}) \subseteq S(\mathcal{T})$ then she asserts that she believes that when the situation $v \in C(\mathcal{T})$ is subjected to a control which changes the probability distribution on its floret the result of that manipulation will be felt by the controlled floret and no other. So in particular, the distribution of the remaining florets will be the same as they would be were the process in idle. This leads to the following definition, see Thwaites *et al.* (2010).

Definition 8.9. The DM's historic event tree $\mathcal{T}$ is said to be *causal on the set of situations* $C(\mathcal{T}) \subseteq S(\mathcal{T})$ into $C'(\mathcal{T}) \subseteq V(\mathcal{T})$ under control D if she asserts that for any subset $V_c \subseteq C(\mathcal{T})$, on reaching a situation $v \in V_c \subseteq C(\mathcal{T})$ a unit is forced under D with probability one to pass along an edge $(v, v'(v))$ where $v'(v) \in C'(\mathcal{T})$ is some child of v in $\mathcal{T}$ then all situations $v \notin V_c$ in the controlled system will be parallel to their corresponding situations in the idle system. Call $\mathcal{T}$ *causal* if $C(\mathcal{T}) = S(\mathcal{T})$ and $C'(\mathcal{T}) = V(\mathcal{T})$.

In the allergenic example it could be plausibly argued that the event tree $\mathcal{T}$ given in Chapter 2 (page 56) was causal on $\{v_4\}$ when $C'(\mathcal{T}) = V(\mathcal{T})$. We have already described the effects of wounding. If we artificially ensure that the compound is not applied to someone who has a natural wound – and thus force the unit to pass along the edge to v_1, then the DM could also plausibly suggest that all developments subsequent from v_1 on would also be parallel to their idle counterparts. Notice that this type of hypothesis could also be applied to conjectures about what might have happened to a unit whose history was already known. For example it could be made to make counterfactual inferences (Imbens and Rubin, 1997; Rubin, 1978): in the example above about what might have happened to someone who became sensitised when their scalp was wounded had they been prevented from being treated because they exhibited this wound.

The implications of causal trees are often fierce. On the other hand, because they are so strong, if they can be made then very strong deductions are also possible in the light of them. This therefore makes them useful. The weakest case is when $C(\mathcal{T}) = \{v\}$ and $C'(\mathcal{T}) = \{v'\}$ where v' is a child of v. Here we consider the single control which is to control the units so that if it arrives in situation v it is forced to pass along to v'. The causal assumption then implies:

(1) The manipulation will have no effect on the past leading to that unit, nor on the DM's beliefs about what would happen to units on whom the controls are not active.
(2) The effect of the control is as intended and applied perfectly.
(3) The effect on the unit after the control under any possible unfolding of history is as it would be for someone naturally finding themselves at v'.

The first point demands that the effect of the control is local to each unit. In particular the development of a unit when there is control in place that the unit *might* have received the treatment had their history not precluded them, will be the same as the development if that control had not been available to anyone. Note for example that such a hypothesis would not be tenable if the population's behaviour could be affected by envy or peer pressure. The second point is an important practical issue for a decision analysis. For example if the control is to evacuate a population after a nuclear accident then we know that there will not be full compliance. Some people will hide or refuse to go despite the threat. Furthermore whether or not someone decides whether or not to comply may well be linked to their vulnerability to the threat, invalidating the third point. The third can be even more fragile an assumption. For example it assumes that there is no placebo effect associated with a treatment regime. All these issues demonstrate that the DM should make any causal assertions with great care.

On the other hand they can be compelling. Return to the elicitation of the probability that two fragments of matching glass are found on the coat of a given guilty suspect seen standing 10 m from a pane smashed by a brick thrown at a particular velocity. We have argued that this could be plausibly considered as a further replicate of a set of experimental units where counts of the number of glass fragments on coats of units standing 10 m from a smashed pane were taken. The argument in this example is particularly compelling because there is a general belief within our cultural world view that the *mechanism* governing the probability distribution of smashed glass falling on a coat from a given distance and at a given velocity is something that does not depend on anything other than certain physical attributes of the incidence – here the distance from the pane and the velocity of the brick – and certainly not the identity of the unit or the type of coat. In particular we believe that whether the suspect was forced to stand 10 m from a pane when the brick was thrown or if he happened to stand at this distance when the brick was thrown would not change the distribution of fragments of glass on his coat. Note that it is not always possible to identify the probability of what might happen when a decision is made to force a unit into a given situation with one where that unit naturally arrives at that situation.

As a general principle it is extremely important in most scenarios to describe *how* the control D is going to be enacted for the DM's causal hypothesis for it to be possible for the assertion behind a causal model be satisfactorily appraised. Second even historic event trees are often not causal. This is because various parts of the process are not explicitly represented in the model. Here it would be difficult for the DM to credibly assert that her event tree is causal on the situation v_2 in the allergenic example whose emanating edges label whether or not the dermis becomes inflamed. If this inflammation is artificially induced then this clearly does not necessarily mean the unit will become sensitised. There may well be a hidden cause present, affecting the sensitiser which means that sensitisation in the controlled unit will not arise from the same process as in the idle system: see later. It often happens that a statistical model of an idle system does not need to be specific about generating processes that are essential to the development of plausible arguments about how the system will behave under certain controls (Smith and Figueroa-Quiroz, 2007). Note that if trees are not historic then all these problems multiply: a criticism of some types of counterfactual models as raised by Dawid (2000). Other examples can be found in Riccomagno and Smith (2005) and for good introductions to this underlying debate see Pearl (2000); Riccomagno and Smith (2005).

8.3 DAGs and causality

8.3.1 Some definitions

We saw in the last chapter that the BN was a very useful and compact framework within which to express collections of irrelevance statements. There is a causal analogue of these frameworks and it is possible to give a more detailed discussion of these relatively well-developed systems below.

The graph of a causal Bayesian network (CBN) Glymour and Cooper (1999); Pearl (1995, 2000); Spirtes *et al.* (1993) defined below provides a framework within which collections of local causal hypotheses can be pasted together into a causal composite. The CBN assumes that the uncontrolled system respects a Bayesian network (BN) with a given DAG but makes a further assumption that the data is generated by mechanisms that are intrinsically robust to different frameworks in which they are found in a sense explained below. Recently links between the original causal hypotheses, counterfactual machinery and CBNs have been vigourously debated; see e.g. Pearl (2003); Robins (1986, 1997); Shafer (1996); Spirtes *et al.* (1993). In this section we will briefly review some of the recent advances in this area beginning with causal notions based on DAGs of BNs.

Several authors, notably Glymour and Cooper (1999); Pearl (1995, 2000); Robins (1986); Spirtes *et al.* (1993) and Robins (1997) have discussed the relationship between causality and BNs, developing useful methods for defining causality and investigating its identifiability under various sampling schemes. Such methods usually begin with the hypothesis that a process which may be biological, medical, social or economic, can be randomly sampled from a population in a *idle* state: i.e. one which is not subject to any control by the DM but is simply observed. This observed system is further believed by the DM to respect a collection of conditional independence assumptions consistent with a BN. Furthermore the joint probability mass function that will embellish this BN is assumed to be strictly positive. Recall from Chapter 6 that a BN asserts that for all values $(x_1, \ldots, x_n)$ of the vector $(X_1, \ldots, X_n)$ of n random variables its joint mass function $p(x_1, \ldots, x_n)$ respects the factorisations into conditional densities

$$p(\boldsymbol{x}) = p_1(x_1) \prod_{i=2}^{n} p_i(x_i|\boldsymbol{x}_{Q_i})$$

where $p_i(x_i|\boldsymbol{x}_{R_i}, \boldsymbol{x}_{Q_i}) = p_i(x_i|\boldsymbol{x}_{Q_i})$, embodying the set of $n-1$ conditional independence statements $X_i \amalg \boldsymbol{X}_{R_i}|\boldsymbol{X}_{Q_i}$ given below where we allow both the sets of indices of parents Q_i of X_i and R_i to be empty. Under the assumption that $p(\boldsymbol{x}) > 0$ for all combinations of levels of $\boldsymbol{x}$ this equivalently asserts that the joint mass function respects the set of $n-1$ statements (7.5) we obtain from a new simplified factorisation formula of (7.4).

Let $p(x_1, x_2, \ldots, x_{j-1}, x_{j+1}, \ldots, x_n||x_j)$ denote the probability distribution of the remaining random variables in the system given when the DM controls the variable X_j so that it takes a value x_j. Note from the discussion in the last section that it is necessary for the DM to stipulate exactly how she envisages doing this manipulation. When the DM believes that

$p(\boldsymbol{x}) > 0$ for all combinations of levels of $\boldsymbol{x}$, Pearl and others now suggest that the joint mass function $p(x_1, x_2, \ldots, x_{j-1}, x_{j+1}, \ldots, x_n || x_j)$ of the other variables in the BN given that the DM controls the variable X_j so that it takes a value x_j should be given by the equations

$$p(x_2, x_3, \ldots, x_n || x_1) = \prod_{i=2}^{n} p_i(x_i | \boldsymbol{x}_{Q_i}) = \frac{p(\boldsymbol{x})}{p_1(x_1)}$$

and for $j = 2, 3, \ldots, n$

$$p(x_1, x_2, \ldots, x_{j-1}, x_{j+1}, \ldots, x_n || x_j) = p_1(x_1) \prod_{i=2, i \neq j}^{n} p_i(x_i | \boldsymbol{x}_{Q_i}) = \frac{p(\boldsymbol{x})}{p_j(x_j | \boldsymbol{x}_{Q_j})}. \quad (8.3)$$

But where does this formula come from and when is it appropriate or at least plausible for the DM to use it?

There is certainly one scenario where it would be the automatic choice. Suppose that the DM has in front of her a network of n simulators. The first simulator gives an output x_1 whilst the ith simulator $i = 2, 3, \ldots, n$ gives an output x_i on the basis of inputs $\boldsymbol{x}_{Q_i}$ with the probability mass function $p_i(x_i | \boldsymbol{x}_{Q_i})$. We can represent this network by a directed graph $\mathcal{G}$ whose vertices are $(X_1, \ldots, X_n)$ and which has an edge from X_j to X_i if and only if x_j is a component of $\boldsymbol{x}_{Q_i}$. Assume in this network that $\mathcal{G}$ is acyclic and so a DAG. Moreover assume that each simulator has a randomisation device which in particular ensures that for any possible set of inputs $\boldsymbol{x}_{Q_i}$ it can return any output x_i with positive probability. Then it can be clearly seen that the output variables will all respect the conditional independences expressed in the BN with DAG $\mathcal{G}$ if all the randomisations of the mechanisms are mutually independent of each other. For by definition the DM knows that X_i can only depend on $(\boldsymbol{x}_{R_i}, \boldsymbol{x}_{Q_i})$ through the value of the input vector $\boldsymbol{x}_{Q_i}$, implying that $X_i \amalg \boldsymbol{X}_{R_i} | \boldsymbol{X}_{Q_i}$, $i = 2, 3, \ldots, n$. A collection of draws from this network of stochastic simulators will therefore constitute a random draw from a population whose distribution has mass function $p(\boldsymbol{x})$ given above.

However if this really is the world view of the DM then she can also make deductions about what would happen were she to control the value of the jth simulator to force it to take the value $\widehat{x}_j$, $j = 1, 2, \ldots, n$. At least in this setting there is an obvious meaning to give to this control. The DM simply discards the jth simulator and replaces it with one that returns the value $X_j = \widehat{x}_j$. The value $\widehat{x}_j$ is subsequently used for the value used in all subsequent simulators i where $x_j \in \boldsymbol{x}_{Q_i}$ i.e. x_j is an input variable for the ith simulator.

Definition 8.10. A discrete *causal Bayesian network* (CBN) with DAG $\mathcal{G}$ on a set of measurement variables $\{X_1, \ldots, X_n\}$ is valid if two conditions are met. First when not subject to control the BN with DAG $\mathcal{G}$ is valid. Second the DM has a control available which can be identified with her setting X_j to each one of its possible values x_j for each $j = 1, 2, \ldots, n$, when she believes that the effects of this control will change the joint mass function across her remaining variables $X_1, X_2, \ldots, X_{j-1}, X_{j+1}, \ldots, X_n$ so that it satisfies equation (8.3).

This is the direct analogue of the causal event tree described in the last section. Clearly if a DM believes that her measurements described by a CBN with DAG $\mathcal{G}$ is valid she is making a much stronger assertion than if she were simply to believe that the BN with DAG $\mathcal{G}$ is valid. It is interesting to note that this hypothesis can be expressed graphically. Suppose a DM believes that a valid BN with DAG $\mathcal{G} = (V(\mathcal{G}), E(\mathcal{G}))$ where $V(\mathcal{G})$ is the set of vertices of $\mathcal{G}$ and $E(\mathcal{G})$ its set of edges is also a CBN. It follows that she also believes that a valid DAG for the problem when any measurement X_j is manipulated to any one of its values $\widehat{x}_j$ is $\mathcal{G}_j$ where the vertex set $V(\mathcal{G}_j) = V(\mathcal{G})$ and the edge set $E(\mathcal{G}_j) = E(\mathcal{G}) \backslash E_j(\mathcal{G})$ where $E_j(\mathcal{G})$ are the set of edges in $E(\mathcal{G})$ into X_j. All the conditional distributions on parent configurations will be shared, but with the manipulated variable X_j having its value held to $\widehat{x}_j$. This relationship is depicted below on a DAG $\mathcal{G}$ where X_2 is the manipulated variable.

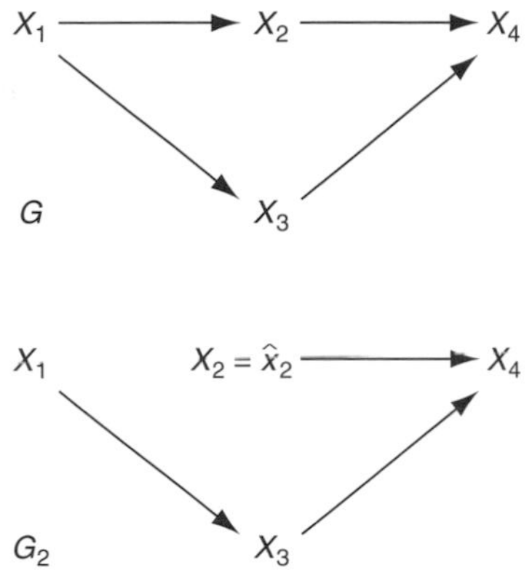

Although the CBN hypothesis is a very strong one to hold we have illustrated above however that there is at least one scenario when its use is compelling. This is one where she is focusing not on a real process but on a simulated process. These are an important subset of decision problems in their own right. For example it is often only possible to begin to examine the effects of various strategies in a complex scenario by simulating these with the use of computer models with stochastic devices that mirror the uncertainty of their outcomes. The network of simulators used to examine the potential consequences of a given type of nuclear disaster at a given plant when certain countermeasures are planned is one example of this type of computer simulation. The examination of the global effect of removing all emissions of a certain type are another.

Of course any rational person would assume that the results of such experiments when projected into any real scenario were extremely speculative. They are clearly subject to possible gross distortions because of features and relationships missed in the computer models of the system, because of such problems as only a partial understanding of the underlying science, lack of observations and the necessary approximations needed to produce a framework where calculations are feasible in real time. But making these calculations is still helpful to the DM who is trying to come to a better understanding of the general implications of applying certain policies. Indeed in scenarios like those illustrated above it might be the *only* possible way of groping towards such a partial understanding. At least if the DM states that her predictions are based on believing a CBN then the assumptions behind her analyses are explicit and later brought into question by an auditor.

It is worth making two technical remarks at this point.

(1) One implication of the argument above is that the distribution of any variable X_i is unaffected by the manipulation of a variable X_j whenever X_i is not downstream of X_j in $\mathcal{G}$ (or is not a descendant of X_j in $\mathcal{G}$). This is because using the analogy above, the value of X_i can be simulated before resolving the value of X_j. This is entirely consistent with the idea that if something happens before a manipulation then that manipulation cannot cause it to change: a rather obvious assertion which is present in most formulations of causality. However the second assumption that causal manipulation of X_k acts like conditioning for its children is more contentious and substantive: see below. In particular it implies that the fact of manipulation does not in itself affect a change in the mechanism of the system. Thus for example if the manipulation is a medical treatment then we preclude the possibility of a placebo effect: i.e. that the health of a patient improves simply because he is being shown attention by being treated. These issues link directly to the three points of the last section.

(2) The concept of a cause as described above probably should be more properly called an "average cause". For example it does not necessarily address the causal effect of a treatment on a *particular individual*. Rather it gives a probabilistic prediction of the effect of a *treatment regime* on a given population of patients all of whose responses respect the CBN $\mathcal{G}$. This distinguishes it from the potential outcome approach to causality which makes predictions about *what would have happened* to a distinguished patient in the sample: see Glymour and Cooper (1999); Rubin (1978).

Note here that the fibre evidence BN of the last chapter could plausibly be argued to be causal.

8.3.2 The total cause

On the basis of the formula (8.3) of a CBN we can immediately write down the marginal mass function of a variable X_k given the variable X_j is manipulated to a value $\widetilde{x}_j$ as the sum of all joint values consistent with each event $\{X_k = x_k\}$ in (8.3). Thus, by henceforth assuming that all primitive probabilities are nonzero

$$p(x_k||x_j) = \sum_{i \neq j,k} \prod_{i=1, i \neq j}^{n} p_i(x_i|\boldsymbol{x}_{Q_i}) = \sum_{i \neq j,k} \frac{p(\boldsymbol{x})}{p_j(x_j|\boldsymbol{x}_{Q_j})}$$

where $p(x_k||x_j)$ is called *the total cause of* x_j on x_k. So provided our simulator network analogy holds for the DM we can immediately predict how an individual variable of interest will respond to manipulation. Note that the probability mass function of X_k after *manipulating* X_j so that it takes a value $\widetilde{x}_j$ is not always the same as the probability mass function of X_k after *observing* X_j takes a value $\widetilde{x}_j$ as illustrated in the following example. Henceforth, as for event trees we call the joint distribution of variables of an unmanipulated/uncontrolled network of simulators the *idle* distribution.

Example 8.11. On various occasions a DM runs a teaching agency. She is uncertain whether or not an individual has assaulted children in the past ($\{X_1 = 1\}$ or $\{X_1 = 0\}$), whether he has moved house in the last five years ($\{X_2 = 1\}$ or $\{X_2 = 0\}$), is on an offender register or not ($\{X_3 = 1\}$ or $\{X_3 = 0\}$) and whether or not he fails child protection clearance ($\{X_4 = 1\}$ or $\{X_4 = 0\}$). The DM believes that someone tempted to assault is more likely to have moved house recently and more likely to be on the offender register. However if she were to learn whether or not a person has assaulted children in the past then learning about whether he had recently moved house or not would not change her probability about whether or not he was on an offender register. Whether or not he obtains clearance will depend only on whether or not he is on an offender register. Following this reasoning the DAG of her BN is

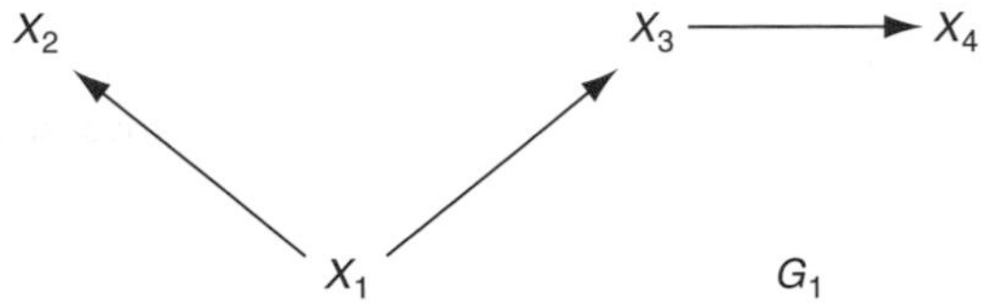

Note that this is equivalent in the idle system to

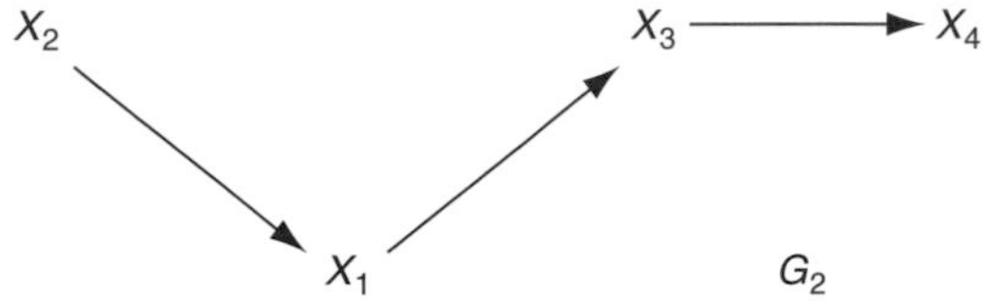

because both DAGs have the same pattern. But in this setting the DM might also reason that $\mathcal{G}_1$ was causal. This would imply, for example, that if the man was prevented from assault or enticed to assault – where these two controls are precisely defined – then her conditional distributions on $X_2|X_1 = x_1$ and $X_3|X_1 = x_1, x_1 = 0, 1$ would be the same as they were in the idle system, where no such prevention or enticement took place. Another consequence of the DM's assertion is that if he was removed from the offender register or unfairly added to it, then $X_4|X_3 = x_3, x_3 = 0, 1$ would be the same as in the idle system as would be the case if he was given one in error $X_3|X_2 = 1$. In this manipulative sense the direction of the arrows in $\mathcal{G}_1$ appear consistent with a generating set of simulators and their associated causal order. On the other hand $\mathcal{G}_2$ is not consistent with a plausible causal order. For example it suggests that forcing someone to move house would tend to cause the man to assault children. Note therefore that CBNs with different DAGs say different things even when the DAG of their BNs have the same pattern.

Finally suppose for the sake of argument

$$P(X_1 = 1) = 0.5,$$
$$P(X_2 = 1|X_1 = 0) = 0.2,$$

$$P(X_2 = 1|X_1 = 1) = 0.6,$$

$$P(X_3 = 1|X_1 = 0) = 0.2,$$

$$P(X_3 = 1|X_1 = 1) = 0.6,$$

$$P(X_4 = 1|X_3 = 0) = 0.9,$$

$$P(X_4 = 1|X_3 = 1) = 0.1.$$

Then for example by Bayes rule, the probability of someone being guilty of assault given they are observed to be on the register

$$P(X_1 = 1|X_3 = 1) = \frac{0.6 \times 0.5}{0.6 \times 0.5 + 0.2 \times 0.5} = \frac{3}{4}$$

whilst the total cause of someone being guilty after he has been (wrongly) added to the registers is

$$P(X_1 = 1||X_3 = 1) = P(X_1 = 1) = \frac{1}{2}.$$

Thus we have demonstrated that we cannot automatically identify the effects of a control with the effects of conditioning on an observation. These are often governed by different formulae.

8.3.3 Identifying a cause in a CBN

Let us assume that a DM believes the simulator analogy accurately describes the relationships between the variables she actually observes in her process and that $\mathcal{G}$ is therefore the DAG of a valid CBN on variables $\boldsymbol{X} = \{X_1, X_2, \ldots, X_n\}$. Suppose that the DM has available to her an enormous population of unmanipulated units that respect $\mathcal{G}$ from which she can pay to take a random sample. For the purposes of the decision analysis she is only interested in the potential effect on X_k – the attribute of her utility function of manipulating the variable X_j to different values it can take. Which of the variables in $\{X_1, X_2, \ldots, X_n\}$ in addition to X_k and X_j will she need to observe on my units before she can accurately calculate the total cause of X_j on X_k?

Obviously with a large sample I will be able to accurately estimate $\{\pi_i(x_i|\boldsymbol{q}_i) : 1 \leq i \leq n\}$ using, for example the product Dirichlet priors as described earlier, or simply by maximum likelihood if the sample is big enough. But could I get away with observing fewer measurements on each unit? Henceforth the variables we plan to see will be called *manifest* and denoted by $\{X_j, X_k, \boldsymbol{X}_Z\}$ where the set $Z \subseteq \{X_1, X_2, \ldots, X_n\} \backslash \{X_j, X_k\}$: possibly the empty set. Variables in $\{X_1, X_2, \ldots, X_n\}$ but not in $Z \cup \{X_j, X_k\}$ are called *hidden*. When vertices of a CBN are manifest they are coloured black and when missing coloured white.

The topology of the CBN $\mathcal{G}$ can tell us when a set of manifest allows us to calculate a total cause from the idle system: i.e. when $\{X_j, X_k, \boldsymbol{Z}\}$ *identify* the total cause of X_j on

X_k. For example suppose we observe the values of all parents in the parent set $\mathbf{Q}_j$ of the manipulated variable X_j so that Z contains $\mathbf{Q}_j$. Then

$$\begin{aligned} p(x_k||x_j) &= \sum_{i \neq j,k} \sum_{i \neq j,k} \frac{p(\boldsymbol{x})}{p_j(x_j|\boldsymbol{x}_{Q_j})} \\ &= \sum_{i:X_i \in Z} \left(\frac{\sum_{i \neq j,k,X_i \notin Z} p(x_1, x_2, \ldots, x_n)}{p_j(x_j|\boldsymbol{x}_{Q_j})} \right) \\ &= \sum_{i:X_i \in Z} \left(\frac{p(\boldsymbol{x}_Z, x_j, x_k)}{p_j(x_j|\boldsymbol{x}_{Q_j})} \right) \end{aligned}$$

is the marginal mass function of the variables whose values we sample, and $p_j(x_j|\boldsymbol{x}_{Q_j})$ are the conditional probabilities of $x_j|\boldsymbol{x}_{Q_j}$ in the idle system which can be calculated from $p(\boldsymbol{x}_Z, x_j, x_k)$. It is clear therefore that we will be able to identify $p(x_k||x_j)$ from a large enough sample of $\{X_j, X_k, \boldsymbol{X}_{Q_j}\}$.

However it is not always possible to identify the total cause from manifest variables. Thus consider the complete graph $\mathcal{G}$ given below where X_1 is a hidden cause and we observe the $\{X_2, X_3\}$ margin whose cell probabilities are

$$p(x_2, x_3) = \sum_{x_1} p(x_1, x_2, x_3) = \sum_{x_1} \pi(x_3|x_1, x_2)\pi(x_2|x_1)\pi(x_1).$$

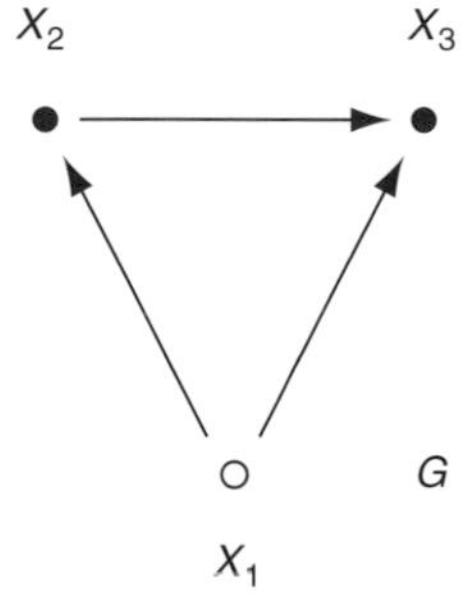

(8.4)

We are interested in the values of the total cause of X_j on X_k. Our formula tells us that this is given by

$$p(x_3, x_1||x_2) = \frac{1}{p(x_2|x_1)} p(x_1, x_2, x_3)$$

whence

$$p(x_3||x_2) = \sum_{x_1} \frac{1}{p(x_2|x_1)} p(x_1, x_2, x_3)$$

if $p(x_2|x_1) > 0, x_1 \in \mathbb{X}_1, x_2 \in \mathbb{X}_2$. Clearly without further conditions, we cannot calculate $p(x_3||x_2)$ from our observed margin $\sum_{x_1} p(x_1, x_2, x_3)$ since, because we do not observe X_1, the weights

$$\{p(x_2|x_1)^{-1} : x_1 \in \mathbb{X}_1, x_2 \in \mathbb{X}_2\}$$

in this sum are completely arbitrary positive numbers satisfying

$$\sum_{x_1} p(x_2|x_1)p(x_1) = p(x_2)$$

for unknown probabilities $\{p(x_1) : x_1 \in \mathbb{X}_1\}$.

What graphical conditions can we impose on $p(x_3||x_2)$ so that it can be written as an explicit function of $\{p(x_2, x_3) : x_2 \in \mathbb{X}_2, x_3 \in \mathbb{X}_3\}$? It is shown in Dawid (2002a) that for this to be so either the edge (X_1, X_2) in $\mathcal{G}$ needs to be missing or the edge (X_1, X_2) needs to be missing.

Thus if $X_2 \amalg X_1$ then

$$p(x_2, x_3) = \sum_{x_1} p(x_3|x_1, x_2)p(x_2)p(x_1) = p(x_2)\sum_{x_1} p(x_3|x_1, x_2)p(x_1) = p(x_2)p(x_3||x_2)$$

so

$$p(x_3||x_2) = \frac{p(x_2, x_3)}{p(x_2)} = p(x_3|x_2)$$

which can be estimated directly from the idle system. Similarly if $X_3 \amalg X_1|X_2$ then

$$p(x_2, x_3) = \sum_{x_1} p(x_3|x_2)p(x_2|x_1)p(x_1) = p(x_3|x_2)\sum_{x_1} p(x_2|x_1)p(x_1) = p(x_2)p(x_3||x_2)$$

so

$$p(x_3||x_2) = \frac{p(x_2, x_3)}{p(x_2)} = p(x_3|x_2)$$

and we can again identify the total cause.

But when there is an unobserved cause of both X_2 and X_3 we cannot in general expect to learn the effect on X_3 of manipulating X_2 just by observing a sample of the $\{X_1, X_2\}$ margin. The best we can hope to do is to obtain certain bounds for this effect: see Pearl (2000). The strength of the effect on X_3 of manipulating X_2 will always be confounded with the effect on X_3 of the unobserved cause X_1 which the data averages over.

8.3.4 Pearl's Backdoor theorem

Suppose the DM observes the cause X_j, the effect X_k together with the vector $\boldsymbol{X}_Z$ whose components lie in $\{X_1, X_2, \ldots, X_n\}\backslash\{X_j, X_k\}$. Suppose the DAG $\mathcal{G}$ is valid for the DM's

CBN. Then there is a sufficient condition for determining whether the total cause of X_j on X_k is identified.

Henceforth suppose X_j is an ancestor of X_k (otherwise trivially $p(x_k||x_j) = p(x_k)$ which is clearly observed). In most examples as applied to a decision analysis X_k can be thought of as an attribute of the DM's utility function whilst X_j is a variable whose value can be set in a decision rule. For example if the DM were searching for efficacious treatments, X_j might be a quantity of medicine given and X_k the speed of recovery of the patient. Here $X_j, X_k, \boldsymbol{X}_Z$ are the variables whose readings she has available to her.

Definition 8.12. A subvector $\boldsymbol{X}_Z$ satisfies the *backdoor criterion* relative to (X_j, X_k) if

(1) no element in Z is a descendant of X_j in $\mathcal{G}$;
(2) the variables in $Z \cup \{X_j\}$ separate X_k from the set of parents Q_j of X_j in the undirected version of the moralised graph of the ancestral set of $X_j, X_k, \boldsymbol{X}_{Q_j} \boldsymbol{X}_Z$ in $\mathcal{G}$.

Thus these conditions demand that none of the elements of $\boldsymbol{X}_Z$ could be affected by the manipulation of X_j and that $\boldsymbol{X}_Z$ separates the hidden causes on X_j from the hidden causes on X_k in the sense above.

Note that for example the set of parents $\boldsymbol{X}_{Q_j}$ of X_j satisfy the backdoor criterion. More technically choose a compatible ordering of variables in $\mathcal{G}$ that lists all variables in $\boldsymbol{X}_Z$ before X_j – condition 1 ensures we can do this. Then we can read directly from the BN that

$$X_j \amalg \boldsymbol{X}_Z | \boldsymbol{X}_{Q_j}$$

where $\boldsymbol{X}_{Q_j}$ denotes the subvector of $\boldsymbol{X}$ that are parents of X_j Also condition 2 implies, directly by the d-separation theorem, that

$$X_k \amalg X_{Q_j} \mid X_Z, X_j$$

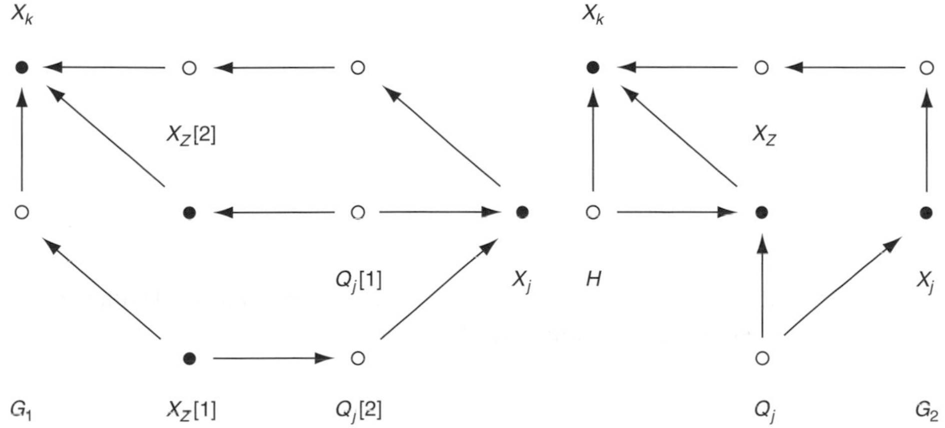

Two illustrations of the second condition are given above. Note that $\boldsymbol{X}_Z = \{X_Z[1], X_Z[2]\}$ satisfies the backdoor criterion in $\mathcal{G}_1$. However $\boldsymbol{X}_Z = X_Z$ satisfies the first condition but not the second: since the path $(\boldsymbol{X}_{Q_j}, H, X_k)$ appearing in the undirected version of the moralised ancestral graph of $\{X_j, X_k, \boldsymbol{X}_{Q_j}, X_Z\}$ in $\mathcal{G}_2$ does not separate $\boldsymbol{X}_{Q_j}$ from X_k. So $\boldsymbol{X}_Z$ is not a backdoor set for (X_j, X_k) in $\mathcal{G}_2$.

We now have the following theorem.

Theorem 8.13 (Backdoor theorem). *If Z satisfies the backdoor criterion relative to $\{X_j, X_k\}$ then the total cause of X_j on X_k is identifiable from $\{\boldsymbol{X}_Z, X_j, X_k\}$ and given by the formula*

$$p(x_k||x_j) = \sum_{\boldsymbol{x}_Z} p(x_k|x_j, \boldsymbol{x}_Z)p(\boldsymbol{x}_Z).$$

Proof See e.g. Pearl (2000). □

8.3.5 A more surprising result

The form of the backdoor theorem is somewhat expected and the formula above has been used for a very long time. However even in simple circumstances, the formulae for the probability of an attribute if subjected to a control may not be simple. For example consider the BN below.

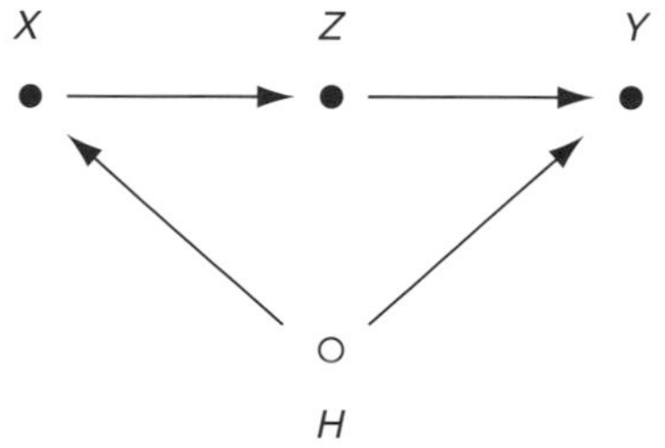

Is the total cause $p(y||x)$ identified from observing the (X, Y, Z) margin? The answer is yes but the related formula is weird!

First note that

$$p(x, y, z, h) = p(h)p(x|h)p(z|x)p(y|z, h)$$

and

$$p(y||x) = \sum_{u,z} p(h)p(z|x)p(y|z, h) \tag{8.5}$$

$$= \sum_{z} p(z|x)\left(\sum_{u} p(h)p(y|z, h)\right). \tag{8.6}$$

Next note that

$$\sum_{u} p(h)p(y|z,h) = \sum_{x'}\sum_{u} p(h|x')p(x')p(y|z,h)$$
$$= \sum_{x'}\sum_{u} p(y|z,h)p(h|x')p(x').$$

But from the DAG above $Z \amalg H|X$ and $Y \amalg X|Z,H$ so in particular

$$p(h|x') = p(h|x',z),$$
$$p(y|h,z) = p(y|h,z,x').$$

It follows that

$$\sum_{h} p(h)p(y|z,h) = \sum_{x}\left(\sum_{h} p(y|h,z,x')p(h|z,x')\right)p(x')$$
$$= \sum_{x} p(y|z,x')p(x')$$

by the rule of total probability. Substituting into (8.5) gives us

$$p(y||x) = \sum_{z}\left(p(z|x)\sum_{x'} p(y|x',z)p(x')\right).$$

Note that all the terms in this equation can be obtained from the mass function $p(x,y,z)$ on the X,Y,Z margin.

In his book (Pearl, 2000) Pearl gives various ways of using the DAG of a valid CBN to guide appropriate substitutions. Graphical conditions which are necessary and sufficient for most forms of causal manipulation of CBN to be identified have now been derived: for example see Tian (2008). In Riccomagno and Smith (2004) we demonstrate how ideas from computer algebra can be used to guide substitutions of this type when additional non-graphical information is also available.

The principle demonstrated above is that even if the DM can only see a subset of the variables in her hypothesised causal system it is still quite often possible for her to make logical deductions about what might happen on the basis of certain controls she might make. These deductions can be quite subtle but can be derived. Furthermore an explanation of the relevant formula comes automatically from the mathematical steps used in each of the steps of its derivation. These mathematical steps can be explained qualitatively simply in terms of irrelevance relationships and the shared parallel situations that the DM posits exist between the idle and controlled systems.

8.4 Time series models*

8.4.1 Time series entirely structured by conditional independence

The final class of structural models that needs to be considered – albeit briefly – are time series models. It is not unusual for variables in a decision problems to have an underlying conditional independence structure which is repeated over and over again in time. The distributions of the different explanatory variables can drift or even change systematically in time but their underlying conditional independence relationships remain immutable. The statistical model needed to describe such a process is the time series. Time series models were some of the first to be studied with a view to their conditional independence structure.

The Markov chain, on observations $Y_t, t = 1, 2, \ldots, n$ is defined by the set of conditional independence statements $Y_t \amalg (Y_1, \ldots, Y_{t-2}) \,|Y_{t-1}, t = 3, 4, \ldots, n$. By definition this has DAG

$$Y_1 \to Y_2 \to \cdots Y_{t-1} \to Y_t \to Y_{t+1} \to \cdots \to Y_n$$

It is now easy to prove some basic properties of this chain. Thus directly from the pattern of this graph we see that one which reverses all the directions of the edges – i.e. reverses time – is equivalent to this model. Furthermore d-separation gives us directly that for $t = 3, 4, \ldots, n-2$ writing $\widehat{\boldsymbol{Y}}_t = (Y_1, \ldots, Y_{t-2}, Y_{t+2}, \ldots, Y_n)$

$$Y_t \amalg \widehat{\boldsymbol{Y}}_t | (Y_{t-1}, Y_{t+1}).$$

These two results are well known in the theory of stochastic processes but notice that they can be proved graphically with virtually no effort using the results above.

More Bayesian and more general is the dynamic linear model (see e.g. Durbin and Koopman (2001); West and Harrison (1997)). This is defined on a sequence of vectors $\boldsymbol{Y}_t, t = 1, 2, \ldots, n$. Let $\boldsymbol{Y}^T \triangleq (\boldsymbol{Y}_1, \boldsymbol{Y}_2, \ldots, \boldsymbol{Y}_T)$ and $\boldsymbol{\theta}^T \triangleq (\boldsymbol{\theta}_1, \boldsymbol{\theta}_2, \ldots, \boldsymbol{\theta}_T)$. Here the time series is described through introducing a vector of explanatory states $\boldsymbol{\theta}_t, t = 1, 2, \ldots, n$ at each time and proposing that the following set of conditional independences hold.

$$\boldsymbol{Y}_t \amalg \boldsymbol{Y}^{t-1}, \boldsymbol{\theta}^{t-1} | \boldsymbol{\theta}_t, t = 1, 2, \ldots, n$$

$$\boldsymbol{\theta}_t \amalg \boldsymbol{Y}^{t-1}, \boldsymbol{\theta}^{t-2} | \boldsymbol{\theta}_{t-1}, t = 2, 3, \ldots, n$$

These are exactly the conditional independences of a valid BN whose DAG is

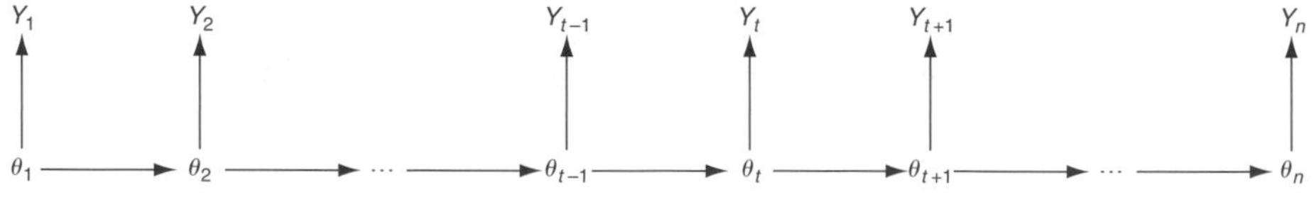

Again various useful non-distributional results can be proved simply by evoking the d-separation theorem that are much more obscure when proved in other ways.

More structure can be represented by separating the components of the states and observation vector so that we have a hierarchical structure on each slice of time. Thus the simple two time slice dynamic model 2TDM (Dean and Kanazawa, 1988; Koeller and Lerner, 1999) with bivariate states $\boldsymbol{\theta}_t = (\theta_{1,t}, \theta_{2,t})$, $t = 1, 2, \ldots, n$ and univariate observations given below with the property

$$Y_t \amalg \boldsymbol{Y}^{t-1}, \boldsymbol{\theta}^{t-1}, \theta_{2,t} | \theta_{1,t}, \; t = 1, 2, \ldots, n,$$

$$\theta_{1,t} \amalg \boldsymbol{Y}^{t-1}, \boldsymbol{\theta}^{t-2} | \boldsymbol{\theta}_{t-1}, \; t = 2, 3, \ldots, n,$$

$$\theta_{2,t} \amalg \boldsymbol{Y}^{t-1}, \boldsymbol{\theta}^{t-2}, \theta_{1,t-1} | \theta_{2,t-1}, \; t = 2, 3, \ldots, n$$

is depicted below, or more conventionally simply by the third, fourth and fifth column of this BN.

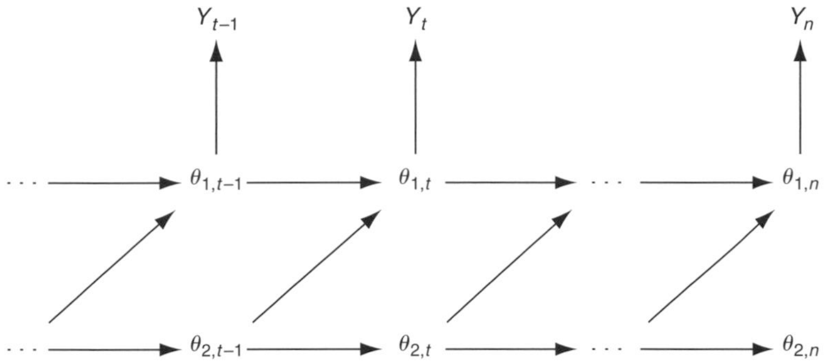

There are many examples of the use of somewhat more complicated two-time slice models than the one illustrated above (for formal definitions and further illustrations of this model class, which is a subclass of the multivariate DLM see e.g. Koeller and Lerner (1999)).

Another useful graphical time series model is the multiregression dynamic model (MDM), Queen and Smith (1992, 1993). This has a different conditional independence structure where the probabilities or regression parameters are linked directly to parents of each component in the time series, and it is these vectors of parameters which are each given their own independent dynamic process. Like the 2TBN the parent configuration of a variable at time t depends only on relationships between variables with a time index t and variables at time $t - 1$ and furthermore these dependences are the same for all time. This means like the the 2TBN the dependences in the full BN can be summarised by a graph depicting dependences on adjacent time frames.

An MDM must have the special property that the parent set of an observation variable can include only its own current state vector and components of the observation vector at the same time listed before it. Furthermore the only parent of a component state vector is the previous state vector. An example of such an MDM for a simple problem where the underlying process has three observations at each time is given below.

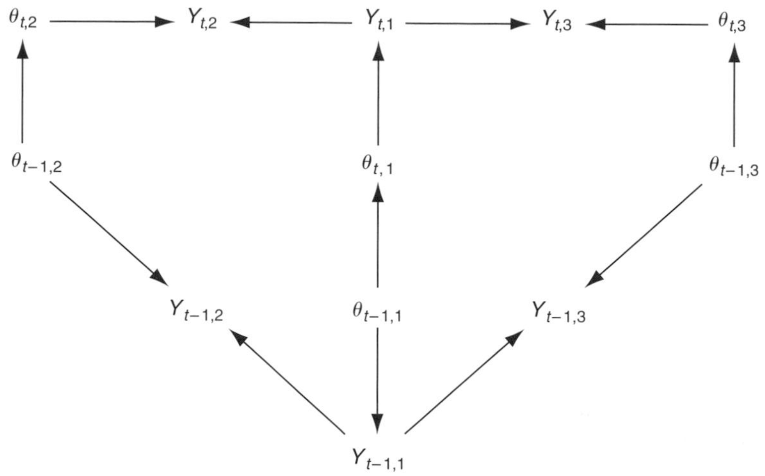

The reasons that this is a useful class is first that all the state vectors remain independent after sampling. Second the close relationship between this model and the linear model means that there are often convenient conjugate families enabling the time series states and recursion on moments of the observables also to be given explicitly.

There is another way to address the challenges to all but the most homogeneous dynamically evolving processes that their structure can quickly become undermined so that they become rather opaque to the user and computationally challenging. Here *power steady models* Ibrahim and Chen (2000); Peterka (1981); Smith (1979a, 1990, 1992) use the idea of increasing the temperature of a joint density at each time step by demanding that

$$p(\boldsymbol{\theta}_t|\mathbf{y}^{t-1}) \propto \{p(\boldsymbol{\theta}_{t-1}|\mathbf{y}^{t-1})\}^k \tag{8.7}$$

for some $0 < k \leq 1$ where the proportionality constant is uniquely determined because $\int p(\boldsymbol{\theta}_t|\mathbf{y}^{t-1})d\boldsymbol{\theta}_t = 1$. Note that although the evolution (8.7) is not fully specified over the whole state space it is enough to give prequential distributions – see Chapter 4. In particular it specifies unambiguously the joint mass function or density $p(\mathbf{y}^T)$ of the observations $\boldsymbol{Y}^T$ up to any future time T, since

$$p(\mathbf{y}^T) = \prod_{t=2}^{T} p(\mathbf{y}_t|\mathbf{y}^{t-1})p(\mathbf{y}_1)$$

where

$$p(\mathbf{y}_1) = \int p(\mathbf{y}_1|\boldsymbol{\theta}_1)p(\boldsymbol{\theta}_1)d\boldsymbol{\theta}_1,$$

$$p(\mathbf{y}_t|\mathbf{y}^{t-1}) = \int p(\mathbf{y}_t|\boldsymbol{\theta}_t)p(\boldsymbol{\theta}_t|\mathbf{y}^{t-1})d\boldsymbol{\theta}_t.$$

It follows that those parts of the joint distributions of the states not determined in the evolution are not identifiable. Their posterior density will be the same as their prior density whatever is observed, and will not influence the future at least of the series observed.

These evolutions have a number of advantages. They give the same steady state recurrences as the conventional Steady DLM West and Harrison (1997) but usually also admit conjugate evolutions when the analogous non-time varying problem has this property, with statistics replaced by familiar exponentially weighted moving average analogues which makes them accessible and interpretable to many users. Logical constraints will be preserved with time as well as all independences and many conditional independences existing from the previous time slice. Finally the evolution can be characterised as an invariant decision based on or using linear shrinkages of either Kullback–Leibler distances or local De Robertis distances discussed below.

These methods have been employed in a number of applications Queen *et al.* (1994); Smith (1983), and Rigat and Smith (2009). Their drawbacks are that they sacrifice the property that states separate the present from the past observations. This means that in particular the evolution is dependent on the time interval on which the process is defined. In odd situations it can produce an improper posterior. It is also rather inflexible: only allowing modelling of processes where the passage of time only induces more uncertainty, with current judgements otherwise being retained.

The variety of the small subset of models discussed above illustrates that there are a myriad of different ways of doing this, each method suited to a different genre of application: e.g. Dahlhaus and Eichler (2003); Didelez (2008); Eichler (2006, 2007); Nodelman *et al.* (2002) and Caines *et al.* (2002), for some of the many other classes. It is therefore often necessary to customise dynamic models to the problem at hand. Needless to say problems of stochastic control where decision rules chosen *feedback* into the dynamics of the process itself are even more complex and resist a simple taxonomy: see reviews of this general area and discussion of their implementation in two different settings in for example Bensoussen (1992); Bersekas (1987); Whittle (1990).

However the basic message that can be taken from this brief review is that many useful classes of time series models can be seen as particular dynamic generalisations of certain graphical models: often BNs.

8.5 Summary

There is an alternative representation of a discrete decision problem to the decision tree which is the influence diagram. This ID can also represent problems involving continuous or mixtures of continuous and discrete variables. A wide number of problems can be efficiently represented in such a form which also provides a fast framework for calculation. We introduced causal trees and causal BNs that allow any hypotheses represented in an event tree or BN to be extended so that they address the DM's beliefs about what might happen were the system controlled in some way. In particular we demonstrated how, under some strong assumptions, formulae could be derived. Time series models could also be built using the same types of framework although these tend to be less generic and often need to be customised to the problem at hand.

So we have seen in Chapters 2 and 7 and in this chapter that a wide variety of decision problems can be decomposed, using an appropriate graphical framework, into manageable components where probabilities can be elicited and utility maximising strategies can be identified quickly and feasibly. Furthermore the frameworks also help the DM to explain why the policy she proposes is a good one.

However ideally the DM would like to give statistical support for the many probability assignments she has given in her analysis and in this way be able to defend as strongly as possible any decision she commits to. It remains in the last chapter to demonstrate that the credence decompositions of the event tree, causal event tree, BN and CBN all provide excellent frameworks for doing this. Furthermore in many circumstances this activity does not significantly complicate the analysis.

8.6 Exercises

8.1 The Coventry Road Race has many competitors. It is suspected that some of the competitors will take a performance-enhancing drug before the race. These drugs are not without side effects and can cause nausea which in turn will have an adverse effect on a racer's performance. Let $\{X_1 = 1\}$ if a competitor takes a performance-enhancing drug and $\{X_1 = 0\}$ otherwise and let $\{X_2 = 1\}$ whenever that person feels nausea with $\{X_2 = 0\}$ otherwise. Let X_3 be a runner's recorded time, measured in seconds. The racers are subject to a random drugs test. Let $\{X_4 = 1\}$ if a competitor is chosen for a test and the test show positive and let $\{X_4 = 0\}$ if that competitor is not chosen or he is tested but shows negative. A match official tells you she believes that the DAG below is a causal Bayesian network (CBN).

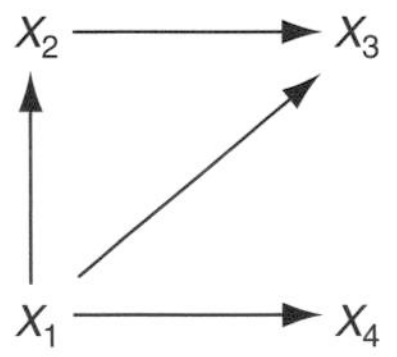

Carefully describe what this causal Bayesian network asserts when all joint probabilities of different configurations of (x_1, x_2, x_3, x_4) are strictly positive. In this context do you see any ambiguity in the definition of any of the potential manipulations? Write down the formula for the total cause of X_2 on X_3. Show that the total cause is not in general the same as $p(x_3|x_2)$. Give two different additional conditional independence statements that allow us to identify the total cause of X_2 on X_3 with $p(x_3|x_2)$, in each case demonstrating why this is so.

8.2 Students sometimes cheat in exams and bring into the exam hall illicit material. Such a strategy is not without risks however, sometimes causing a level of stress of being found out far outweighing the potential benefits of using the introduced material or being caught. Let $\{X_1 = 1\}$ if a student decides to illicitly take a copy of material she believes might be useful into an exam and $\{X_1 = 0\}$ if she brings in no such material. Let $\{X_2 = 1\}$ whenever her chosen course of action introduces debilitating stress reactions and $\{X_2 = 0\}$ otherwise. Let X_3 denote the student's mark in the exam. So as not to disturb other students a suspicious candidate is checked for illicit materials brought into the exam hall only at the end of the exam. Let $\{X_4 = 1\}$ if a student is checked and is found to have brought in illicit material and set $\{X_4 = 0\}$ otherwise. The chief examiner believes that the DAG below is a causal Bayesian network (CBN).

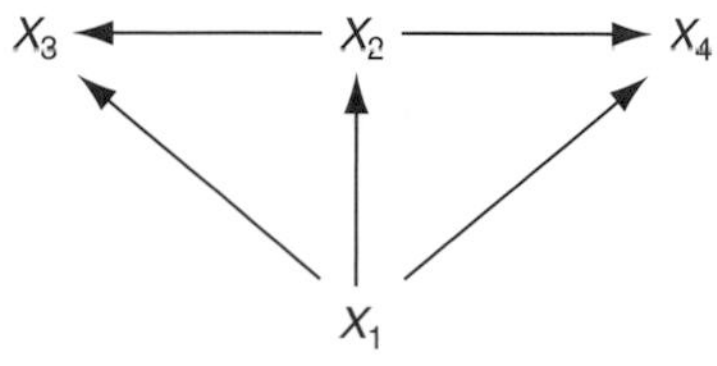

a) Carefully describe what this CBN asserts when all joint probabilities of different configurations of (x_1, x_2, x_3, x_4) are strictly positive. In this context do you see any ambiguity in the definition of any of the potential manipulations? Give one reason why a different examiner might not want to believe the formula for $p(x_2, x_3, x_4 || x_1 = 1)$.

b) Write down the formula for the total cause of X_2 on X_4. Show that this total cause is not in general the same as $p(x_4|x_2)$. Give two different additional conditional independence statements that allow the total cause of X_2 on X_4 with $p(x_4|x_2)$ to be identified, in each case demonstrating algebraically why this is so. Explain these two assumptions in the context given above. In a general CBN G whose vertices are $\{X_1, X_2, \ldots, X_n\}$ suppose the cause X_j, the effect X_k – X_j is an ancestor of X_k in G – and the set $Z \subseteq \{X_1, X_2, \ldots, X_n\} \backslash \{X_j, X_k\}$ are all observed but that the values of the remaining variables remains hidden. When does a set $Z \subseteq \{X_1, X_2, \ldots, X_n\} \backslash \{X_j, X_k\}$ satisfy the backdoor criterion relative to (X_j, X_k)? Use the backdoor theorem on the CBN above, and for each pair of variables (X_j, X_k), $1 \leq j < k \leq 4$ list the subsets Z satisfying the backdoor criterion.

8.3 Use the d-separation theorem to prove that the component state processes of an MDM remain independent after sampling.

9
Multidimensional learning

9.1 Introduction

Drawing together data relevant to different parts of a complex model is a challenge for a number of reasons. First even if the prior density has a simple and interpretable form before any sampling has taken place, sampling may well introduce dependences across large sections of the model. If this happens then the salient features needed for inference can become much more difficult to calculate and this can be critical. Even more of a problem is when the values of certain variables remain unsampled.

However sometimes this is not the case. It is not unusual for the DM to be able to assume that different functions of the data sets she has at hand inform only certain factors in the credence decomposition she chooses. However the circumstances when such assumptions are transparent – or failing that plausible – are closely linked to how sampling schemes, observational studies and experiments are designed. In the last chapter we focused on decision models that could be structured round a BN. We showed how the decomposition of a problem into smaller explanatory components not only made a dependence structure more explicit but also provided a framework for the fast propagation of evidence using local structure in the large joint probability space. Now hierarchical models have been a bedrock of Bayesian modelling for some time and these are usually expressible as a BN. Bayesian models of this type exploit the conditional independence structure/density factorisation/credence decomposition of a model to enable accommodation of survey and experimental information *locally* into the uncertainty model.

In the next section we begin by describing the hierarchical model and how data vectors can in principle be drawn in to inform the process under study using a BN. The practical validity of drawing information from different data sets into a multifaceted problem depends on positive answers to the next two questions:

(1) Do the different sources of data really inform only certain aspects of the problem at hand?
(2) Can the densities of probabilities or parameters in the experimental data be directly related to the distributions of probabilities/parameters in the current instance of interest?

In the second section of this chapter we present some common types of data structure and discuss the extent to which the second question can be answered affirmatively. Whatever the credence decomposition – for example whether it is based on a BN or an event tree – the answer to the first question depends on whether or not the likelihood exhibits appropriate separation properties in a sense formalised below. We illustrate various circumstances when it is secure or plausible for the DM to assume this type of separation. We will focus this section first on event tree estimation and then proceed to show how the case of modular estimation of the probabilities in a discrete BN flows from the separation exhibited in its corresponding tree representation.

In many practical implementations of large decision analyses certain compromises need to be made. Feasibility and speed of the necessary calculations, the acceptability and transparency and the faithfulness of the formal model to the DM's actual beliefs often have to be weighed against each other. When the analysis of at least some components of the model is supported by large and relevant data sets it is natural to hope that inferences are robust to the impacts of such compromises. For example the DM might hope that the impact of mis-specified priors on parameters would not be too great.

Certainly in the simple parametric models discussed in Chapter 5 this appeared to be the case. However the story is actually a lot more complicated. Under standard parametric models such as hierarchical models some features of the subjective prior are overwritten whilst others endure. It is important for both an analyst and an auditor to develop an awareness of what features of a subjective prior can have a critical influence on inferences even after extensive sampling because then she will understand the limits to which different DMs with the same utility function might come to different conclusions even when they accept the same evidence. This chapter ends with such a discussion.

9.1.1 Hierarchical models

One reason the BN has become such a popular tool over the last thirty years is that it can be used as a framework for depicting many of the widely used models used for Bayesian inference. We now use these semantic to describe and analyse some common classes of statistical model. For example we noted in Chapter 5 that eight binary variables $\{Y_i : i = 1, 2, 3, \ldots, 8\}$ in an exchangeable sequence implies the conditional independence $\amalg_{i=1}^{\infty} Y_i | \theta$ where θ, a random variable taking values between zero and one, could be interpreted as the probability. It is easy to check using the d-separation theorem that these conditional independence models can be expressed as the BN below.

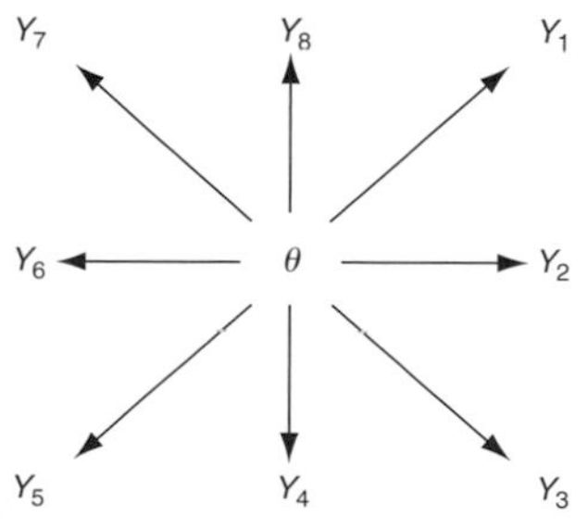

More generally the Bayesian model has a BN specified by

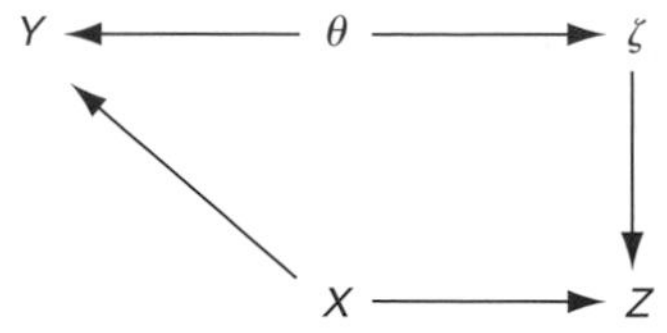

which by d-separation can be used to prove that $Z \amalg \boldsymbol{Y} \mid (\boldsymbol{X}, \boldsymbol{\theta})$ – a property we proved using factorisation formulae in Chapter 5. These sorts of observations suggested separating out some of the components of the various vectors in this depiction to obtain a more detailed insight into the underlying dependence structure of a given model.

Models for Bayesian inference usually focus on the faithful representation of the relationship between $\boldsymbol{\theta}, \boldsymbol{X}$ and $\boldsymbol{Y}$. The most obvious way to introduce certain types of structured dependences transparently is to introduce a new explanatory random vector $\boldsymbol{\phi}$ into the model. Perhaps the most widely studied class of Bayesian models is the hierarchical model or multilevel model (Gelman and Hill, 2007). Here the parameters or the unobserved parameter random variables associated with each observation are defined to have a structured relationship between themselves. A simple example of one such structure is given below.

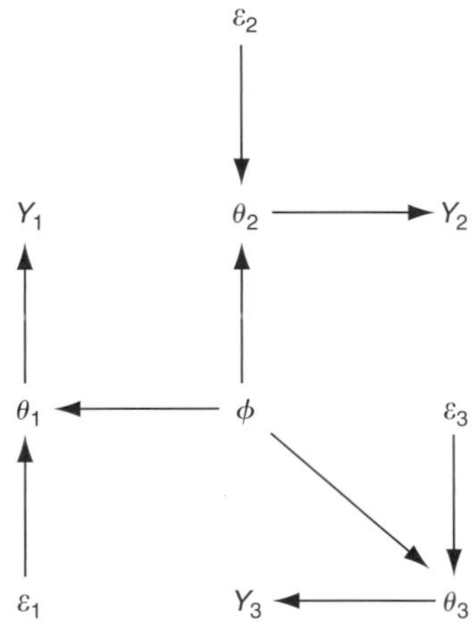

Here the distribution of each observation depends on its own single parameter, for example its mean. These parameters are dependent but on a common shared random variable here denoted by ϕ. Thus for example, we could set

$$\theta_i = \phi + \varepsilon_i$$

where $\amalg_{i=1}^{3} \varepsilon_i$. Here the random variables Y_i are in the first level of the hierarchy, θ_i in the second and ϕ and ε_i, $i = 1, 2, 3$ in the third. Such a hierarchy makes it possible for structural qualitative information to be built in the statistical model at its inception. Thus in the example above ϕ represents the *common* variation in the means of the process. It simply encodes the belief that any prior dependence between the means of the observation is explained by their relationship to the shared variable ϕ. In any given context this sort of qualitative structuring – elicited in much the same way as we described in the water company problem described in Chapter 7 – can be accommodated before detailed elicitation about appropriate settings of distributions takes place.

Hierarchical models are expressively extremely powerful. Furthermore this BN framework can be elicited without reference to particular distributions. So for example software works with a given elicited BN to allow the user to specify a distribution of $Y_i|\theta_i \ \ \varepsilon_i$, $i = 1, 2, 3$ and ϕ customised to her given context once the qualitative framework of the BN is agreed. If this does not admit closure under sampling, the software calculates good approximations to the posterior distributions that can then be fed back to the user.

The analyst needs to be aware that there is an associated risk when building models in this way. It used to be thought that models could be made more and more flexible simply by increasing the number of levels of the hierarchy. This is only true in a restrictive sense. Thus let $\boldsymbol{Y}$ denote the vector of observations, $\boldsymbol{\theta}$ denote the second level parameters and $\boldsymbol{\phi}$ the remaining parameters in the model. Then by definition $\boldsymbol{Y} \amalg \boldsymbol{\phi}|\boldsymbol{\theta}$ or equivalently, by the symmetry property of conditional independence $\boldsymbol{\phi} \amalg \boldsymbol{Y}|\boldsymbol{\theta}$. This in turn means that

$$p\left(\boldsymbol{\phi}|\boldsymbol{\theta},\boldsymbol{y}\right) = p\left(\boldsymbol{\phi}|\boldsymbol{\theta}\right).$$

It follows that when the joint density $p\left(\boldsymbol{\phi},\boldsymbol{\theta}\right)$ is elicited whatever densities that are given a priori by the DM concerning $p\left(\boldsymbol{\phi}|\boldsymbol{\theta}\right)$ will remain unchanged however big a sample of data we collect. This property can go unnoticed in an analysis because $p\left(\boldsymbol{\phi},\boldsymbol{\theta}\right)$ is usually elicited consistently with the "causal" order of the parameters in the explanation. In this case the factorisation $p\left(\boldsymbol{\phi},\boldsymbol{\theta}\right) = p\left(\boldsymbol{\theta}|\boldsymbol{\phi}\right) p(\boldsymbol{\phi})$ has this property and both terms on the right-hand side will change in response to the data. However different candidate distributional families chosen for $p\left(\boldsymbol{\theta}|\boldsymbol{\phi}\right)$ and $p(\boldsymbol{\phi})$ give rise to very different prior densities $p\left(\boldsymbol{\phi}|\boldsymbol{\theta}\right)$.

In Chapter 5 a statistical decision model was formulated so – given any known covariates $\boldsymbol{X}$ – all learning happened through the impact data had on the distribution of the vector of parameters $\boldsymbol{\zeta}$ associated with the problem faced. It follows that unless $\boldsymbol{\zeta} \amalg \boldsymbol{\phi}|\boldsymbol{\theta},\boldsymbol{X}$ then the DM's inferences may depend critically on how the expert set up his prior on $(\boldsymbol{\phi},\boldsymbol{\theta},\boldsymbol{X})$. This dependence will not simply dissolve because a large relevant data set is accommodated into the analysis and the effect of the prior will be strong.

On the other hand if inference need only depend upon $\boldsymbol{\theta}$ then by expressing $p\left(\boldsymbol{\theta}\right) = \int p\left(\boldsymbol{\theta}|\boldsymbol{\phi}\right) p(\boldsymbol{\phi})d\boldsymbol{\phi}$ we have used structural information to perform a credence decomposition. We have argued above that this useful in order to put more credible information into the prior. Also in Smith and Rigat (2009) we have proved that this decomposition can help make the inference more robust to certain sorts of mis-specification of $p\left(\boldsymbol{\theta}|\boldsymbol{\phi}\right)$. So in this sense the introduction of new levels of the parameters into the model can be helpful. For some other observations about sensitivity to prior specification in hierarchical models and stability of second level parameters see Schervish (1995) and recent results on the effect of these structural issues on the convergence of numerical methods in Papaspiliopoulos and Roberts (2009).

9.2 Separation, orthogonality and independence

9.2.1 Separability of a likelihood

Over the years statisticians have needed to design experiments or surveys whose results apply to domains that are as universal as possible. We argued in Chapter 5 that if the DM needs to accommodate experimental information then she needs to believe that the parameters of the sampling distribution was linked appropriately to the parameters of the distribution of the events of interest and addressed decision analysis. Suppose the DM believes this to be so. She has seen the results of various different experiments each giving information about the parameters in different components of her model. So ideally she would like to use data informative to different components in her credence decomposition, use Bayes rule to accommodate this information to transform her prior beliefs about each component into her posterior beliefs and then aggregate up these more refined beliefs into the relevant composite she needs to evaluate the expected utilities under the different decisions open to her.

There are substantive conditions and assumptions she needs to make before this sort of combination rule is appropriate. However happily there are also many circumstances when such a rule of combination is formally sound. In this section we will discuss when this is so and illustrate how such information can be integrated into to a moderately sized decision analysis.

Some use of data to inform a decision problem is of course straightforward. For example in the forensic example of Chapter 5 recall that the components in the vector corresponded to the probability that a person in a given age and social class category had a certain number of fragments of glass on their clothing. A juror might plausibly accept that these probabilities were appropriate to a suspect in court whose age and social class matched those of the experiment. Similarly, well-designed randomised longitudinal surveys of cancers developed in children exposed to known amounts of radiation could be brought to bear to address estimates of these effects on populations in a particular nuclear accident.

On the other hand experiments often need to be designed and assumptions made about different populations and extrapolations made from the subjects in the experiments or surveys to the case of interest. We will show that one property we need for this local accommodation of information to be valid is that the likelihood of the evidence from all the experiments is separable.

Definition 9.1. A likelihood $l(\boldsymbol{\theta}|\boldsymbol{y},\boldsymbol{x})$, strictly positive for all values of $\boldsymbol{\theta}$ is *separable* in $\boldsymbol{\theta}$, $\boldsymbol{\theta} = (\boldsymbol{\theta}_1, \boldsymbol{\theta}_2, \ldots, \boldsymbol{\theta}_q) \in \Theta$, $\Theta = \Theta_1 \times \Theta_2 \times \cdots \times \Theta_m$, $q \leq m$ at a value $\boldsymbol{x}$ of covariates when $\log l(\boldsymbol{\theta}|\boldsymbol{y},\boldsymbol{x})$ can be written in the form

$$\log l(\boldsymbol{\theta}|\boldsymbol{y},\boldsymbol{x}) = \sum_{j=1}^{q} \log l_j(\boldsymbol{\theta}_j|\boldsymbol{y},\boldsymbol{x})$$

where $\log l_j(\boldsymbol{\theta}_j|\boldsymbol{y},\boldsymbol{x})$ is a function of $\boldsymbol{\theta}$ only through the subvector $\boldsymbol{\theta}_j \in \Theta_j$, $1 \leq j \leq m$.

Suppose the Bayesian DM genuinely believes that her prior probabilities over the different components of her parameter space are a priori independent and $p(\boldsymbol{\theta}|\boldsymbol{x}) > 0$ for all possible values of $(\boldsymbol{\theta}, \boldsymbol{x})$. This implies that her prior density has a product form and

$$p(\boldsymbol{\theta}|\boldsymbol{x}) = \prod_{j=1}^{q} p_j(\boldsymbol{\theta}_j|\boldsymbol{x}) \Leftrightarrow \log p(\boldsymbol{\theta}|\boldsymbol{x}) = \sum_{j=1}^{q} \log p_j(\boldsymbol{\theta}_j|\boldsymbol{x}).$$

Bayes rule now implies that

$$\log p(\boldsymbol{\theta}|\boldsymbol{x},\boldsymbol{y}) + \log p(\boldsymbol{y}|\boldsymbol{x}) = \log p(\boldsymbol{y}|\boldsymbol{\theta},\boldsymbol{x}) + \log p(\boldsymbol{\theta}|\boldsymbol{x})$$

$$\log p(\boldsymbol{\theta}|\boldsymbol{x},\boldsymbol{y}) = c(\boldsymbol{x},\boldsymbol{y}) + \sum_{)j=1}^{q} \log l_j(\boldsymbol{\theta}_j|\boldsymbol{x},\boldsymbol{y}) + \sum_{j=1}^{q} \log p_j(\boldsymbol{\theta}_j|\boldsymbol{x})$$

for some constant $c(\boldsymbol{x},\boldsymbol{y})$ not depending on $\boldsymbol{\theta}$. But since, also by Bayes rule,

$$\log p_j(\boldsymbol{\theta}|\boldsymbol{x},\boldsymbol{y}) = c_j(\boldsymbol{x},\boldsymbol{y}) + \log l_j(\boldsymbol{\theta}_j|\boldsymbol{x},\boldsymbol{y}) + \log p_j(\boldsymbol{\theta}_j|\boldsymbol{x})$$

for some constant $c_j(\boldsymbol{x},\boldsymbol{y})$ not depending on $\boldsymbol{\theta}, j = 1, 2, \ldots, q$

$$\log p(\boldsymbol{\theta}|\boldsymbol{x},\boldsymbol{y}) = c'(\boldsymbol{x},\boldsymbol{y}) + \sum_{j=1}^{q} \log p_j(\boldsymbol{\theta}_j|\mathrm{y},\boldsymbol{x})$$

for some constant $c'(\boldsymbol{x},\boldsymbol{y})$ not depending on $\boldsymbol{\theta}$. It follows that

$$p(\theta|\boldsymbol{x}) \propto \prod_{j=1}^{q} p_j(\boldsymbol{\theta}_j|\boldsymbol{x},\boldsymbol{y}) \Leftrightarrow p(\theta|\boldsymbol{x},\boldsymbol{y}) = \prod_{j=1}^{q} p_j(\boldsymbol{\theta}_j|\boldsymbol{x},\boldsymbol{y})$$

since all these densities must integrate to unity. So if the data input by the DM has a separable likelihood over the components of its parameters and she has a prior which makes these components independent then the component parameters are independent after the accommodation of the evidence. This means that we can safely update the densities parameter components separately and then aggregate these to obtain the DM's full probability distribution $p(\theta|\boldsymbol{x},\boldsymbol{y})$. This is a precious property even when the parameter space of the problem is even moderately large. It not only facilitates fast computation but also more importantly allows the DM legitimately to explain her learning in terms of the components of the problem. Note that the separation of the likelihood can be written in terms of conditional independence as the condition

$$\coprod_{j=1}^{q} \boldsymbol{\theta}_j|\boldsymbol{X} = \boldsymbol{x} \Leftrightarrow \coprod_{j=1}^{q} \boldsymbol{\theta}_j|\boldsymbol{X} = \boldsymbol{x}, \boldsymbol{Y}.$$

The most obvious situation where likelihoods are separable is when the distribution of the jth data set only depends on the parameter vector $\boldsymbol{\theta}_j$, $j = 1, 2, \ldots, q$. and all these experiments are independent of each other. However there are many other experiments exhibiting separable likelihoods where observations can be highly dependent on each other. One important example is when the sampling distribution of each observation takes a chain form so that the density or mass function $p_i(\boldsymbol{y}_i|\boldsymbol{x}_i,\boldsymbol{\theta})$ respects the *chain factorisation*

$$p_i(\boldsymbol{y}_i|\boldsymbol{x}_i,\boldsymbol{\theta}) = p_{i,j}(\boldsymbol{y}_{i,1}|\boldsymbol{x}_i,\boldsymbol{\theta}_1)\prod_{j=2}^{q} p_{i,j}(\boldsymbol{y}_{i,j}|\boldsymbol{y}_{i,1},\boldsymbol{y}_{i,2},\ldots,\boldsymbol{y}_{i,j-1}\boldsymbol{x}_i,\boldsymbol{\theta}_j)$$

where $q \geq 2, \boldsymbol{y}_i = (\boldsymbol{y}_{i,1},\boldsymbol{y}_{i,2},\ldots,\boldsymbol{y}_{i,q}), \boldsymbol{x}_i = (\boldsymbol{x}_{i,1},\boldsymbol{x}_{i,2},\ldots,\boldsymbol{x}_{i,q})$ and each term in the factorisation above is a function of its listed arguments. Thus we construct the conditional sample distributions in a recursive sequential order where each component vector of observations is only allowed to depend on the value of their covariates, previous listed responses and parameters that only appear in that conditional. On observing a random sample of n such variables it then follows that we can write

$$l(\boldsymbol{\theta}|\boldsymbol{y},\boldsymbol{x}) = \prod_{j=1}^{q} l_j(\boldsymbol{\theta}_j|\boldsymbol{y},\boldsymbol{x})$$

where

$$l_1(\boldsymbol{\theta}_1|\boldsymbol{y},\boldsymbol{x}) = \prod_{i=1}^{n} p_{i,1}(\boldsymbol{y}_{1,1}|\boldsymbol{x}_i,\boldsymbol{\theta}_1)$$

and for $j = 2, 3, \ldots, q$

$$l_j(\boldsymbol{\theta}_j|\boldsymbol{y},\boldsymbol{x}) = \prod_{i=1}^{n} p_{i,j}(\boldsymbol{y}_{i,j}|\boldsymbol{y}_{i,1},\boldsymbol{y}_{i,2},\ldots,\boldsymbol{y}_{i,j-1},\boldsymbol{x}_{i,},\boldsymbol{\theta}_j).$$

It follows that whenever it is valid for the DM to believe that $\amalg_{j=1}^{q}\boldsymbol{\theta}_j$ are a priori independent they will remain so after sampling.

There are two important corollaries to this observation. First learning can be modularised. Suppose that the DM believes the different components of $\boldsymbol{\theta}$ are independent a priori. She can then *delegate* her learning to her trusted expert associated with each component vector. That trusted expert can use $l_j(\boldsymbol{\theta}_j|\boldsymbol{y},\boldsymbol{x})$ to update his prior – that would have been adopted by the DM a priori – and simply report his *posterior* density over these parameters to the DM who can then recombine her beliefs about the system using her prior credence decomposition.

Second suppose the DM performs a subsidiary experiment where a new set of vectors of observations $\{\boldsymbol{Y}'_{i,j} : i = 1,2,\ldots,n'\}$ are taken at respective points $\{(\boldsymbol{y}'_{i,1},\boldsymbol{y}'_{i,2},\ldots,\boldsymbol{y}'_{i,j-1},\boldsymbol{x}'_{i,}) : i = 1,2,\ldots,n'\}$ *controlled* to take these values. Then under the assumption that this factorisation is causal in the sense described in the last chapter it will follow that

$$l_j(\boldsymbol{\theta}_j|\boldsymbol{y},\boldsymbol{x},\boldsymbol{y}',\boldsymbol{x}') = l_j(\boldsymbol{\theta}_j|\boldsymbol{y},\boldsymbol{x})l_j(\boldsymbol{\theta}_j|\boldsymbol{y}',\boldsymbol{x}').$$

So sampling data and experimental data can be seamlessly combined in this modular way if a model is causal and the independence results still apply. Here is a simple example of a model from this class.

Example 9.2 (A multiregression model). This family consists of a set of linear models with conjugate priors as discussed in Chapter 5 as defined in n random vectors $\boldsymbol{Y}(x) = \{\boldsymbol{Y}_{i,j}(x) : i = 1,2,\ldots,n, j = 1,2,3\}$ on three levels where

$$\psi_{i,1}(x_i) = \theta_{1,1} + \theta_{1,2}x_i,$$
$$\psi_{i,2}(x_i) = \theta_{2,1} + \theta_{2,2}x_i^2 + \theta_{2,3}y_{i,1},$$
$$\psi_{i,3}(x_i) = \theta_{3,1} + \theta_{3,2}\sin x_i + \theta_{3,3}y_{i,1}^3$$

and respective conditional precisions $\theta_{1,3} = \sigma_1^2$, $\theta_{2,4} = \sigma_2^2$ and $\theta_{3,4} = \sigma_3^2 y_{i,1}^2$. Suppose we put the usual normal inverse gamma prior on the variables. Then if the vectors $\boldsymbol{\theta}_1 \triangleq (\theta_{1,1}, \theta_{1,2}, \theta_{1,3}), \boldsymbol{\theta}_2 \triangleq (\theta_{2,1}, \theta_{2,2}, \theta_{2,3}, \theta_{2,4}), \boldsymbol{\theta}_3 \triangleq (\theta_{3,1}, \theta_{3,2}, \theta_{3,3}, \theta_{3,4})$ are all a priori independent, this will be so a posteriori. The DAG of a valid BN of this process and the first observation is given below.

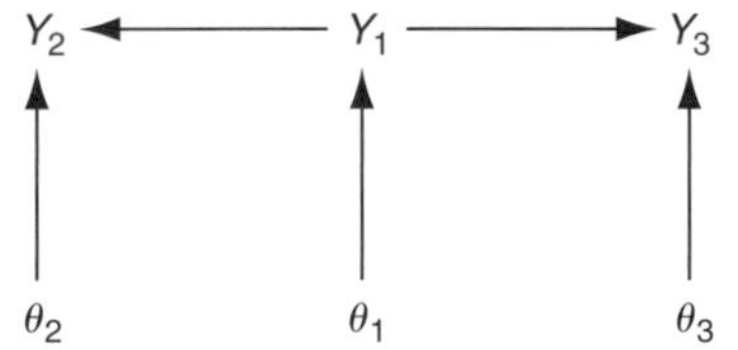

Furthermore each will have a conjugate normal inverse gamma posterior, whose hyperparameters are updated using the recurrences given in Chapter 5. You are asked to calculate these recurrences explicitly in Exercise 9.3.

It is worth pointing out that a subclass of this genre of models, called recursive structural models, has been studied widely by econometricians. A good introduction to Bayesian analyses of these models is given in Lancaster (2004) and some excellent practical Bayesian analyses of models in this class concerning marketing decision modeling using structural models in Rossi *et al.* (2005).

9.2.2 Conditional separability

Note that another very useful property is conditional separability.

Definition 9.3. A likelihood is *separable* in $\boldsymbol{\theta} = (\boldsymbol{\theta}_1, \boldsymbol{\theta}_2, \ldots, \boldsymbol{\theta}_q)$ conditional on $\boldsymbol{\theta}_{q+1}$, $\boldsymbol{\theta} = (\boldsymbol{\theta}_1, \boldsymbol{\theta}_2, \ldots, \boldsymbol{\theta}_q, \boldsymbol{\theta}_{q+1})$, $\Theta = \Theta_1 \times \Theta_2 \times \cdots \times \Theta_m \times \Theta_{m+1}$, $q \leq m$ at a value $\boldsymbol{x}$ of covariates when $\log l(\boldsymbol{\theta}|\boldsymbol{y},\boldsymbol{x})$ can be written in the form

$$\log l(\theta|\boldsymbol{y},\boldsymbol{x}) = \sum_{j=1}^{q} \log l_j(\boldsymbol{\theta}_j,\boldsymbol{\theta}_{q+1}|\boldsymbol{y},\boldsymbol{x})$$

where $\log l_j(\boldsymbol{\theta}_j,\boldsymbol{\theta}_{q+1}|\boldsymbol{y},\boldsymbol{x})$ is a function of $\boldsymbol{\theta}$ only through the subvector $\boldsymbol{\theta}_j \in \Theta_j$ and $\boldsymbol{\theta}_{q+1}$, $1 \leq j \leq m$.

You are asked to prove in Exercise 9.1 below that if the likelihood of an experiment is separable conditional on $\boldsymbol{\theta}_{q+1}$ at $\boldsymbol{x}$ then

$$\coprod_{j=1}^{q} \boldsymbol{\theta}_j|\boldsymbol{\theta}_{q+1}, \boldsymbol{X} = \boldsymbol{x} \Leftrightarrow \coprod_{j=1}^{q} \boldsymbol{\theta}_j|\boldsymbol{\theta}_{q+1}, \boldsymbol{X} = \boldsymbol{x}, \boldsymbol{Y}.$$

This property is also useful since if the parameter ζ is such that

$$\zeta \amalg \{\theta_k : 1 \leq k \neq j \leq q\}|\boldsymbol{\theta}_{q+1}, \boldsymbol{\theta}_j, \boldsymbol{X} = \boldsymbol{x}$$

so that ζ only depends on $(\boldsymbol{\theta}_j, \boldsymbol{\theta}_{q+1})$ and the known covariates, then a posteriori beliefs about $(\boldsymbol{\theta}_j|\boldsymbol{\theta}_{q+1})$ are more stable in the sense that they will only depend on the term $l_j(\boldsymbol{\theta}_j,\boldsymbol{\theta}_{q+1}|\boldsymbol{y},\boldsymbol{x})$ and sampling does not destroy the prior conditional independence $\coprod_{j=1}^{q} \boldsymbol{\theta}_j|\boldsymbol{\theta}_{q+1}, \boldsymbol{X} = \boldsymbol{x}$ as, for example, expressed in a BN. It follows in particular that it is more secure to assign an enduring meaning to these parameter vectors. This sort of argument is fleshed out below.

9.2.3 Separability, experiments and the Gaussian regression prior

The separability properties of a likelihood were first used in the design and analysis of experiments under the term orthogonality. This provides a useful introduction to the use of this idea in conjunction with decomposition of large-scale models. So consider the following example.

Example 9.4. The DM agrees with the experimentalist that it is reasonable to believe that the logarithm $Y(\boldsymbol{x})$ of the time it takes to complete a machine task is approximately normally distributed with mean $\mu(\boldsymbol{x},\boldsymbol{\theta})$ and variance σ^2 where

$$\mu(\boldsymbol{x},\boldsymbol{\theta}) = \theta_1 + x_2\theta_2 + x_3\theta_3.$$

In a well-designed experiment the experience and training of the employee doing the task was measured by x_2 in a way that was universal to these types of task, whilst x_3 measures the quality of equipment. With this parameterisation θ_1 reflects the actual complexity of the task in question, θ_2 whether the operative has attended a training course and θ_3 the quality of the machinery used by the operative. The experiment from which the DM wishes to draw her information consisted of n replicates providing data $\{(y_i(\boldsymbol{x}_i), \boldsymbol{x}_i) : i = 1, 2, \ldots, n\}$ on tasks of the same complexity where units were chosen so that

$$\sum_{i=1}^{n} x_{i1} = \sum_{i=1}^{n} x_{i,2} = 0.$$

Note that the design matrix $A = \{a_{i,j} : i = 1, 2, \ldots, n, j = 1, 2, 3\}$ as defined in the linear model example in Chapter 5 is such that $a_{i,1} = 1$, $a_{i,2} = x_{i,2}$, $a_{i,3} = x_{i,3}$. The DM calculates her posterior distribution as described in the analysis of the Gaussian linear model of Chapter 5. She needs to assess the effect θ_2 of the decision to send a new employee on the same training course investigated in the experiment. She knows that both the complexity θ_1 of her task and the quality of her machinery θ_3 are completely different from those in the experiment. However she believes that the linear model of the experiment and her task have the same incremental additive effect so that the value θ_2 in the problem she faces can be identified with this parameter in the experiment.

In the example above suppose that the DM believes that $\theta_2 \amalg (\theta_1, \theta_3) \,|\theta_4$ and a new observation Z of the same experiment but where the design is not necessarily orthogonal is given by the BN

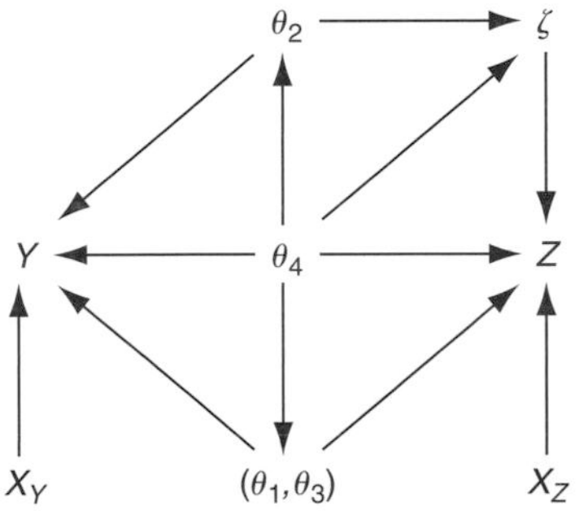

In Example 1.9 below you are asked to use the d-separation theorem to check that after sampling we cannot conclude that $\theta_2 \amalg (\theta_1, \theta_3) \,|\boldsymbol{Y}, \boldsymbol{X}, \theta_4$. Note that the alternative prior with $(\theta_1, \theta_2, \theta_3) \amalg \theta_4$ simply omits the edges (θ_4, θ_2) and $(\theta_4, (\theta_1, \theta_3))$ from this graph. Then even with this simplification it still cannot be concluded that $\theta_2 \amalg (\theta_1, \theta_3) \,|\boldsymbol{Y}, \boldsymbol{X}, \boldsymbol{\theta}_4$. So the sampling destroys the conditional independence structure of this BN and hence its explanatory power.

However the conditional independence above *will* hold if given the precision θ_4 the likelihood is separable in the three components $(\theta_1, \theta_2, \theta_3)$. In the example above we note that the loglikelihood of (5.19) satisfies

$$2 \log l(\boldsymbol{\theta}|\boldsymbol{y}, A) = n \log \theta_4 - \theta_4 (\boldsymbol{y} - A\boldsymbol{\theta}_1)^T (\boldsymbol{y} - A\boldsymbol{\theta}_1).$$

Therefore given θ_4 this separates in $(\theta_1, \theta_2, \theta_3)$ if and only if there are no cross product terms in quadratic form on the second half of this expression: i.e. if $A^T A$ is diagonal. We can conclude that if the DM, experimenter and auditor all believed that $\theta_1, \theta_2, \theta_3$ were mutually independent conditional on θ_4 a priori then having accommodated the data they would still all believe that θ_2 was independent of their beliefs about (θ_1, θ_3) conditional on θ_4. This has two advantages. The first is computational. The posterior distribution of $\theta_2|\theta_4$ is simplified. This is not very important in this simple context but in analogous much larger problems this can be critical.

The second advantage is more subtle but more important from an inferential point of view. This conditional independence both prior and posterior to this experiment θ_2 can be given a stand alone label of "the effect of training". The independence above of θ_2 from (θ_1, θ_3) given the precision θ_4 both a priori and a posteriori means that it is a logical to give its distribution without qualifying it with (θ_1, θ_3), the complexity and machine effects. If the DM were subsequently told more about (θ_1, θ_3) then this would have no effect on their beliefs about θ_2 given θ_4.

Designed experiments like these usually control the covariates of the experimental units. It follows that they are most directly informative about the probable effect on an attribute when a DM controls the covariates of a given unit or population of units. In the example above this was to predict the effect of sending an employee on a training course. However with the appropriate causal hypotheses these inferences can also be used to make deductions about what might happen when employees volunteered themselves to go on a training course. So for example if all parties believed the DAG above was a CBN then the same deductions about a volunteering employee will be the same as one who is sent.

9.3 Estimating probabilities on trees

9.3.1 Estimating probabilities on trees with no symmetries

In Chapter 2 we saw various examples where trees were used to represent a decision problem. Decision trees need probabilities on the edges of their chance nodes. These could obviously be simply added as a single probability elicited either from the DM herself or from an expert she trusts. However we have already argued that the DM will often feel more secure if, as far as is possible, experimental information is accommodated into these judgements. We have illustrated above how in a given court case sampling information from a designed experiment can be used to determine the probability that an innocent person had glass on their clothing and hence – if the suspect matches can be thought of as a random draw from this population – the probability that glass is present if he is innocent. Information from many independent experiments of this type will allow the DM to support and formally accommodate this type of experimental evidence into her judgements. Because each experiment is independent of the others the likelihood will separate. Then, provided the DM believes that different vectors of probabilities labelling the emanating edges of each situation in the tree are a priori independent of one another they will be independent a posteriori, ensuring the advantages discussed above.

However in other sampling scenarios our data is observational. Assume there is sampling information in the form of a vector of observations from random draws of units from a population containing our unit of interest. Further suppose that the DM wants to measure the effect of a particular educational programme on a child drawn from a particular population of exchangeable children. Our evidence will simply then be the performance of other children in that population. In this scenario if the random sample of observed units respecting the distribution of the tree is ancestral – in a sense defined below – then the likelihood also separates. So in this framework learning can be made local.

For simplicity in this book we will restrict our attention to examples where the DM believes that the development of every member of the population containing the unit of interest and the sampled units can be faithfully represented by the same tree $\mathcal{T}$ with various shared but uncertain edge probabilities. Then by definition the probability $p(y_i|\boldsymbol{\theta})$ of observing the path $(v_0, v_{1,i}, \ldots, v_{k(i),i})$ in any given member of this population is

$$p(y_i|\boldsymbol{\theta}) = \prod_{j=1}^{k(i)} \theta_{v_{j-1},v_j}. \tag{9.1}$$

Now suppose the tree is sampled so that it satisfies the following condition.

Definition 9.5. Units whose evolution is governed by the tree $\mathcal{T}$ are said to constitute a *Poisson ancestral sample* if

(1) the evolution of all sampled units is independent given the vector of probabilities $\boldsymbol{\theta}$ given by the edge probabilities of $\mathcal{T}$ as given in (9.1);
(2) each unit k is observed until it reaches a *terminating vertex* $v(k)$ after which it remains unseen;
(3) the terminating vertex $v(k)$ of any unit is independent of its evolution of the unit after that point.

The second and third condition corresponds to a particular instance of the missing at random (MAR) hypothesis Little and Rubin (2002) and ensures that the likelihood takes a particularly simple form. Note that this is ancestral sampling in the sense that we implicitly learn the passage of parents to children taken by a unit before it reaches its terminating vertex.

Obviously if the terminating vertex is informative about how a unit will subsequently evolve then this information should be formally accommodated into a Bayesian analysis. Thus for example suppose the units are patients who have all received a treatment and are being followed up to monitor their recovery. If a patient absents herself the DM might well believe that this could indicate that she believes she is fully recovered. It would then follow that the absence of this record would indicate that the full recovery of the patient would be higher than for someone with the same history who continues to allow herself to be monitored. In this sort of circumstance the third assumption would be insecure. Notice that if the full evolution of each unit to the leaf of the tree is always observed then the second condition and third conditions are automatically satisfied.

On observing an ancestral sample $\boldsymbol{y} = (y_1, y_2, \ldots, y_n)$ of the parts of the evolution of n units the assumptions above ensure that the sample mass function $p(\boldsymbol{y}|\boldsymbol{\theta})$ of our data is simply given by the product of the probabilities of the units. So

$$p(\mathbf{y}|\boldsymbol{\theta}) = \prod_{i=1}^{n} \prod_{j=1}^{k(i)} \theta_{v_{j-1},v_j} = \prod_{v \in V(\mathcal{T})\backslash\{v_0\}} \theta_{u,v}^{y(u,v)} \tag{9.2}$$

where $y(u, v)$ are the number of units in the sample passing from u to v along the path of its observed development. Let the subset of vertices $V(u) \triangleq \{v \in V(\mathcal{T}) : (u, v) \in E(\mathcal{T})\}$ – the set of edges connected from u to v and $\boldsymbol{\theta}_u \triangleq \{\theta_v : (u, v) \in E(\mathcal{T})\}$ – i.e. the vector of probabilities labelling the set of edges emanating from u. Then it is easily checked that (9.2) can be rearranged as

$$p(\mathbf{y}|\boldsymbol{\theta}) = \prod_{u \in S(\mathcal{T})} l_u(\mathbf{y}_u|\boldsymbol{\theta}_u) \tag{9.3}$$

where, for each situation $u \in S(\mathcal{T})$, $l_u(\mathbf{y}_u|\boldsymbol{\theta}_u)$ is a function only of

$$\mathbf{y}_u \triangleq \{y[i] = y(u, v) \text{ for some } v \in V(\mathcal{T})\}$$

and $\boldsymbol{\theta}_u$ is defined above. Here

$$l_u(\mathbf{y}_u|\boldsymbol{\theta}_u) = \prod_{v \in V(\mathcal{T})\backslash\{v_0\}} \theta_{u,v}^{y(u,v)}.$$

Note that $l_u(\mathbf{y}_u|\boldsymbol{\theta}_u)$ is a multinomial likelihood on the vector $\boldsymbol{\theta}_u$ of probabilities where by definition all the probabilities that are components $\theta_{u,v}$ of $\boldsymbol{\theta}_u$ must sum to unity – the unit must develop somewhere – so

$$\sum_{v \in V(u)} \theta_{u,v} = 1 \text{ where, for each } v \in V(u), \theta_{u,v} \geq 0$$

and that $l_u(\mathbf{y}_u|\boldsymbol{\theta}_u) = 1$ if no children of u are observed.

The notation in the equations above – which is based on a labelling of a tree – is necessarily rather opaque. However their implications are easily stated. Recall that the set of situations $S(\mathcal{T}) \subset V(\mathcal{T})$ of a tree $V(\mathcal{T})$ is the set of its non-leaf vertices. If an ancestral random sample is taken, then the sample mass function $p(\mathbf{y}|\boldsymbol{\theta})$ – and hence by definition the likelihood $l(\boldsymbol{\theta}|\mathbf{y})$ – separates in the vectors $\boldsymbol{\theta}_u$, where $\boldsymbol{\theta}_u$ is the vector of probabilities notating the edges emanating from the situation u. It follows that if the DM believed that $\amalg_{u \in S(\mathcal{T})} \boldsymbol{\theta}_u$, i.e. that learning the values of some subset of these vectors of edge probabilities gave no useful information about the rest, so that

$$p(\boldsymbol{\theta}) = \prod_{u \in S(\mathcal{T})} p_u(\boldsymbol{\theta}_u) \tag{9.4}$$

then this will remain true after sampling, i.e. $\amalg_{u \in S(\mathcal{T})} \boldsymbol{\theta}_u | \boldsymbol{Y}$. This is very useful practically for two reasons. The first advantage is a computational one. Because of the separation of the likelihood, the posterior density $p(\boldsymbol{\theta}|\mathbf{y})$ can be written

$$p(\boldsymbol{\theta}|\mathbf{y}) = \prod_{u \in S(\mathcal{T})} p_u(\boldsymbol{\theta}_u|\mathbf{y}_u) \tag{9.5}$$

where the posterior density $p_u(\boldsymbol{\theta}_u|\mathbf{y}_u)$ can be calculated by the formula

$$p_u(\boldsymbol{\theta}_u|\mathbf{y}_u) \propto l_u(\mathbf{y}_u|\boldsymbol{\theta}_u)p_u(\boldsymbol{\theta}_u|\mathbf{y}_u).$$

Since $l_u(\mathbf{y}_u|\boldsymbol{\theta}_u)$ is a multinomial likelihood on $\boldsymbol{\theta}_u$ if we can choose $p_u(\boldsymbol{\theta}_u)$ from one of the families of densities closed under multinomial sampling discussed in Chapter 5 then the posterior density of $p(\boldsymbol{\theta}|\mathbf{y})$ can also be calculated in closed form simply by using (9.5). In particular if we were to use a Dirichlet $D(\boldsymbol{\alpha}_u^0)$ as the prior density $p_u(\boldsymbol{\theta}_u)$ of $\boldsymbol{\theta}_u, u \in S(\mathcal{T})$, then the posterior density $p_u(\boldsymbol{\theta}_u|\mathbf{y}_u)$ is Dirichlet $D(\boldsymbol{\alpha}_u^+)$ where $\boldsymbol{\alpha}_u^+ = \boldsymbol{\alpha}_u^0 + \mathbf{y}_u$. So even when the tree is enormous the prior to posterior densities, marginal likelihoods and predictive densities can all be calculated using (9.5) as simple linear relationships between the hyperparameters of the different situation probability vectors.

The second advantage of the analysis given above is a modelling and inferential one. Suppose the DM is happy that the event tree faithfully expresses the dependence structure of units in the study and the unit of interest and that the edge probabilities emanating from different situations are independent of one another a priori. Then from (9.4) the DM's prior beliefs can be expressed as a single framework $\mathcal{T}$ together with local information on certain marginal mass functions on vertices of the tree – here $\{p_u(\boldsymbol{\theta}_u) : u \in S(\mathcal{T})\}$. Having accommodated information from the observational study the framework $\mathcal{T}$ is still faithful to her beliefs and the results of the experiment just requires new marginal posterior densities $\{p_u(\boldsymbol{\theta}_u|\mathbf{y}_u) : u \in S(\mathcal{T})\}$ to be substituted for their prior analogues. This is the simplest class of models where learning new information simply involves retaining a graphical structure and modifying some local features.

In Exercise 9.4 you are asked to prove that non-ancestral sampling can lead to a likelihood that is no longer separable so these two convenient properties are lost.

9.3.2 Estimating probabilities in staged event trees

Quite often it is unrealistic to assume the DM believes all the vectors $\{\boldsymbol{\theta}_u : u \in S(\mathcal{T})\}$ are independent of each other. More usually some of these vectors of probabilities will be equal. For example in the running examples in Chapter 2 we argued that many of the sets of probabilities linked with the edges emanating from one situation were equal to those emanating from another. However there is a set of models accommodating these sorts of beliefs which still gives rise to a separable likelihood under ancestral sampling. Let $\mathbb{U} = \{U_1, U_2, \ldots, U_m\}$ be a partition of the situations $S(\mathcal{T})$ of $\mathcal{T}$ into *stages* U_i, $i = 1, 2, \ldots, m$ such that if $v, v' \in U$ then $\boldsymbol{\theta}_v = \boldsymbol{\theta}_{v'} \triangleq \boldsymbol{\theta}_U = (\theta_{U,1}, \theta_{U,2}, \ldots, \theta_{U,r(U)})$ and

$$\amalg_{U \in \mathbb{U}} \boldsymbol{\theta}_U.$$

Definition 9.6. A *staged tree* is an event tree $\mathcal{T}$ together with a partition $\mathbb{U} = \{U_1, U_2, \ldots, U_m\}$ such that such if $v, v' \in U$ then $\boldsymbol{\theta}_v = \boldsymbol{\theta}_{v'} \triangleq \boldsymbol{\theta}_U$ and

$$\amalg_{U \in \mathbb{U}} \boldsymbol{\theta}_U .$$

Thus in a staged tree the DM states a partition of its situations. The DM believes that situations in the same set in this partition are parallel to each other: i.e. that the vectors of edge probabilities emanating from situations in the same stage (with an appropriate labelling of these edges) can be identified (see Chapter 2). On the other hand probability vectors in different stages will all be independent of each other. Note that the analyst can elicit a staged tree qualitatively before eliciting the prior density. The DM just needs to state which probabilities she believes will be different and which seem unconnected. Of course she may not be prepared to make this stark distinction, but surprisingly there are many scenarios when it is appropriate to do this.

If sampling is ancestral with respect to T then directly from (9.3) we have that

$$p(\boldsymbol{y}|\boldsymbol{\theta}) = \prod_{U \in \mathbb{U}} l_U(\boldsymbol{y}_U|\boldsymbol{\theta}_U)$$

where, for each situation $u \in S(\mathcal{T})$, $l_U(\boldsymbol{y}_U|\boldsymbol{\theta}_U)$ is a function only of

$$\boldsymbol{y}_U \triangleq (y(U,1), y(U,2), \ldots, y(U, r(U))$$

where $y(U,j)$ are the number of observed units passing to a situation $u \in U$ and then along an edge labelled by the probability θ_{Uj} in $\mathcal{T}$ where $\boldsymbol{\theta}_U$ is defined above and

$$l_U(\boldsymbol{y}_U|\boldsymbol{\theta}_U) = \prod_{j=1}^{r(U)} \theta_{U,j}^{y(U,j)} .$$

Again $l_U(\boldsymbol{y}_U|\boldsymbol{\theta}_U)$ can be seen to be a multinomial likelihood on the vector $\boldsymbol{\theta}_U$ of probabilities where by definition all the probabilities that are components $\theta_{U,j}$ of $\boldsymbol{\theta}_U$ must sum to unity so

$$\sum_{j=1}^{r(U)} \theta_{U,j} = 1 \text{ where, for } \theta_{U,j} \geq 0\, , j = 1, 2, \ldots, r(U)$$

and $l_U(\boldsymbol{y}_U|\boldsymbol{\theta}_U) = 1$ if no unit passes to a situation $u \in U$. Thus we again have the property that if an ancestral random sample is taken, then the sample mass function $p(\boldsymbol{y}|\boldsymbol{\theta})$ – and hence by definition any likelihood $l(\boldsymbol{\theta}|\boldsymbol{y})$ – separates, this time in the vectors $\boldsymbol{\theta}_U$, where $\boldsymbol{\theta}_U$ is the vector of probabilities notating the edges emanating from a situation $u \in U$. So if the DM believed that $\amalg_{u \in S(\mathcal{T})} \boldsymbol{\theta}_u$, i.e. all the values of some subset of these vectors of edge probabilities gave no useful information about the rest, so that

$$p(\boldsymbol{\theta}) = \prod_{U \in \mathbb{U}} p_U(\boldsymbol{\theta}_U) \tag{9.6}$$

then this will remain true after sampling, i.e. $\amalg_{U\in\mathbb{U}}\boldsymbol{\theta}_U|\boldsymbol{Y}$ and $p(\boldsymbol{\theta}|\boldsymbol{y})$ can be written

$$p(\boldsymbol{\theta}|\boldsymbol{y}) = \prod_{U\in\mathbb{U}} p_U(\boldsymbol{\theta}_U|\boldsymbol{y}_U) \tag{9.7}$$

where the posterior density $p_U(\boldsymbol{\theta}_U|\boldsymbol{y}_U)$ can be calculated by the formula

$$p_U(\boldsymbol{\theta}_U|\boldsymbol{y}_U) \propto l_U(\boldsymbol{y}_U|\boldsymbol{\theta}_U)p_U(\boldsymbol{\theta}_U|\boldsymbol{y}_U).$$

Furthermore since $l_U(\boldsymbol{y}_U|\boldsymbol{\theta}_U)$ is a multinomial likelihood on $\boldsymbol{\theta}_U$ if we can choose $p_U(\boldsymbol{\theta}_U)$ from one of the families of densities closed under multinomial sampling discussed in Chapter 5 then the posterior density of $p(\boldsymbol{\theta}|\boldsymbol{y})$ can also be calculated in closed form. In particular if we were to use a Dirichlet $D(\boldsymbol{\alpha}_U^0)$ as the prior density $p_U(\boldsymbol{\theta}_U)$ of $\boldsymbol{\theta}_U$, $U \in \mathbb{U}$, then the posterior density $p_U(\boldsymbol{\theta}_U|\boldsymbol{y}_U)$ is Dirichlet $D(\boldsymbol{\alpha}_U^+)$ where

$$\boldsymbol{\alpha}_U^+ = \boldsymbol{\alpha}_U^0 + \boldsymbol{y}_U. \tag{9.8}$$

So the staged tree gives another example where with ancestral sampling a prior to posterior analysis is closed under sampling and where the components of the factorisation in (9.7) concern only features of the problem concerning each individual stage in the partition $\mathbb{U}$ and what has been observed to happen to units directly after arriving at that stage. Later in this chapter we will see that the popular finite discrete Bayesian network is a special case of a staged tree where both the tree and the partition of stages takes a particular form.

9.3.3 Sampling populations of units with different trees

Sometimes a decision analysis has to use information from units not just drawn at random from a population respecting different trees. We have already considered causal models where units can be either manipulated or just observed. Each subpopulation associated with each staged tree, under the ancestrality assumption above, will have a likelihood which separates and is a monomial in the probabilities and so supports a conjugate analysis. If the DM believes that all probabilities appearing in the florets of parallel situations in any tree are independent then again her learning will be modular. In particular under the hypothesis of a tree being causal evidence from designed experiments can be applied to a sampled unit, evidence from a sample survey can be applied to estimate the effect of a control or sample evidence can be combined together simply by multiplying their respective likelihoods. Note that we have already used this argument on the elicitation of the probability of two fragments of matching glass being found on the coat of a given guilty suspect seen standing 10 m from a pane smashed by a brick thrown at a particular velocity.

Although we illustrate below that it is by no means automatic that a sampling scheme is ancestral they commonly are. When they are, focus can centre on inferences about each stage floret in turn in the tree decomposition and the inferences then a posteriori can be pieced together using the prior decomposition a posteriori. Moreover in the last section of this chapter we will see that this will usually provide a good approximation of the appropriate inference when sample sizes are large even when the floret probability vectors are not a priori independent.

9.4 Estimating probabilities in Bayesian networks

9.4.1 Introduction

The conjugate estimation of the probability distributions in a discrete BN actually follows directly from the results given in the last section because it can always be represented as a particular kind of staged tree. To first illustrate this consider the following very simple example.

Example 9.7. Suppose W_1, W_2 and W_3 are three different weather features that can occur on any one day in August and suppose we believe the simple DAG. is valid

$$\mathcal{G} = W_1 \longrightarrow W_2 \longrightarrow W_3$$

If these variables are all binary then the state space of this BN are the root to leaf paths $\{\lambda_1, \lambda_2, \ldots, \lambda_8\}$ that can be represented by the event tree

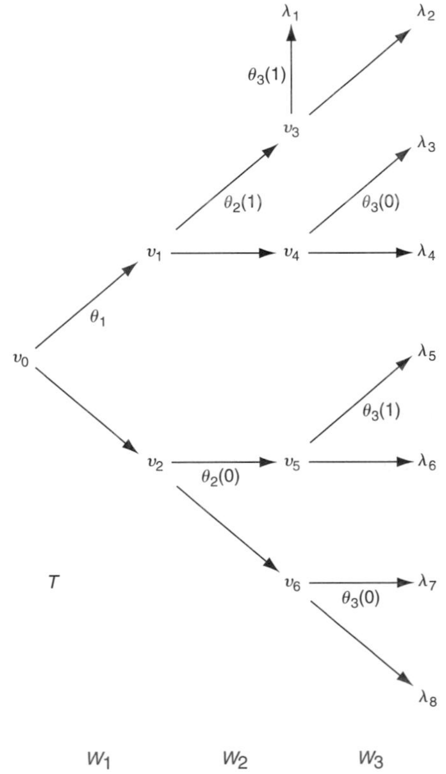

The conditional independence in the tree $W_3 \amalg W_1 | W_2$ give us a staged tree. Thus to complete this model we would need to specify the probabilities $\theta_1 = P(W_1 = 1)$, $\boldsymbol{\theta}_2 = (\theta_{2|0}, \theta_{2|1})$ and $\boldsymbol{\theta}_3 = (\theta_{3|0}, \theta_{3|1})$ where

$$\theta_{2|0} = P(W_2 = 1|W_1 = 0), \quad \theta_{2|1} = P(W_2 = 1|W_1 = 1),$$
$$\theta_{3|0} = P(W_3 = 1|W_2 = 0), \quad \theta_{3|1} = P(W_3 = 1|W_2 = 1)$$

so that the stages are given by $\{\{v_0\}, \{v_1\}, \{v_2\}, \{v_3, v_5\}, \{v_4, v_6\}\}$ where edge probabilities in the two nontrivial stages are associated as shown on the tree above. For a Bayesian analysis of this example the DM needs to first specify a joint prior distribution over the vector of five random variables $\boldsymbol{\theta} = (\theta_1, \theta_{2|0}, \theta_{2|1}, \theta_{3|0}, \theta_{3|1})$, and then update this in the light of records of each new unit as it arrives. The assumption that the vectors $\boldsymbol{\theta}_1, \boldsymbol{\theta}_2, \boldsymbol{\theta}_3$ associated with the conditional distributions W_1, W_2 and W_3 respectively are mutually independent, is sometimes called the *global independence* assumption: see e.g. Cowell *et al.* (1999); Lauritzen and Speigelhalter (1988). The assumption that the components of θ_2 and θ_3 are independent is sometimes called the *local independence assumption*. If the DM has a prior distribution over parameters which satisfies both these assumptions then it is easy to check from the semi-graphoid properties that all the components of $\boldsymbol{\theta}$ are mutually independent of each other. This prior assumption is the one we used in the staged tree but applied to this special case. Although this prior assumption is not always appropriate – see below – in a later section it is shown that inferences are often very robust to violations of this assumption.

In the next example we notice that the stages of this staged tree correspond to the different configurations parents of each child variable in the BN can take. Thus there is one (null) configuration of parents of W_1 corresponding to the root of the event tree, one for each configuration of W_1 which is the parent of W_2 in $\mathcal{G}$ and the configuration of the parent $W_2 = 1$ equates to the two situations $\{v_3, v_5\}$ whose emanating edges are embellished with the probabilities of this configuration. We can similarly equate $W_2 = 0$ to the two situations $\{v_4, v_6\}$. The stages forming these partitions are given by the slightly more complicated scenario described below.

Example 9.8. Patients suffering viral infection A exhibit the four symptoms

X_1 : taking values $1 =$ normal temperature, $2 =$ raised temperature, $3 =$ high temperature;

X_2 : taking values $0 =$ no headache, $1 =$ headache;

X_3 : taking values $0 =$ no aching limbs, $1 =$ aching limbs;

X_4 : taking values $0 =$ no dizziness, $1 =$ dizziness

that are believed to respect a BN whose DAG is given by

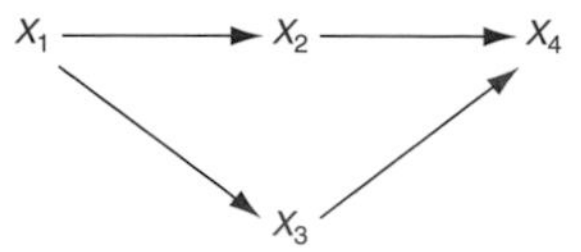

The event tree of this BN takes situations in an order consistent with their indexing

$$\{v_0, v_1, v_2, v_3, v_{10}, v_{20}, v_{30}, v_{11}, v_{21}, v_{31},$$
$$v_{100}, v_{200}, v_{300}, v_{101}, v_{201}, v_{301}, v_{110}, v_{210}, v_{310}, v_{111}, v_{211}, v_{311}\}$$

and whose leaves are

$$\{v_{1000}, v_{2000}, v_{3000}, v_{1010}, v_{2010}, v_{3010}, v_{1100}, v_{2100}, v_{3100}, v_{1110}, v_{2110}, v_{3110},$$
$$v_{1001}, v_{2001}, v_{3001}, v_{1011}, v_{2011}, v_{3011}, v_{1101}, v_{2101}, v_{3101}, v_{1111}, v_{2111}, v_{3111}\}.$$

Here the root vertex of the tree is v_0 and where, for example, v_{201} denotes the situation associated with the event $\{X_1 = 2, X_2 = 0, X_3 = 1\}$. Note the stages form the following partition of the situations

$$\{u_1 \triangleq \{v_0\}, u_{2|1} \triangleq \{v_1\}, u_{2|2} \triangleq \{v_2\}, u_{2|3} \triangleq \{v_3\},$$
$$u_{3|1.} \triangleq \{v_{10}, v_{11}\}, u_{3|2.} \triangleq \{v_{20}, v_{21}\}, u_{3|3.} \triangleq \{v_{30}, v_{31}\},$$
$$u_{4|.00} \triangleq \{v_{100}, v_{200}, v_{300}\}, u_{4|.01} \triangleq \{v_{101}, v_{201}, v_{301}\},$$
$$u_{4|.10} \triangleq \{v_{110}, v_{210}, v_{310}\}, u_{4|.11} \triangleq \{v_{111}, v_{211}, v_{311}\}\}.$$

Thus for example since $X_4 \amalg X_1 | X_2, X_3$ in particular

$$P(X_4 = 1|X_1 = 1, X_2 = 0, X_3 = 0) = P(X_4 = 1|X_1 = 2, X_2 = 0, X_3 = 0)$$
$$= P(X_4 = 3|X_1 = 1, X_2 = 0, X_3 = 0)$$

i.e. v_{100}, v_{200} and v_{300} all lie in the same stage $u_{4|.00}$, the stage associated with the random variable X_4 associated with the parental configuration of $X_2 = 0, X_3 = 0$.

In general all BNs on random variables taking finite discrete sets of values are a subset of staged trees where the tree is drawn is some arbitrary order and the stages of this tree correspond to each of the possible configurations of values of the parents of each vertex variable of the DAG of the BN. This means that we can translate the results concerning the separation properties of staged trees and apply them directly to Bayes estimation of the probabilities in a discrete BN. In particular if the DM believes that the probabilities emanating from situations in a given stage are Dirichlet and these Dirichlet random vectors are all mutually independent of each other then after ancestral sampling they will remain independent Dirichlets posterior to sampling with parameters updating in an obvious linear way.

Thus consider the updating of the probabilities using the observed symptoms of a random sample of patients suffering from the virus in the last example. For simplicity assume that all symptoms on all patients in the sample are observed so that sampling is complete and the ancestrality condition is trivially satisfied. Let the vectors of probabilities associated with the eleven stages be associated with the four random variables. These are the three vectors

$\boldsymbol{\theta}_1$ of probabilities associated with the values X_1 can take; $\boldsymbol{\theta}_2 = (\theta_{2|1}, \theta_{2|2}, \theta_{2|3})$ is associated with X_2 where $\theta_{2|x_1} = P(X_2 = 1|X_1 = x_1)$, $x_1 = 1, 2, 3$; $\boldsymbol{\theta}_3 = (\theta_{2|1.}, \theta_{2|2.}, \theta_{2|3.})$ is associated with X_3 where $\theta_{3|x_1.} = P(X_3 = 1|X_1 = x_1)$, $x_1 = 1, 2, 3$. Finally let $\boldsymbol{\theta}_2 = (\theta_{4|.00}, \theta_{4|.01}, \theta_{4|.10}, \theta_{4|.11})$ be the probabilities associated with different parental configurations of X_4 where $\theta_{4|.x_2x_3} = P(X_4 = 1|X_2 = x_2, X_3 = x_3)$, $x_2, x_3 = 0, 1$.

Suppose the DM believes that all these eleven parameters are a priori mutually independent – this is the local and global independence property. Moreover suppose the DM's prior density over $\boldsymbol{\theta}_1$ is Dirichlet $D(\alpha_{11}^0, \alpha_{12}^0, \alpha_{13}^0)$; $\theta_{2|x_1}$ has a beta $Be(\alpha_{2|x_1}^0, \beta_{2|x_1}^0)$ prior distribution, $x_1 = 1, 2, 3$; $\theta_{3|x_1.}$ has a beta $Be(\alpha_{3|x_1.}^0, \beta_{3|x_1.}^0)$ prior distribution, $x_1 = 1, 2, 3$; and $\theta_{4|.x_2x_3}$ has a beta $Be(\alpha_{4|.x_2x_3}^0, \beta_{2|.x_2x_3}^0)$ prior, $x_2, x_3 = 0, 1$.

Then on observing this complete data set the DM's posterior joint density exhibits the same independences it did a priori and the posterior distribution over $\boldsymbol{\theta}_1$ is Dirichlet $D(\alpha_{11}^+, \alpha_{12}^+, \alpha_{13}^+)$; $\theta_{2|x_1}$ has a beta $Be(\alpha_{2|x_1}^+, \beta_{2|x_1}^+)$ posterior, $x_1 = 1, 2, 3$; $\theta_{3|x_1.}$ has a beta $Be(\alpha_{3|x_1.}^+, \beta_{3|x_1.}^+)$ prior distribution, $x_1 = 1, 2, 3$; and $\theta_{4|.x_2x_3}$ has a beta $Be(\alpha_{4|.x_2x_3}^+, \beta_{4|.x_2x_3}^+)$ posterior, $x_2, x_3 = 0, 1$.

Here, by (9.8) the posterior parameters of the vector $\boldsymbol{\theta}_1$ are linked to the data

$$\alpha_{1x_1}^+ = \alpha_{1x_1}^0 + y_{x_1}$$

where y_{x_1} are the number of units in the sample taking the value $x_1, x_1 = 1, 2, 3$. Thus we use the obvious Dirichlet updating formula described in Chapter 5 to estimate the probability of ranges of temperature a patient might have. Equation (9.8) translates into recurrence of the conditional probabilities associated with X_2 satisfying

$$\alpha_{2|x_1}^+ = \alpha_{2|x_1}^0 + y_{x_1,1}, \quad \beta_{2|x_1}^+ = \beta_{2|x_1}^0 + y_{x_1,0}$$

where $y_{x_1,1}$ denote the number in the sample for which $X_1 = x_1$ and go on to exhibit a headache ($X_2 = 1$) and $y_{x_1,0}$ the number of these people who do not. Note again this is exactly analogous to the beta parameter updating equation given in Chapter 5 but simply applied to this conditional probability in the obvious way. Similarly the hyperparameters of components of the $\boldsymbol{\theta}_3$ vector update via

$$\alpha_{3|x_1.}^+ = \alpha_{3|x_1}^0 + y_{x_1,\ldots,1}, \quad \beta_{3|x_1.}^+ = \beta_{3|x_1.}^0 + y_{x_1,\ldots,0}$$

where $y_{x_1,\ldots,x_3}$ denote the number in the sample for which $X_1 = x_1$ and proceed to exhibit $X_3 = x_3$. Finally the hyperparameters of components of the $\boldsymbol{\theta}_4$ vector are given by the recurrences

$$\alpha_{4|.x_2x_3}^+ = \alpha_{4|.x_2x_3}^0 + y_{.x_2x_31}, \quad \beta_{4|.x_2x_3}^+ = \beta_{4|.x_2x_3}^0 + y_{.x_2x_30}$$

where $y_{.x_2x_3x_4}$ are the numbers of those in the sample for which $X_2 = x_2$ and $X_3 = x_3$ and $X_4 = x_4$, $x_2, x_3, x_4 = 0, 1$.

This conjugate updating is totally general and works for any BN. Furthermore it continues to apply if the sampling is ancestral to a particular tree. So under a good sampling scheme and expedient prior assumptions the updating of the parameters of the different variables in a BN can be updated variable by variable, drawing information in an obvious way from the sample. Thus the only data used to update a conditional probability vector are the units satisfying those configurations of values of parents corresponding to the conditioning event and the proportion of those taking the different values associated with that random variable. This modularity property of learning, applicable to both trees and BNs is absolutely critical for the efficient and quick learning of large systems. It not only allows various parts of the model to be updated in parallel but it also allow us to modularise the learning, making different agents responsible for different parts of the system, confident that the whole system can be recomposed in a coherent fashion. More details of how these ideas can be characterised and implemented can be found in for example, Heckerman (1998); Speigelhalter and Lauritzen (1990) and references therein. Details of how these learning systems can remain coherent in settings where agents associated in the estimation of parts of the system can only communicate locally are given in Xiang (2002). For a good review of analogous methods of Bayesian estimation of probabilities in CBNs see for example Heckerman (2007).

9.5 Technical issues about structured learning*

9.5.1 A simple three-variable example

When each individual unit in a random sample respecting a given BN is sampled ancestrally – i.e. there is a tree taking variables in a total order compatible with the DAG of the BN – then its sample distribution will just be a product of the probabilities associated with the variables in the BN. It follows that a likelihood which is proportional to the product of these products over the different units will be separable. On the other hand if there is even one unit for which this is not the case then the likelihood will be a polynomial but not a monomial in the conditional probabilities of that particular BN. This in turn means that if the DM believes a priori that all the conditional probabilities in the tree are mutually independent, sampling will induce dependences between at least some of them and the modularity property discussed above starts to be eroded.

Sometimes there is a partial solution to this problem. Thus consider the first example of the last section where the DM observes just the last two weather conditions W_2 and W_3 but W_1 is not observed for any unit. We note that a Markov equivalent BN – i.e. one sharing the same pattern and so a logically equivalent set of conditional independent statements – is

$$W_1 \longleftarrow W_2 \rightarrow W_3$$

and we note that we now have an ancestral sample for each unit. (Draw the tree beginning with W_2 then add W_3 situations and then W_1.). However this tree will be parameterised differently: $\theta_2' = P(W_{2'} = 1)$, $\boldsymbol{\theta}_1' = (\theta_{1|0}', \theta_{1|1}')$ and $\boldsymbol{\theta}_3 = (\theta_{3|0}, \theta_{3|1})$ where

$$\theta_{1|0}' = P(W_1 = 1|W_2 = 0), \quad \theta_{1|1}' = P(W_1 = 1|W_2 = 1),$$
$$\theta_{3|0} = P(W_3 = 1|W_2 = 0), \quad \theta_{3|1} = P(W_3 = 1|W_2 = 1).$$

If the DM is content to assume that all these parameters are independent a priori then they will also be independent a posteriori. Note that under this reparameterisation the posterior joint distribution of the parameter vector $\boldsymbol{\theta}'_1 \triangleq (\theta'_{1|0}, \theta'_{1|1})$ associated with the distribution of W_1 conditional on W_2 is identical to the prior joint distribution of this vector. We will see in the final section of this chapter that provided that the prior distributions are chosen to be mutually smooth, with large data sets whether we use a prior exhibiting local and globally independent in the original parameterisation or in the new one, posterior distributions will be close in variation distance. So even if the DM's beliefs dictate prior independence in the original parameterisation then posterior separation of the conditional probabilities in the new parameterisation will hold approximately, with all the advantages of transparency of interpretation this brings.

9.5.2 Priors invariant to an equivalence class of BNs

The comments above provoke the following question. Suppose the DM believes only that a particular set of conditional independence conditions hold. If this is so, then her beliefs about the joint density of the conditional probability parameters should not depend on the particular choice of BN in a given equivalence class. So is it possible for her to believe a priori that all parameters are mutually independent whatever equivalent parameterisation she uses?

The answer to this question is affirmative and was proved for the class of decomposable BNs in Dawid and Lauritzen (1993) and more generally in Daneshkhah and Smith (2004). It uses the well-known property of the Dirichlet distribution that if the joint probabilities of the finite discrete random vector $\boldsymbol{X} = (X_1, X_2, \ldots, X_k)$ have a Dirichlet distribution then the probabilities of X_1 and $X_i|X_1, \ldots, X_{i-1}$, $i = 1, 2, \ldots, k$, each also have a Dirichlet density and furthermore that these parameter vectors are mutually independent of each other. Clearly this continues to be so after a reindexing of the components of $\boldsymbol{X}$ since the Dirichlet family is closed under such reindexing.

A sketch proof of the result when the BN is decomposable is straightforward. Give each joint distribution of the cell probabilities of the joint table of the vector of random variables of $\big(X_1(c), X_2(c), \ldots, X_{k(c)}(c)\big)$ of each clique c of the decomposable BN a Dirichlet density. Ensure that the margins over the separator vector over the two cliques containing them – which from the property of the Dirichlet given above will itself be Dirichlet – agree. This is always possible because the parameters of the separator Dirichlet distribution are just the sums of the Dirichlet parameters in either of the Dirichlets on the cliques, so consistency is ensured by demanding these sums of clique hyper-parameters are identical. By equation (7.18) this gives a prior density over the space of all probabilities. Furthermore the property that each clique has a Dirichlet distribution ensures in particular that the conditional probabilities associated with a clique margin taking conditional probability vectors in any order associated with a factorisation compatible with the DAG are also all Dirichlet distributed. Again by the properties of the Dirichlet distribution, each of these vectors of conditional probabilities associated with a particular factorisation will be mutually independent of each other. This family of densities is called the hyper-Dirichlet family (Dawid and Lauritzen, 1993).

Example 9.9. Consider the weather example with three binary variables. This is a decomposable BN with two cliques $c_1 = (W_1, W_2)$ and $c_2 = (W_3, W_2)$. Suppose that the DM wants to set up a prior over the parameters of this model which has the invariance properties discussed above, and gives a Dirichlet $D(\alpha)$ distribution to the four probabilities in c_1 and a Dirichlet $D(\boldsymbol{\beta})$ distribution to the four probabilities in c_2 as below where for example $\beta_{01} = P(W_3 = 0, W_2 = 1)$.

c_1	α_{ij}	$\alpha_{\cdot 0}$	$\alpha_{\cdot 1}$	W_1	c_2	β_{ij}	$\beta_{\cdot 0}$	$\beta_{\cdot 1}$	W_3
	$i \backslash j$	0	1			$i \backslash j$	0	1	
$\alpha_{0\cdot}$	0	3	7	10	$\beta_{0\cdot}$	0	10	4	14
$\alpha_{1\cdot}$	1	9	1	10	$\beta_{1\cdot}$	1	2	4	6
W_2		12	8	20	W_2		12	8	20

The consistency condition we need to ensure that we have the same distribution on the probabilities of the separator W_2 of these two is that

$$\alpha_{\cdot 0} \triangleq \alpha_{00} + \alpha_{10} = P(W_2 = 0) \triangleq 1 - \theta_2' = \beta_{00} + \beta_{10} = \beta_{\cdot 0},$$

$$\alpha_{\cdot 1} \triangleq \alpha_{01} + \alpha_{11} = P(W_2 = 1) \triangleq \theta_2' = \beta_{01} + \beta_{11} = \beta_{\cdot 1}.$$

The properties of the Dirichlet distribution tell us that $\theta_2' \sim Be(\alpha_{\cdot 1}, \alpha_{\cdot 0})$ which in the table above is $Be(8, 12)$. The parameters of the BN $W_1 \to W_2 \to W_3$ are all independent beta distributed with $\theta_1 \sim Be(\alpha_{1\cdot}, \alpha_{0\cdot}) = Be(10, 10)$

$$\theta_{2|0} \sim Be(\alpha_{01}, \alpha_{00}) = Be(7, 3), \quad \theta_{2|1} \sim Be(\alpha_{11}, \alpha_{10}) = Be(1, 9),$$

$$\theta_{3|0} \sim Be(\beta_{10\cdot}, \beta_{00}) = Be(2, 10), \quad \theta_{3|1} \sim Be(\beta_{11}, \beta_{01}) = Be(4, 4),$$

whilst the parameters associated with $W_1|W_2$ associated with the alternative parameterisation are

$$\theta_{1|0}' \sim Be(\alpha_{10}, \alpha_{00}) = Be(9, 3), \quad \theta_{1|1}' \sim Be(\alpha_{11}, \alpha_{01}) = Be(1, 7).$$

There are various points to notice from this example. First the setting of the hyperparameters of the clique probability tables when these are integers sum like counts on a contingency table and satisfy exactly the consistency constraints on the margins of these tables. This has led their sum (here 20) to be called the effective sample size of this prior. We have already discussed in simpler scenarios the elicitation device of setting the values of prior hyperparameters to reflect the strength of evidence associated with a comparable sample. This is sometimes a convenient way of setting the values of the hyperparameters in practice.

Second it has been shown by Geiger and Heckerman (1997) that the demand for parameter independence over all equivalent decomposable BNs characterises the hyperdirichlet family of distributions: i.e. this is the *only* family of densities over discrete distributions with this property. So if the DM wants to choose a model which is invariant to the change in order of conditioning discussed above then she is forced to choose a candidate from this class. This interesting property is used to help characterise priors for model selection between BNs: see Heckerman (1998).

Third, there are some scenarios – for example in certain contexts of model selection – where the family of hyperdirichlet distributions are appropriate for the hyperparameters of a BN. However it must be remembered that they are also very restrictive. Consider the situation where the BN is proposed where $X \to Y$ and X indexes a disease category whilst Y indexes different collections of symptoms. Then it is common for a DM to be much less certain of the relative disease probabilities for X – these may well depend strongly on unobserved or fast-changing circumstances – than the probabilities of Y given disease X that are much less dependent on the underlying environment. In contexts like these the DM will want to have a smaller number of data equivalents, as measured by the sum of the hyperparameters of the disease Dirichlet, than the number of data equivalent pieces of information of the priors over the symptoms – as measured by the sum of the hyperparameters of the symptom given each diseases Dirichlet. But the hyperdirichlet forces the DM – under this measure of her uncertainty – to be much surer of the disease probabilities! In this quite common scenario the DM should not choose from a hyperdirichlet. Notice however it is elementary to find a conjugate product of densities with more uncertainty on the disease probabilities than the symptoms given disease probabilities but with both the marginal and all conditional densities all Dirichlet: it is just that their hyperparameters will not satisfy the summation constraints that need to be imposed on the hyperdirichlet in order for the invariance properties to be satisfied.

Third as a consequence of results presented in the last section of this chapter that for random samples giving very large counts in each parent–child configuration however the prior is set, within certain regularity conditions, the posterior distributions will not heavily depend on the default choice of a hyperdirichlet. However this is not so when the analysis is used for model selection. In fact even within the hyperdirichlet class model selection is highly sensitive to how the priors on the hyperparameters are set: see Freeman and Smith (2010); Silander *et al.* (2007); Steck (2008); Steck and Jaakkola (2002).

9.5.3 Non-separable missingness, asymptotic unidentifiability and ambiguity*

When data is systematically missing in a way that cannot be transformed into ancestral sampling we can regularly induce dependences between the densities of probabilities that are really hard to understand and explain to a DM. Sadly many interesting problems of inference associated with latent class analyses – models well studied by psychologists – phylogenetic models – studied by evolutionary biologists and Markov switching models – used for example in speech recognition, are just some of the many examples of real models exhibiting these difficulties. The characteristic of such a model is that an intermediate state through which a unit proceeds from one part of a process to another is not observed. The path it took cannot then be determined however large the sample of the end points. For these types of model the choice of subjective prior usually critically determines the deductions being made. The simplest example of when sampling induces complex dependences can again be illustrated using the first example of the last section but where the value of (W_1, W_3) is seen in all units but the value of W_2 – determining the path taken from W_1 to W_3 – is never observed.

In sampling structures like these a property called aliasing always raises its head. This property is not simply linked to conditional independence but to group symmetries within the joint probability model. This means that even if the structure of a model was a priori specified as a simple set of conditional independences between the variables in the BN, sampling will induce other more complicated structure. This structure – unlike conditional independence structure – depends critically on the number of levels the hidden (intermediate) variables can take. But even when the variables are all binary – as in our weather example – the structure of the likelihood is non-trivial and generally exhibits maximum likelihood estimates that are not only non-unique but also line segments. The posterior density of the probabilities is therefore of a complicated analytic form and very prior dependent.

The situation becomes much worse when the hidden variable can take more than two levels. In (Mond *et al.*, 2003; Smith and Croft, 2003; Robins *et al.*, 2003) we showed that when W_1 and W_3 each have four levels and W_2 has three levels, it is not unusual for the limit as $n \to \infty$ of the observed likelihood on $(\theta_1, \boldsymbol{\theta}_2, \boldsymbol{\theta}_3)$ associated with a very large data set to have as many as 48 global maxima, each corresponding to a very different but equally likely explanation of the data. Note that this phenomenon does *not* go away as the sample size increases.

These types of issue may seem obscure but are actually very important to understand well. For example in phylogenetic trees (Settimi and Smith, 2000), describing the evolution of one species to another, we typically only have genetic information about species that are currently alive. If evolution is represented by a rooted tree where the root is a hypothesised common ancestor and each species is related to a four-level marker, all we have data about marker values on are the species associated with leaves of the tree. This type of tree is simply a more complicated version of the weather example. So we know that any inferences that can be made about the pedigree will tend to present sets of possible interpretations all quite different and all equally supported by the data observed.

Because conjugacy is often lost, missing data problems need to be analysed numerically. The more routine methods based on Metropolis–Hastings algorithms or Markow chain Monte Carlo methods appear to work quite well provided that about 80% of the data on each unit is not missing and there is no node only sparsely informed. But if this is not so then – as in phylogenetic models – numerical methods need to be customised using an awareness of the underlying geometry of the problem if the solution converged to is not going to be one of many equally good alternatives that are undiscovered.

9.6 Robustness of inference given copious data*

9.6.1 Introduction

An auditor will often be prepared to agree the likelihood of this well-designed experiment. On the other hand he might want to choose a different prior density than the one the DM herself has used. Indeed the DM herself may be concerned that the inevitable elicitation errors in her prior, perhaps inherited from a remote expert, might influence her posterior. When the sample is large and informative, can she at least formally demonstrate that in this scenario she will discover the appropriate posterior distribution of this vector and that a different plausible prior would give very similar conclusions a posteriori, for any decision analysis she may want to perform? Illustrations of conjugate prior to posterior analyses given in Chapter 5 suggested this might be so.

One common way to address this issue is for the analyst to perform a numerical *sensitivity analysis*. Other plausible candidate priors the DM or analyst might use could be encoded and the posterior density checked to see whether the optimal policies change much in the light of these alternatives. It is wise to perform such sensitivity analyses as a matter of course. But even when such numerical sensitivity analyses appear to demonstrate a robustness to her prior the thoughtful DM or appraiser may still be rightly concerned that different perturbations they did not check, perhaps ones outside a known parametric family, might give rise to big changes in the optimal decision even though the sample sizes of supporting experiments are large. If this were so then it would be a big concern.

For the purposes of this section let f_0 and f_n denote respectively the *functioning prior* and *functioning posterior* – i.e. the one the DM actually proposes to use – and let g_0 and g_n denote respectively the *genuine prior* and *genuine posterior* – i.e. the one the DM if she thought much harder or the one the auditor would use. All densities are assumed to be over a finite parameter vector $\boldsymbol{\theta} \in \Theta \subseteq \mathbb{R}^m$. Here an observed sample of n observations is denoted by $\boldsymbol{y}_n = (y_1, y_2, \ldots, y_n)$, $n \geq 1$. Assume a sequence of *observed* sample densities $\{p(\boldsymbol{y}_n|\boldsymbol{\theta})\}_{n\geq 1}$ are all continuous on Θ and the experiment is such that both the DM and auditor can agree about this conditional density.

Recall that in Chapter 3 we argued that if the difference between the posterior densities is measured by *variation distance* $d_V(f_n^*, g_n^*) \triangleq \int |f_n^*(\boldsymbol{\zeta}) - g_n^*(\boldsymbol{\zeta})| d\boldsymbol{\zeta}$ then the expected utilities associated with the same utility function and the same class of decisions will be uniformly close for the two densities, whatever the bounded utility – by definition depending only on $(\boldsymbol{Y}, \boldsymbol{Z}, \boldsymbol{\theta}, \boldsymbol{\zeta})$ through the vectors of uncertain quantities $(\boldsymbol{Z}, \boldsymbol{\zeta})$ provided that $d_V(f_n^*, g_n^*)$ is small, where $\boldsymbol{Z}, \boldsymbol{\zeta} \amalg \boldsymbol{Y}|\boldsymbol{\theta}$. It follows that if we can show that $d_V(f_n^*, g_n^*)$ gets progressively smaller as $n \to \infty$ then provided the sample is chosen large enough, whatever the DM's utility function whether f_n^* or g_n^* is used will make no substantive difference to the DM's or auditor's evaluation of the efficacy of different decisions. So closeness in variation will be sufficient to ensure the robustness and hence the persuasiveness of the DM's analysis.

The robustness of posterior densities to the mis-specification of prior densities has now been widely formally studied and there is an extensive literature on this topic to which I cannot possibly do justice in this short section. However some results are key to appreciating those features of the prior that fade quickly as data is accommodated and those features that endure. The broad conclusion we can make is that probabilities needed for a typical decision analysis are extremely robust to prior mis-specification when data is truly informative about $\boldsymbol{\theta}$ and when $p(\boldsymbol{\zeta}|\boldsymbol{\theta})$ is agreed by all parties, with some important caveats that are discussed below.

We saw earlier in this chapter that even when data sets become progressively larger it is not necessarily the case that the likelihood gives progressively better information about parameters in the system. For example we saw that conditional densities of lower level parameters in a unit hierarchy were not observable and that it was impossible to learn about certain features associated with the dependence between state random vectors in a state space model from sampling. So suppose we are outside such situations and it is possible to learn about the parameter vector the DM needs with progressive accuracy. Here throughout we condition implicitly on the known covariates $\boldsymbol{X}$.

Assume then that with the prior we are using, the posterior distribution concentrates its probability mass on an ever-smaller neighbourhood of the parameter vector. In this section we focus our attention on these circumstances. Let

$$B(\boldsymbol{m}, \rho) \triangleq \{\boldsymbol{\theta} : \|\boldsymbol{\theta} - \boldsymbol{m}\| < \rho\}$$

for $\boldsymbol{x} = (x_1, x_2, \ldots, x_r)$ and $\|\boldsymbol{x}\| = \left(\sum_{i=1}^{p} x_i^2\right)^{1/2}$ so that $B(\boldsymbol{m}, \rho)$ denotes the open ball centred at $\boldsymbol{m}$ and with (small) radius ρ.

Definition 9.10. We say that f_n *concentrates* its mass on $\boldsymbol{m}_n$ as $n \to \infty$ if for each $\delta > 0$ there exists a sequence of location vectors $\boldsymbol{m}_n$ and sets $A_n(\delta) = \{\boldsymbol{\theta} : |\boldsymbol{\theta} - \boldsymbol{m}_n| \leq \delta\}$ having the property that

$$\int_{\boldsymbol{\theta} \notin A_n} f_n(\boldsymbol{\theta}) d\boldsymbol{\theta} \triangleq \alpha_n \to 0$$

as $n \to \infty$.

Suppose the density $p(\boldsymbol{\zeta}|\boldsymbol{\theta})$ is sufficiently smooth to have the property that for all $\boldsymbol{\theta} \in B(\boldsymbol{m}, \rho)$

$$|p(\boldsymbol{\zeta}|\boldsymbol{\theta}) - p(\boldsymbol{\zeta}|\boldsymbol{\theta} = \boldsymbol{m})| \leq \delta(\boldsymbol{m}, \rho)$$

where

$$\sup_{\boldsymbol{m} \in \boldsymbol{\Theta}} \delta(\boldsymbol{m}, \rho) \leq \delta(\rho)$$

and where $\delta(\rho) \to 0$ as $\rho \to 0$. Noting that the marginal density $f_n^*(\boldsymbol{\zeta})$ after seeing $\boldsymbol{y}_n$ of the parameters of interest can be calculated using the formula

$$f_n^*(\boldsymbol{\zeta}) = \int p(\boldsymbol{\zeta}|\boldsymbol{\theta}) f_n(\boldsymbol{\theta}) d\boldsymbol{\theta}$$

then the difference between the functioning posterior density $f_n^*(\boldsymbol{\zeta})$ and the density obtained by simply plugging in an estimate $\boldsymbol{m}$ of $\boldsymbol{\theta}$ into $p(\boldsymbol{\zeta}|\boldsymbol{\theta})$

$$\begin{aligned}\left|f_n^*(\boldsymbol{\zeta}) - p(\boldsymbol{\zeta}|\boldsymbol{\theta}=\boldsymbol{m})\right| &= \left|\int [p(\boldsymbol{\zeta}|\boldsymbol{\theta}) - p(\boldsymbol{\zeta}|\boldsymbol{\theta}=\boldsymbol{m})] f_n(\boldsymbol{\theta})d\boldsymbol{\theta}\right| \\ &\le \int |p(\boldsymbol{\zeta}|\boldsymbol{\theta}) - p(\boldsymbol{\zeta}|\boldsymbol{\theta}=\boldsymbol{m})| f_n(\boldsymbol{\theta})d\boldsymbol{\theta} \\ &\le \delta(\rho)\int f_n(\boldsymbol{\theta})d\boldsymbol{\theta} = \delta(\rho).\end{aligned}$$

This is interesting on its own account. Thus assume the expert has strong information about $\boldsymbol{\theta}$ but simply tells the estimate $\boldsymbol{m}_n$ of $\boldsymbol{\theta}$ – often a probability – to the DM not his full density. He nevertheless can assure her that f_n concentrates its mass on $\boldsymbol{m}_n$ and that n is large enough to make both ρ_n and $\delta(\rho_n)$ negligibly small. Then under the conditions above the DM can simply approximate her posterior density $f_n^*(\boldsymbol{\zeta})$ by the plug-in estimate $p(\boldsymbol{\zeta}|\boldsymbol{\theta}=\boldsymbol{m}_n)$ and that this approximation will enable her to identify decision rules which give almost the highest utility. The practical implications of this is that if any expert system delivers estimates for all its parameters – say a vector of probabilities – that have concentrated on a point estimate then that point estimate is probably all the DM needs to perform her analysis.

Secondly the simple inequality above can be used in another context. The most common results that have been proved about Bayesian robustness take the following form. Suppose data really does come from one of the sample densities $\{p(\boldsymbol{y}|\boldsymbol{\theta})\}_{n\ge1}$ here the one where $\boldsymbol{\theta}=\boldsymbol{\theta}_0$ and $\boldsymbol{\theta}_0$ is not on the boundary of Θ. Also assume that a consistent estimator of $\boldsymbol{\theta}$ exists; i.e. there is a function of the data that tends almost surely to the true value $\boldsymbol{\theta}_0$ of $\boldsymbol{\theta}$. So for example if y were a probability and $\boldsymbol{y}$ the indicator variables on a random sequence of coin tosses where θ is the probability of a head then it is well known that the sample proportion $\overline{Y} \triangleq n^{-1}\sum_{i=1}^{n} Y_i$ is a consistent estimator of θ.

Under such conditions, in various senses made explicit in Schervish (1995), the posterior density converges almost surely to $\boldsymbol{\theta}_0$ whatever the prior. A different strong convergence result concerning posterior densities is proved in Ghosh and Ramamoorthi (2003), p18.

These are very useful theoretical results for the decision analyst. By letting $\boldsymbol{m} \triangleq \boldsymbol{\theta}_0$ then under the consistency conditions discussed above and extending notation in the obvious way,

$$\left|f_n^*(\boldsymbol{\zeta}) - g_n^*(\boldsymbol{\zeta})\right| \le \left|f_n^*(\boldsymbol{\zeta}) - p(\boldsymbol{\zeta}|\boldsymbol{\theta}_0)\right| + \left|g_n^*(\boldsymbol{\zeta}) - p(\boldsymbol{\zeta}|\boldsymbol{\theta}_0)\right| \le 2\delta(\rho)$$

so that the functioning and genuine analyses will give approximately the same expected utilities for each decision for large enough n.

Interestingly the least robust scenario is when the DM believes that $\boldsymbol{\zeta} = \boldsymbol{\theta}$ where the DM's utility function may not be smooth. Here if the DM simply needs to predict the next observation in an exchangeable sequence then convergence happens in an even stronger sense than the one given above, see Blackwell and Dubins (1962). For this reason we focus our discussion on this most problematic case when the DM needs to learn directly about some property of $\zeta = \boldsymbol{\theta}$ itself to determine her expected utility. You are asked to check in an exercise that in the contexts we describe above $d_V(f_n^*, g_n^*) \leq d_V(f_n, g_n)$, so the bounds we obtain below apply also to the cases discussed above, albeit rather coarsely.

Exactly how large does n need to be for two Bayesian posterior densities to be close and what happens if the sample density has been mis-specified by the expert? Whilst accepting that the mis-specification will mislead her inferences, it would be helpful for the DM to know that she will be mislead in almost the same way whatever her prior. At least then both she and the auditor will come to approximately the same conclusion using their different priors even if this conclusion is wrong.

9.6.2 A Bayesian learns nothing about smoothness

Using the notation above, then using Bayes rule (5.10) the genuine posterior density $g_n(\boldsymbol{\theta}) \triangleq g(\boldsymbol{\theta}|\boldsymbol{y}_n)$ and functioning posterior density $f_n(\boldsymbol{\theta}) \triangleq f_0(\boldsymbol{\theta}|\boldsymbol{y}_n)$ after n observations will be given respectively by

$$\log g_n(\boldsymbol{\theta}) = \log g_0(\boldsymbol{\theta}) + \log p(\boldsymbol{y}_n|\boldsymbol{\theta}) - \log p_g(\boldsymbol{y}_n),$$
$$\log f_n(\boldsymbol{\theta}) = \log f_0(\boldsymbol{\theta}) + \log p(\boldsymbol{y}_n|\boldsymbol{\theta}) - \log p_f(\boldsymbol{y}_n) \qquad (9.9)$$

where $p_g(\boldsymbol{y}_n)$ and $p_f(\boldsymbol{y}_n)$ are the predictive densities/mass functions of $\boldsymbol{y}_n$ using the genuine and functioning prior respectively. Let the *local log density ratio distance* $d_A^L(f, g)$ over a set $A \subseteq \Theta$ between two densities f and g be defined by

$$d_A^L(f, g) \triangleq \sup_{\boldsymbol{\theta},\boldsymbol{\phi}\in A} \{\log f(\boldsymbol{\theta}) - \log f(\boldsymbol{\phi}) + \log g(\boldsymbol{\phi}) - \log g(\boldsymbol{\theta})\}.$$

Note that these distances are easy to interpret. For example if f_n, g_n, f_0, g_0 are all continuous and A is closed and bounded then $d_A^L(f_n, g_n)$ is simply the maximum value of $\log f_n(\boldsymbol{\theta}) - \log g_n(\boldsymbol{\theta})$ in A subtracted from the minimum value of this function. Now by subtracting the equations (9.9) from each other gives that

$$\log g_n(\boldsymbol{\theta}) - \log f_n(\boldsymbol{\theta}) = \log g_0(\boldsymbol{\theta}) - \log f_0(\boldsymbol{\theta}) + \Delta(\boldsymbol{y}_n) \qquad (9.10)$$

where, by definition, $\Delta(\boldsymbol{y}_n) = \left(\log p_f(\boldsymbol{y}_n) - \log p_g(\boldsymbol{y}_n)\right)$ is not a function of $\boldsymbol{\theta}$. So by subtracting (9.10) evaluated at a point $\boldsymbol{\theta} \in A$ from (9.10) evaluated at a different point $\boldsymbol{\phi} \in A$ allows us to deduce that, for any $A \subseteq \Theta$

$$d_A^L(f_n, g_n) = d_A^L(f_0, g_0).$$

Thus posterior distances $d_A^L(f_n, g_n)$ remain the same as the prior distances $d_A^L(f_0, g_0)$ whatever we learn from data, however informative our sample is, provided what we observe is not a logical impossibility and could be explained, however surprisingly, as a possible observation from each $\boldsymbol{\theta} \in A$. So any inferences dependent on these distances will also be dependent on how we chose to specify priors: what we put in a priori is what we get out a posteriori. This property was first noted by DeRobertis (1978), Hartigan and De Robertis when $A = \Theta$, and was subsequently characterised in Wasserman (1992) and the local versions of the properties studied more recently in Smith and Daneshkhah (2009) and Smith and Rigat (2009).

9.6.3 Strong robustness to prior mis-specification

Here we will choose the sets $A = B(\boldsymbol{m}, \rho)$ to be small and the DM believes that $d^L_{B(\boldsymbol{m},\rho)}(f, g) \to 0$ as $\rho \to 0$. Without this second condition her approximating functioning posterior may well not approximate her genuine posterior in nearly all standard inferential scenarios (Gustafson and Wasserman, 1995) even when the functioning posterior concentrates its mass on to a ball $B(\boldsymbol{m}, \rho)$ whose radius ρ is arbitrarily small. So at least in the case when $\boldsymbol{\zeta} = \boldsymbol{\theta}$ without these conditions inference might be unstable.

But often the DM is happy to assume that $\log f_0$ and $\log g_0$ are differentiable on $\boldsymbol{\theta} \in B(\boldsymbol{m}, \rho)$ where $\boldsymbol{m}$ is in a part of the parameter space. If for some points in $\boldsymbol{\theta} \in B(\boldsymbol{m}, \rho)$ the DM believes that the derivative of $\log f_0$ is much larger than $\log g_0$ whilst at others the derivative of $\log f_0$ is much smaller than $\log g_0$ – i.e. that the genuine prior could be very lumpy in the way it distributes its mass – then the posterior distance $d^L_{B(\boldsymbol{m},\rho)}(f_n, g_n)$ will be large however small ρ is. It can be shown that in this case if f_n concentrates its mass on $B(\boldsymbol{m}, \rho)$ then the variation distances $d_V(f_n, g_n)$ between f_n and g_n will be large and so Bayes decisions identified using the approximating posterior density f_n may be totally different than those that should be used. So robustness of Bayesian inference about $\boldsymbol{\theta}$ at least is critically dependent on getting the smoothness of the approximating prior in the right ball park.

Let $\boldsymbol{\theta} \in B(\boldsymbol{m}, \rho)$ where $\boldsymbol{m}$ is in a part of the parameter space. Suppose that the DM is happy to assume that within $B(\boldsymbol{m}, \rho)$ both $\log f_0$ and $\log g_0$ are differentiable and that the derivative of each on this set is bounded in modulus value by M. Then it is easy to check that for all $\rho \leq R$ for some small value of R

$$d^L_A(f_n, g_n) = d^L_A(f_0, g_0) \leq 3M\rho.$$

This in turn assures, with a regularity condition that the variation distances between the posterior using the genuine and functioning priors will become increasingly close. Suppose the DM believes that $p_f(\boldsymbol{y}) \leq p_g(\boldsymbol{y})$, i.e. that the marginal likelihood of the functioning prior is no larger than c times larger than the genuine one and the genuine prior will explain the data at least as well as the functioning prior. Then

$$d_V(f_n, g_n) \leq d^R_{A_n(\delta)}(f_0, g_0) + 2\alpha_n(1 + \Lambda) \tag{9.11}$$

where $A_n(\delta)$ $\{A_n(\delta)\}_{n\geq 1}$ are defined as a function of the statistics of our functioning posterior with the property given above and Λ is defined so that

$$\sup_{\boldsymbol{\theta}\in\Theta} \frac{g(\boldsymbol{\theta})}{f(\boldsymbol{\theta})} = \Lambda < \infty \tag{9.12}$$

and is set so that when $g(\boldsymbol{\theta})$ is bounded then this condition requires that the tails of g are no thicker than those of f. This result not only proves the type of robustness we need regardless of mis-specification of the sampling distribution but also enables us to formally bound the variation distance due to prior mis-specification of the prior. These bounds will of course depend on how the DM chooses Λ and M but otherwise just depend on statistics like the functioning posterior variance which will usually be routinely available anyway. The explicit calculation of these bounds is rather context specific and so outside the scope of this book. Many examples of these constructions are found in Smith (2007); Smith and Rigat (2009) and their application to high-dimensional BNs under numerical prior to posterior analyses in Smith and Daneshkhah (2009).

The basic rule is therefore that we usually do not need to worry about mis-specification of a prior when there is a lot of data available informative about the parameters. The times when there may be problems are the following:

(1) When the genuine prior is very rough compared to our choice of functioning – a case not often met in practice except for non-parametric formulations; see e.g. Ghosh and Ramamoorthi (2003).

(2) When the data lies in the remote tail of the functioning prior – or in the case of parameters with bounded support at the edge of the parameter space – so that in this neighbourhood $d^L_{A_n}(f_0, g_0)$ might be very large if the tail of the prior densities have different characteristics. Thus the DM might be happy to assume that

$$\sup_{\boldsymbol{\theta},\boldsymbol{\phi}\in A} |f_0(\boldsymbol{\theta}) - f_0(\boldsymbol{\phi})| + \sup_{\boldsymbol{\theta},\boldsymbol{\phi}\in A} |g_0(\boldsymbol{\theta}) - g_0(\boldsymbol{\phi})|$$

is small for small neighbourhoods A_n. This is however not sufficient to assure $d^L_{A_n}(f_0, g_0)$ is small only if $f_0(\boldsymbol{\theta})$ is bounded away from zero because otherwise both $\log f_0(\boldsymbol{\theta})$ and $\log g_0(\boldsymbol{\theta})$ are very different large negative values. It is well known – see e.g. Dawid (1973); O'Hagan (1979); Smith (2007) – that prior densities with different tail characteristics respond to outliers in completely different ways. This should not worry a DM too much because if her prior were so mis-specified, by using a diagnostic she would detect this and probably want to re-evaluate her whole analysis in any case.

(3) The functioning prior has too tight tails so that a prior permanently dominates the data when the data is surprising. This encourages the expert in charge of a probabilistic expert system to be conservative and whenever plausible to choose a prior density with heavy tails, seen by many as good Bayesian practice anyway; see for example O'Hagan and Forster (2004).

So despite robustness to prior specification measured by posterior variation distance being one of the strongest forms of robustness we could reasonably demand it hold, in most decision analytic scenarios we could reasonably expect it to. Less stringent forms of stability – see French and Rios Insua (2000) for examples of these – obviously allow convergence with even fewer prior conditions. If data is informative then in many formal senses a decision analysis is usually robust to prior settings. The analyst should therefore concentrate his attention on ensuring that the credence decomposition used is faithful to the DM and focus much of his energy on eliciting beliefs about those features in the model where there is sparse empirical evidence.

9.7 Summary

In complex applications there are often straightforward and elegant ways of formally including data from experiments, surveys and experimental studies into the probabilistic evaluation of features of the problem related to the distribution of a utility function under various decisions. The preservation after sampling of the DM's credence decompositions can make the online accommodation of data very fast. This is especially the case when a prior can be chosen which is closed under sampling. Then not only will these computations be quick but also the analysis will be able to provide a transparent narrative enabling the DM to explain the effects of the data she has used on her current probabilistic beliefs.

As we stated in Chapter 5, there are many circumstances it will be necessary for the DM – or the expert whose analyses are adopted by the DM – to calculate her posterior distributions numerically. However there is widely available software and open-access code which allows her to do this. The algorithms used to make these calculations make use of the credence decomposition of the problem to speed up their calculations and to preserve the types of modularity discussed here.

It is impossible to learn anything from data about certain properties of a distribution supporting a Bayesian decision analysis. Properties such as the smoothness of distributions on probabilities, discussed in the section above, the distributions of lower level parameters in a hierarchical model, discussed in the first section of this chapter, or the joint distribution of two first level parameters given data and their margins, discussed in the last section of Chapter 8, are all invariant to learning using Bayes rule. So in this sense Bayesian inference and Bayesian decision modelling relies on appropriate prior densities if it is to give a faithful representation of *all* uncertain quantities in the system. On the other hand an expected utility is usually only a function of features of this joint distribution that *can* be learned about with increasing accuracy as data support increases.

The analyses, both exact and numerical, tend to be robust to prior mis-specification – at least from the standpoint of a decision analysis – provided that priors do not conflict with the data and the data is informative about the parameters of the model. The main problems we encounter therefore tend to centre on the DM's ability faithfully to represent the relationships between her information sources, her model of process and her utilities. It is the appropriate *structuring* of the problem – as described in these last three chapters – which is the necessary prerequisite for effective decision making in large problems. Therefore the development of effective frameworks that help the DM perform such structuring is one of the key tasks of a decision analyst. If the structuring is faithful to the structure of the problem then a wise decision analysis will normally follow.

9.8 Exercises

9.1 Prove that if the likelihood of an experiment is separable conditional on $\boldsymbol{\theta}_{q+1}$ at $\boldsymbol{x}$ then

$$\coprod_{j=1}^{q} \boldsymbol{\theta}_j | \boldsymbol{\theta}_{q+1}, \boldsymbol{X} = \boldsymbol{x} \Leftrightarrow \coprod_{j=1}^{q} \boldsymbol{\theta}_j | \boldsymbol{\theta}_{q+1}, \boldsymbol{X} = \boldsymbol{x}, \boldsymbol{Y}.$$

9.2 Use the d-separation theorem to check that after sampling we cannot in general conclude from the DAG in Section 2.3 above that $\theta_2 \amalg (\theta_1, \theta_3) \,|\boldsymbol{Y}, \boldsymbol{X}, \theta_4$. Also show that if the alternative prior with $(\theta_1, \theta_2, \theta_3) \amalg \theta_4$ simply omits the edges (θ_4, θ_2) and $(\theta_4, (\theta_1, \theta_3))$ from this graph it is still not possible to conclude that $\theta_2 \amalg (\theta_1, \theta_3) \,|\boldsymbol{Y}, \boldsymbol{X}, \boldsymbol{\theta}_4$.

9.3 Calculate the recurrences for the multiregression model given above.

9.4 You take a random sample from the tree below where you observe only whether or not each unit reaches the leaf v_4. Write down the likelihood of this experiment and prove that this likelihood does not separate the probability parameters of its two florets

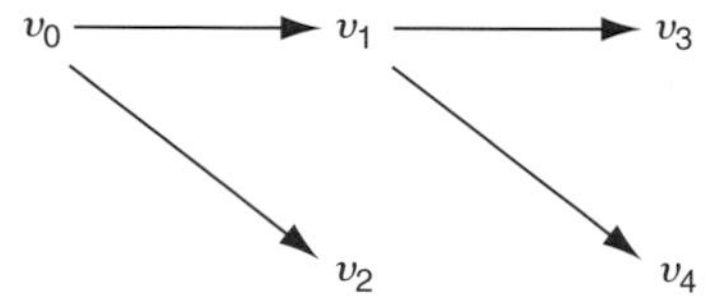

Prove that non-ancestral sampling can lead to a likelihood that is no longer separable so these two convenient properties are lost.

9.5 A valid BN with three binary random variables $W_1 \to W_2 \to W_3$ and associated vectors of parameters $(\theta_1, \boldsymbol{\theta}_2, \boldsymbol{\theta}_3)$ was defined in Example 9.9. Show that if the DM believes that $(\theta_1, \boldsymbol{\theta}_2, \boldsymbol{\theta}_3)$ are globally independent that the following DAG of $(\theta_1, \boldsymbol{\theta}_2, \boldsymbol{\theta}_3, W_1, W_2, W_3)$ is also valid.

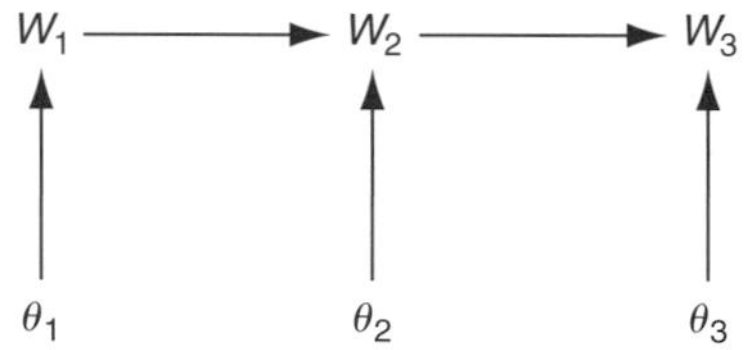

Use the d-separation theorem to prove that $\boldsymbol{\theta}_2$ and $\boldsymbol{\theta}_3$ will become dependent after sampling the value of (W_1, W_3) of a unit respecting this BN. Construct an example where, if $(\widehat{\theta}_1, \widehat{\boldsymbol{\theta}}_2, \widehat{\boldsymbol{\theta}}_3)$ maximises the likelihood then another quite different combination of values $(\widehat{\theta}_1, \widetilde{\boldsymbol{\theta}}_2, \widetilde{\boldsymbol{\theta}}_3)$ also does. What does this tell us about the posterior over these parameters?

9.6 You have collected data on a set of 100 patients. You have recorded whether or not they exhibited each of three symptoms and whether they have an infection $\{Z = 1\}$, or not $\{Z = 0\}$. Say that the binary random variables $Y[i] = 1$ iff the patient exhibited symptom i, $1 \le i \le 3$ and 0 otherwise.

Which conditional independence would the idiot Bayes model assume about the random variables $Z, Y[1], Y[2], Y[3]$? Represent this model as a BN. Let $\theta_Z = P(Z = 1)$ and $\theta_{Y[i]|Z=j} = P(Y[i] = 1|Z = j)$, $i = 1, 2, 3$ and $j = 0, 1$. What does it mean for a BN to exhibit local and global independence? Carefully stating but without proof any results you might need, find the posterior distribution of the seven vectors of these probabilities when you observe the following joint probability table of 100 observations of $(Z, Y[1], Y[2], Y[3])$

$y[1], y[2], y[3]$	000	001	010	011	100	101	110	111
$Z = 0$	10	8	6	4	5	2	1	0
$Z = 1$	2	5	5	5	8	12	12	15

and each of the seven probabilities are a priori thought to be independent each with a uniform $Be(1, 1)$ prior density.

9.7 The DAG G of an influence diagram on the binary random variables $\{X[i] : 1 \leq i \leq 9\}$ has directed edges $\{(X[1], X[2]), (X[2], X[3]), (X[3], X[4]), (X[3], X[5]), (X[4], X[5]), (X[4], X[6]), (X[4], X[8]), (X[4], X[9]), (X[5], X[6]), (X[5], X[7]), (X[8], X[9])\}$.

Write down a set of conditional independence statements of the influence diagram with the DAG G. Although the probabilities $P\{X[i] = 1|Q_i\}$, $4 \leq i \leq 9$, are known where Q_i is any configuration of parents in G of $X[i]$, the joint distribution of $\{X[1], X[2], X[3]\}$ is unknown to you. You decide to assign a uniform distribution to the five probabilities $P(X[1] = 1), P(X[2] = 1|X[1] = 0), P(X[2] = 1|X[1] = 1), P(X[3] = 1|X[2] = 0), P(X[3] = 1|X[2] = 1)$. You believe probabilities are locally and globally independent. Use the d-separation theorem to prove that, if you observed $X[1] = X[2] = X[3] = 0$, global independence would be preserved. Also prove that local independence would be preserved.

You now take a random sample of measurements of $(X[1], X[2], X[3])$ on 100 machines where the measurements respect the conditional independence coded in the DAG G. The number $n(x[1], x[2], x[3])$ of machines observed in each configuration is given in the table below.

$(x[1], x[2], x[3])$	$(0,0,0)$	$(0,0,1)$	$(0,1,0)$	$(0,1,1)$	$(1,0,0)$	$(1,0,1)$	$(1,1,0)$	$(1,1,1)$
$n(x[1], x[2], x[3])$	6	20	10	4	12	16	23	9

Assuming local and global independence, state without proof the posterior distribution of this vector of five probabilities. Can you see anything in the data that might cause you to question any conditional independence assumption in this model?

9.8 You take a multinomial sample of size $N = \sum_{i=1}^{5} x_i$ on six categories to obtain a likelihood $l(\boldsymbol{\theta}|\boldsymbol{x})$ from the sample density $p(\boldsymbol{x}|\boldsymbol{\theta})$ where

$$p(\boldsymbol{x}|\boldsymbol{\theta}) = \frac{N!}{x_1!x_2!x_3!x_4!x_5!}\theta_1^{x_1}\theta_2^{x_2}\theta_3^{x_3}\theta_4^{x_4}\theta_5^{x_5}\theta_6^{x_6}$$

and where $\boldsymbol{\theta} = (\theta_1, \theta_2, \theta_3, \theta_4, \theta_5, \theta_6)$, $\theta_i > 0$ for $1 \leq i \leq 6$ and $\sum_{i=1}^{6} \theta_i = 1$. Suppose your prior density $\pi(\boldsymbol{\theta}|\boldsymbol{\alpha}_0)$ on the vector of probabilities $\boldsymbol{\theta}$ is Dirichlet $D(\boldsymbol{\alpha}_0)$ where $\boldsymbol{\alpha}_0 = (\alpha_{0,1}, \alpha_{0,2}, \alpha_{0,3}, \alpha_{0,4}, \alpha_{0,5}, \alpha_{0,6})$, $\alpha_{0,i} > 0$ for $1 \leq i \leq 6$ is only strictly positive when $\boldsymbol{\theta}$ satisfies the constraints above, when

$$\pi(\boldsymbol{\theta}|\alpha_0) = \frac{\Gamma\left(\sum_{j=0}^{6} \alpha_{0,j}\right)}{\prod_{j=0}^{6} \Gamma(\alpha_{0,j})} \theta_1^{\alpha_{0,1}-1} \theta_2^{\alpha_{0,2}-1} \theta_3^{\alpha_{0,3}-1} \theta_4^{\alpha_{0,4}-1} \theta_5^{\alpha_{0,5}-1} \theta_6^{\alpha_{0,6}-1}$$

where $\Gamma(\alpha) = \int_0^\infty u^{\alpha-1} e^{-u} du$, $\alpha > 0$ is the Gamma function with the property that $\Gamma(\alpha) = (\alpha - 1)\Gamma(\alpha - 1)$, $\Gamma(1) = 1$. You learn from a scientist that in fact components of $\boldsymbol{\theta}$ are constrained so that

$$\begin{array}{lll} \theta_1 = \psi_1 \phi_1, & \theta_2 = \psi_2 \phi_2, & \theta_3 = \psi_3(1 - \phi_2), \\ \theta_4 = \psi_1 (1 - \phi_1), & \theta_5 = \psi_2 (1 - \phi_2), & \theta_3 = \psi_3 \phi_2 \end{array}$$

where $\sum_{i=1}^{3} \psi_i = 1$, $\psi_i > 0$, $i = 1, 2, 3$ and $0 < \phi_1, \phi_2 < 1$. Show that within this submodel $\theta_i > 0$ for $1 \leq i \leq 6$ and $\sum_{i=1}^{6} \theta_i = 1$. Write down the likelihood $l(\boldsymbol{\psi}, \phi_1, \phi_2|\boldsymbol{x})$ associated with the multinomial sample above as a function of $\boldsymbol{\psi} = (\psi_1, \psi_2, \psi_3), \phi_1$ and ϕ_2.

Now suppose you choose a family of prior densities $\pi(\boldsymbol{\psi}, \phi_1, \phi_2|\boldsymbol{\beta}_0, \boldsymbol{\lambda}_0, \boldsymbol{\rho}_0)$ where $\boldsymbol{\beta}_0 = (\beta_{0,1}, \beta_{0,2}, \beta_{0,3})$, $\beta_{0,i} > 0$ for $1 \leq i \leq 3$; $\boldsymbol{\lambda}_0 = (\lambda_{0,1}, \lambda_{0,2})$, $\lambda_{0,i} > 0$ for $i = 1, 2$; $\boldsymbol{\rho}_0 = (\rho_{0,1}, \rho_{0,2})$, $\rho_{0,i} > 0$ for $i = 1, 2$ and where $\pi(\boldsymbol{\psi}, \phi_1, \phi_2|\boldsymbol{\beta}_0, \boldsymbol{\lambda}_0, \boldsymbol{\rho}_0)$ can be written in the form

$$\pi(\boldsymbol{\psi}, \phi_1, \phi_2|\boldsymbol{\beta}_0, \boldsymbol{\lambda}_0, \boldsymbol{\rho}_0) = \pi_0(\boldsymbol{\psi}|\boldsymbol{\beta}_0).\pi_1(\phi_1|\boldsymbol{\lambda}_0).\pi_2(\phi_2|\boldsymbol{\rho}_0).$$

where, for $(\boldsymbol{\psi}, \phi_1, \phi_2)$ satisfying $\sum_{i=1}^{3} \psi_i = 1$, $\psi_i > 0$, $i = 1, 2, 3$ and $0 < \phi_1, \phi_2 < 1$,

$$\pi_0(\boldsymbol{\psi}|\boldsymbol{\beta}_0) = \frac{\Gamma(\beta_{0,1} + \beta_{0,2} + \beta_{0,3})}{\Gamma(\beta_{0,1})\Gamma(\beta_{0,2})\Gamma(\beta_{0,3})} \psi_1^{\beta_{0,1}-1} \psi_2^{\beta_{0,2}-1} \psi_3^{\beta_{0,3}-1},$$

$$\pi_1(\boldsymbol{\phi}_1|\boldsymbol{\lambda}_0) = \frac{\Gamma(\lambda_{0,1} + \lambda_{0,2})}{\Gamma(\lambda_{0,1})\Gamma(\lambda_{0,2})} \phi_1^{\lambda_{0,1}-1} (1 - \phi_1)^{\lambda_{0,2}-1},$$

$$\pi_2(\boldsymbol{\phi}_2|\boldsymbol{\lambda}_0) = \frac{\Gamma(\rho_{0,1} + \rho_{0,2})}{\Gamma(\rho_{0,1})\Gamma(\rho_{0,2})} \phi_2^{\rho_{0,1}-1} (1 - \phi_2)^{\rho_{0,2}-1}.$$

Show that this family of priors is closed under sampling to $l(\boldsymbol{\psi}, \phi_1, \phi_2|\boldsymbol{x})$ and calculate $\pi(\boldsymbol{\psi}, \phi_1, \phi_2|\boldsymbol{\beta}_0, \boldsymbol{\lambda}_0, \boldsymbol{\rho}_0, \boldsymbol{x})$ explicitly.

A DM is interested in the proportion θ of particles that contain a particular chemical C and she takes a random sample of N such particles. If the particles have not been contaminated she believes that θ will have a $Be(\alpha_0, \beta_0)$. However she believes that there is a small probability π, $0 \leq \pi \leq 1$, that the sample is contaminated and given this she believes that θ will have a $Be(\alpha_c, \beta_c)$ distribution. It follows that her prior density is drawn from the family of distributions whose densities are given by

$$q(\theta|\alpha_0, \beta_0, \alpha_c, \beta_c, \pi) = \pi p(\theta|\alpha_c, \beta_c) + (1 - \pi)p(\theta|\alpha_0, \beta_0)$$

where $p(\theta|\alpha, \beta)$ is the beta density defined above. Prove that this family is also closed under sampling and calculate the posterior distribution explicitly when you observe that all the particles contain the chemical, i.e. when $x = N$.

Recalling that $\Gamma(t) = (t-1)!$ for $t = 1, 2, 3, \ldots$ prove that the client's posterior odds for the contaminated model tends to ∞ as $N = x \to \infty$ when $(\alpha_0, \beta_0, \alpha_c, \beta_c, \pi, N)$ is $(2, 3, 2, 2, 0, 1, 5)$. Interpret this result.

9.9 Using the notation above prove that $d_V(f_n^*, g_n^*) \leq d_V(f_n, g_n)$.

10
Conclusions

10.1 A summary of what has been demonstrated above

The results and analyses in this book have demonstrated the following points:

- A Bayesian decision analysis delivers a *subjective* but *defensible* representation of a problem that guides wise decision making, and provides a compelling supporting narrative for why the chosen action was taken. By crystallising the reasons behind a chosen action it can be used as a platform for new creative innovative thinking about the problem at hand and so is always *open to re-evaluation and reformulation*.
- A DM can use the framework above to address not only simple decision problems but also highly structured, *high-dimensional* multifaceted problems.
- The DM will usually need *guidance* to identify both the structure of her utility function and an appropriate credence decomposition over the features of the problem she believes might influence her decision. We have seen that it is extremely helpful if these elicitation processes are supported by graphs. Detailed discussions of several of these have been given above but there are many more. Graphs are important because they can not only describe evocatively consensual thinking about underlying processes but also provide a conduit into faithful and computationally feasible probabilistic models.
- The *quantification* of a decision model's utility functions and probabilities will usually be the most contentious and most difficult features to elicit faithfully. However if the underlying credence decomposition and the structure of the utility function have been faithfully elicited then analyses are usually surprisingly *robust* to moderate mis-specifaction of these functions.
- The most compelling decision analyses support as many as possible of its assertions with *hard evidence*: often in terms of the results of designed experiments and sample surveys. Such information can be incorporated into an analysis seamlessly within the Bayesian methodology in ways illustrated above.
- We have also seen how the framework of Bayesian decision theory allows the DM to formally and transparently find good strategies that *balance* the achievement of objectives or sources of evidence which pull decisions in different directions. However it also helps her to automatically identify when any compromise is a poor option. It then guides the DM to choose decisions that strive to attain high scores in one subset of attributes at the cost of any significant gain in the others, or leads her to act as if she believed mainly in a particular subset of the sources of information and largely ignore the rest.

10.2 Other types of decision analyses

In this book I have necessarily focused on an important subset of the types of problem that a DM often faces. These are the ones that I believe to be best served by a full Bayesian analysis as described above and have most experience in. However there are of course many others types of decision problem addressed.

The closest is one where the DM would like to conduct an analysis like the one above but there is no time for a detailed elicitation. I have found that great clarity can often be obtained as soon as the DM's attributes have been elicited and the relevant graphically based credence decomposition has been discovered. Then the optimal decision is sometimes so transparent to the DM that no further quantitative analysis is necessary. This is because the elicitation of the structure of the problem is helpful *in itself*. The embellishments of the structure provided by the full quantificaction of the model just add to the specificity of the conclusions of the analysis.

It is therefore often fruitful to embark on the elicitation of the structure of a decision problem even when the analyst is aware that the full quantification of the model will be impossible because of time constraints. Of course in such cases the analyst may choose a different representation of a problem, perhaps more familiar to the DM, on which to base the analysis. This is indeed widely done see French and Rios Insua (2000) for a good review of alternative methods. One difficulty of using such methods is that their semantics may not be such a so precisely defined, or coherent like the decision tree, ID, BN or CBN so the results of the analysis can be difficult to communicate unambiguously. But in the very early exploratory stages of a decision analysis, not to demand coherence may not be such a bad thing. In common with other decision analysts I believe however that any good analysis needs, however informally, to in some way elicit the DM's values which should then focus any discussion: see e.g. Keeney (1992, 2007).

A second related scenario often encountered is one where the DM needs support to develop a framework which will enable her to structure her thinking over a *range* of problems. The methods discussed above for structuring a specific problem can obviously be used to form such a template. It is often the case that the structure of, for example, trees and BNs usually endure over ranges of problem, albeit with some modifications to each case. Because they admit numerical embellishments when these are needed, these frameworks are especially useful in this type of support.

A third scenario I have not discussed in the book is when several experts give different probabilities to the same event and the DM needs to adopt some combination of these. There is now a wide literature dedicated to the different ways such probabilistic judgements can be combined: see e.g. Bedford and Cooke (2001); Clemen and Winkler (2007) for reviews of some of these and Faria and Smith (1996, 1997) for how one of these – the logarithmic pool – can be applied to a BN with several different collections of experts advising on the distributions of different variables. Different combination rules are appropriate for different decision making environments.

In an environment close to the one described here where the DM adopts a function of these probabilities as her own then it is clear that any good method will utilise any knowledge she has of the way in which information between experts might be shared. For example if two experts come to the same judgements about the distribution of a parameter based on the same set of observations then this has the same weight to the DM as hearing from just one. But if they do this and their sources of information are quite different then the two experts have complementary information and both are useful together. So any method the DM adopts should have the flexibility to treat these two scenarios differently. Codings of the problem sympathetic to these ideas are given in Lindley (1988); Winkler (1986) and Smith and Faria (2000).

Some responsible DMs do not have the language of probability and so cannot take even partial ownership of these methods. When this is the case they shouldn't be used. Sometimes it is clear from initial interviews with the DM that they are trying to construct an alibi for a current commitment. Helping a DM to produce a coherent argument for their current committed choice should not be a role a Bayesian decision analyst should willingly take on. We discussed at the end of Chapter 5 the dangers of misleading a DM by constructing worldviews for the past consistent with what is already known. If this process is undertaken when the DM's first priority is not to be honest then the results of the analysis can be disastrous as well as being ethically dubious.

Once there is no longer a single responsible and acting DM at the hub of the analysis then the attractiveness of the Bayesian paradigm as described here begins to fade. In particular the demand for coherence and a total order on preferences is unlikely to be compelling. Then the analyst should look for other tools to support the DM.

10.2.1 Some concluding remarks

Effective decision analysis is intrinsically subjective. It helps the DM build a defensible view of the world she both believes in and owns. We have encountered various reasons why there is invariably a subjective component to an analysis. We have seen that although data from well-designed experiments, sample surveys, analogous instances and expert judgements can make various features in the model more consensual there will usually be features of a problem which depend on the beliefs about those populations that are believed to be related to the problem being studied. Furthermore a decision analysis requires the DM to relate her beliefs about how the current instance relates to the empirical evidence from related populations and this is nearly always a subjective judgement. Typically the best she can hope for is a compelling argument and rationale for why, on the evidence in front of her and within her limitations of processing this information, she plans to act in a certain way.

The demand for a subjectivist approach to decision analysis does *not* however preclude agreement between all important parties that the model presented by the DM represents a professional attempt to marshal all available evidence that can reasonably be accommodated into a model and which draws conclusions in a mututally acceptable way. The analysis may well achieve broad consensus about its conclusions, even an acceptance that a chosen course is the *only* possible good action. But any such consensus, like any currently accepted scientific theory, is provisional. It should be accepted that future further insights and new evidence will almost certainly improve the deductions and quite often overturn the current consensus completely.

Any DM who goes as far as placing probabilities on a large collection of events will almost inevitably be shown through subsequent analyses to have been flawed in some of her judgements. Because she makes committing statements and owns responsibility for these she can and will be often proved wrong. The boldness of presenting testable statements for someone to disprove goes much further than many non-Bayesian methodologies. But the clarity of communication achieved through the DM's movement towards a description which tries to encompasses all the major features she perceives as critical to a wise choice of decisions, whilst making her vulnerable to criticism, is often a necessary step towards successful decision making. The decision model facilitates her so that she can present her ideas fully to another. In particular relating to others by allowing her ideas to be critiqued also enables her to refine her understanding of the problem she faces, her domain and herself which empowers her to make better and more defensible decisions.

It can be argued that recognising the necessity for subjective judgements in an analysis and embracing this need is an ethical imperative. As Levinas (1969) p219 once wrote concerning the relationship between ethics and reason:

> The will is free to assume the responsibility in whatever sense it likes; it is not free to refuse this responsibility itself; it is not free to ignore the meaningful world into which the face of the Other has introduced it. In the welcoming face the will opens to reason.

I believe that *only* when the DM is prepared to take ownership of and a responsibility for her beliefs and actions within the real world and face others through this ownership can she be truly liberated. The role of the analyst is to be a welcoming face, encouraging and supporting her engagement in this dynamic as she grows in the honest subjective responsibility for her beliefs and actions within the world in which she operates.

References

Aitken, C. and Taroni, F. (2004) "Statistics and the Evaluation of Evidence for Forensic Scientists". 2nd edn. Wiley, Chichester.

Aitken, C. G. G., Taroni, F. and Garbolino, P. (2003) "A graphical model for the evaluation of cross-transfer evidence in DNA profiles". Theoretical Population Biology 63, 179–190.

Andersson, S., Madigan, D. and Pearlmean, M. (1997) "Alternative Markov properties for chain graphs". Scandinavian Journal of Statistics 24, 81–102.

Andrade, J. A. A. and O'Hagan, A. (2006) "Bayesian robustness modelling using regularly varying distributions". Bayesian Analysis 1, 169–188.

Atwell, D. N. and Smith, J. Q. (1991) "A Bayesian forecasting model for sequential bidding". Journal of Forecasting 10, 565–577.

Bedford, T. and Cooke, R. (2001) "Probabilistic Risk Analysis: Foundation and Methods". Cambridge University Press.

Bensoussen, A. (1992) "Stochastic Control of Partially Observable Systems". Cambridge University Press.

Berger, J. O. (1985) "Statistical Decision Theory and Bayesian Analysis". 2nd edn. Springer-Verlag, New York.

Bernardo, J. M. and Smith, A. F. M. (1996) "Bayesian Theory". Wiley, Chichester.

Bersekas, D. P. (1987) "Dynamic Programming". Prentice Hall, Englewood Cliffs.

Blackwell, D. and Dubins, L. (1962) "Merging of opinions with increasing information". Annals of Mathematical Statistics 33, 882–886.

Bonet, B. (2001a) "A calculus for causal relevance". In: Proceedings of the Seventeenth Conference on Uncertainty in Artificial Intelligence (eds. J. Breese, D. Koller). Morgan Kaufmann, San Francisco, pp. 40–47.

Bonet, B. (2001b) "Instrumentality tests revisited". In: Proceedings of the Seventeenth Conference on Uncertainty in Artificial Intelligence (eds. J. Breese, D. Koller). Morgan Kaufmann, San Francisco, pp. 48–54.

Caines, P., Deardon, R. and Wynn, H. (2002) "Conditional orthogonality and conditional stochastic realization". In: New Directions in Mathematical Systems Theory and Optimization, LNCIS 286 (eds. A. Rantzer and C. I. Byrnes). Springer-Verlag, New York, pp. 71–84.

Canning, C., Thompson, E. A. and Skolnick, M. H. (1978) "Probability functions on complex pedigrees". Advances in Applied Probability 10, 26–61.

Carlson, B. W. (1993) "The accuracy of future forecasts and past judgements". Organizational Behaviour and Human Decision Processes 54, 245–276.

Chen, M.-H., Shao, Q.-M. and Ibrahim, J. G. (2000) "Monte Carlo Methods in Bayesian Computation". Springer-Verlag, New York.

Clemen, R. T. and Lichtendahl, K. C. (2002) "Debiasing expert overconfidence: a Bayesian calibration model". Working paper, Duke University, Durham NC.

Clemen, R. T. and Winkler, R. L (2007) "Aggregating probability distributions". In: Advances in Decision Analysis: From Foundations to Applications (eds. W. Edwards, R. F. Miles, Jr. and D. von Winterfeldt). Cambridge University Press, pp. 154–176.

Covaliu, Z. and Oliver R. M. (1995) "Representation and solution of decision problems using sequential decision diagrams". Management Science 41, 1860–1881.

Cowell, R. G., Dawid, A. P., Lauritzen, S. L. and Spiegelhalter, D. J. (1999) "Probabilistic Networks and Expert Systems". Springer-Verlag, New York.

Cox, D. R. and Wermuth, N. (1996) "Multivariate Dependences". Chapman and Hall, London.

Curley, S. P. (2008) "Subjective probability". In: Encyclopedia of Quantitative Risk Analysis and Assessment (eds. E. L. Melnick and B. S. Everitt). Wiley, Chichester, pp. 1725–1734.

Dahlhaus, R. and Eichler, M. (2003) "Causality and graphical models for time series". In: Highly Structured Stochastic Systems (eds. P. Green, N. Hjort and S. Richardson). Oxford University Press, pp. 115–137.

Daneshkhah, A. and Smith, J. Q. (2004) "Multicausal prior families, randomisation and essential graphs". In: Advances in Bayesian Networks, Physica-Verlag, 1-17 and Proceedings of the First European Workshop on Probabilistic Models Cuenca Spain, 25–34.

Dawid, A. P. (1973) "Posterior expectations for large observations". Biometrika 60, 664–667.

Dawid, A. P. (1979) "Conditional independence in statistical theory (with discussion)". Journal of the Royal Statistical Society B 41(1), 1–31.

Dawid, A. P. (1982) "The well calibrated Bayesian (with discussion)". Journal of the American Statistical Association 77, 604–613.

Dawid, A. P. (1992) "Prequential analysis, stochastic complexity and Bayesian inference". In: Bayesian Statistics (eds. J. M. Bernardo *et al.*). Oxford University Press, pp. 109–125.

Dawid, A. P. (2000) "Causality without counterfactuals (with discussion)". Journal of the American Statistical Association 95, 407–448.

Dawid, A. P. (2001) "Separoids: a mathematical framework for conditional independence and irrelevance". Annals of Mathematics and Artificial Intelligence 32, 335–372.

Dawid, A. P. (2002a) "Influence diagrams for causal modelling and inference". International Statistical Reviews 70, 161–189.

Dawid, A. P. (2002b) "Bayes theorem and the weighing of evidence by juries". In: Proceedings of the British Academy, vol. 113 (ed. R. Swinburne). Oxford University Press, pp. 71–90.

Dawid, A. P. (2007) "The geometry of proper scoring rules". Annals of the Institute of Statistical Mathematics 59(1), 77–93.

Dawid, A. P. and Evett, I. W. (1997) "Using a graphical model to assist the evaluation of complicated patterns of evidence". Journal of Forensic Science 42, 226–231.

Dawid, A. P. and Lauritzen, S. (1993) "Hyper-Markov laws in the statistical analysis of decomposable graphical models". Annals of Statistics 21(3), 1272–1317.

Dawid, A. P. and Vovk, V. G. (1999) "Prequential probability: principles and properties". Bernoulli 5, 125–162.

De Finetti, B. (1974) "Theory of Probability, Vol. 1". Wiley, Chichester.
De Finetti, B. (1980) "Foresight, its logical laws, its subjective sources". In: Studies in Subjective Probability (eds. H. E. Kyburg and H. E. Smokler). Dover Publications, New York, pp. 93–158.
De Groot, M. H. (1970) "Optimal Statistical Decisions". McGraw-Hill, New York.
Dean, T. and Kanazawa, K. (1988) "Probabilistic temporal reasoning". Proceedings AAAI-88, 524–528.
Denison, D. G. T., Holmes, C. C., Mallick, B. K. and Smith, A. F. M. (2005) "Bayesian Methods for Non-linear Classification and Regression". Wiley, Chichester.
DeRobertis, L. (1978) "The use of partial prior knowledge in Bayesian inference". Ph. D. dissertation, Yale University.
Didelez, V. (2008) "Graphical models for marked point processes based on local independence". Journal of the Royal Statistical Society, Series B 70, 245–264.
Dodd, L., Moffat, J. and Smith, J. Q. (2006) "Discontinuity in decision making when objectives conflict: a military command decision case study". Journal of the Operational Research Society 57, 643–654.
Dowie, J. (1976) "On the efficiency and equity of betting markets". Economica 43, 139–150.
Drton, M. and Richardson, T. S. (2008) "Binary models for marginal independence". Journal of the Royal Statistical Society B 70(2), 287–310.
Durbin, J. and Koopman, S. J. (2001) "Time Series Analysis by State Space Methods". Oxford University Press.
Edwards, D. (2000) "Introduction to Graphical Modelling". Springer-Verlag, New York.
Eichler, M. (2006) "Graphical modelling of dynamic relationships in multivariate time series". In: Handbook of Time Series Analysis (eds. M. Winterhalder, B. Schelter and J. Timmer). Wiley-VCH, Berlin, pp. 335–372.
Eichler, M. (2007) "Granger-causality and path diagrams for multivariate time series". Journal of Econometrics 137, 334–353.
Evans, M. and Swartz, T. (2000) "Approximating Integrals via Monte Carlo and Deterministic Methods". Oxford University Press.
Faria, A. E. and Smith, J. Q. (1996) "Conditional External Bayesianity in Decomposable Influence Diagrams". In: Bayesian Statistics 5 (eds. J. M. Bernardo, J. O. Berger, A. P. Dawid, and J. Q. Smith). Oxford University Press, pp. 551–560.
Faria, A. E. and Smith, J. Q. (1997) "Conditionally externally Bayesian pooling operators in chain graphs". Annals of Statistics 25(4), 1740–1761.
Feller, W. (1971) "An Introduction to Probability Theory and its Applications, Vol. 2". 2nd edn. John Wiley, New York.
Fine, T. L. (1973) "Theories of Probability: an Examination of Foundations 2". Academic Press, New York.
Flores, M. J. and Gamez, J. A. (2006) "A review on distinct methods and approaches to perform triangulation for Bayesian networks". In: Advances in Probabilistic Graphical Models (eds. J. A. Gamez and A. Salmeron). Springer-Verlag, New York, pp. 127–152.
Freeman, G. and Smith, J. Q. (2010) "Bayesian MAP Selection of Chain Event graphs". Journal of Multivariate Analysis (to appear).
Freeman, G. H., Jacka, S. D., Shaw, J. E. H. and Smith, J. Q. (1996) "Modelling the management of underground water assets". Journal of Applied Statistics 23(2,3), 273–284.
French, S. and Rios Insua, D. (2000) "Statistical Decision Theory". Kendall's Library of Statistics 9. Arnold, London.

French, S., Papamichail, K. N., Ranyard, D. C. and Smith, J. Q. (1995) "Decision support for nuclear emergency response". Proceedings of the Fifth Hellenistic Conference on Informatics, Athens, Vol. 2, pp. 591–600.

French, S., Harrison, M. T. and Ranyard, D. C. (1997) "Event conditional attributer modelling in decision making when there is a threat of a nuclear accident". In: The Practice of Bayesian Analysis (eds. S. French and J. Q. Smith). Arnold, London.

French, S., Papamichail, K. N., Ranyard, D. C. and Smith, J. Q. (1998) "Design of a decision support system for the use in the event of a radiation accident". In: Applied Decision Analysis (eds. F. J. Giron and M. L. Martinez). Kluwer Academic, Dordrecht, pp. 3–18.

French, S., Maule, J. and Papamichail, N. (2009) "Decision Behaviour, Analysis and Support". Cambridge University Press.

Fruthwirth-Schnatter, S. (2006) "Finite Mixture and Markov Switching Models". Springer-Verlag, New York.

Gammerman, D. and Lopez, H. F. (2006) "Markov Chain Monte Carlo". Chapman and Hall, London.

Geiger, D. and Heckerman, D. (1997) "A characterization of the Dirichlet distribution through local and global independence". Annals of Statistics 25, 3, 731–792.

Geiger, D. and Pearl, J. (1990) "On the Logic of Causal Models in Uncertainty in Artificial Intelligence, Vol. 4" (eds. R. D. Shachter, T. S. Lewitt, L. N. Kanal and J. F. Lemmer). North Holland, Amsterdam, pp. 3–14.

Geiger, D. and Pearl, J. (1993) "Logical and algorithmic properties of conditional independence and graphical models". Annals of Statistics 21, 2001–2021.

Gelman, A. and Hill, J. (2007) "Data Analysis using Regression and Multilevel/Hierarchical Models". Cambridge University Press.

Gelman, A., Carlin, J. B., Stern, H. S. and Rubib, D. B. (1995) "Bayesian Data Analysis". Chapman and Hall, London.

Ghosh, J. K. and Ramamoorthi, R. V. (2003) "Bayesian Nonparametrics". Springer-Verlag, New York.

Gigerenzer, G. (2002) "Reckoning with Risk". Penguin, London.

Glymour, D. and Cooper, G. F. (1999) "Computation, Causation, and Discovery". MIT Press, Cambridge, MA.

Goldsein, M. (1985) "Temporal Coherence". In: Bayesian Statistics 2 (eds. J. M. Bernardo *et al.*). Oxford University Press, pp. 189–209.

Goldstein, M. and Rougier, J. C. (2009) "Reified Bayesian modelling and inference for physical systems". Journal of Statistical Planning and Inference 139(3), 1221–1239.

Goldstein, M. and Wooff, D. (2007) "Bayesian Linear Statistic: Theory and Methods". Wiley, Chichester.

Gomez, M. (2004) "Real world applications of influence diagrams". In: Advances in Bayesian Networks (eds. J. A. Gámez, S. Moral, A. S. Cerdan). Springer-Verlag, New York, pp. 161–180.

Goodwin, P. and Wright, G. (2003) "Decision Analysis for Management Judgement". 3rd edn. John Wiley and Sons, Chichester.

Grimmet, G. R. and Stirzaker, D. R. (1982) "Probability and Random Processes". Oxford University Press.

Gustafson, P. and Wasserman, L. (1995) "Local sensitivity diagnostics for Bayesian inference". Annals of Statistics 23, 2153–2167.

Harrison, P. J. and Smith, J. Q. (1979) "Discontinuous decisions and conflict". In: Proceedings of the First International Meeting in Bayesian Statistics, Valencia, Spain, pp. 99–127.

Heard, N. A., Holmes, C. C. and Stephens, D. A. (2006) "A quantitative study of gene regulation involved in the immune response of anopheline mosquitoes: an application of Bayesian hierarchical clustering of curves". Journal of the American Statistical Association 101, 473, 18–29.

Heath, D. and Sudderth, W. (1989) "Coherent inference from improper priors and from finitely additive priors". Annals of Statistics 17, 2, 907–919.

Heckerman D. (1998) "A tutorial to learning with Bayesian networks". In: Learning in Graphical Models (ed. M. I. Jordon). MIT Press, Cambridge, MA, pp. 301–354.

Heckerman, D. (2007) "A Bayesian approach to learning causal networks". In: Advances in Decision Analysis: From Foundations to Applications (eds. W. Edwards, R. F. Miles Jr., D. von Winterfeldt). Cambridge University Press, pp. 202–220.

Hill, B. M. (1988) "De Finnetti's theorem, induction, A(n) or Bayesian nonparametric inference". In: Bayesian Statistics 3 (eds. J. M. Bernardo *et al.*). Oxford University Press, pp. 211–241 (with discussion).

Hoffman, K. and Kunze, R. (1971) "Linear Algebra". 2nd edn. Prentice Hall, New Jersey.

Hora, S. C. (2007) "Eliciting probabilities from experts". In: Advances in Decision analysis: From Foundations to Applications (eds. W. Edwards, R. F. Miles Jr., D. von Winterfeldt). Cambridge University Press, pp. 129–153.

Howard, R. A. (1988) "Decision analysis: practice and promise". Management Science 34(6), 679–695.

Howard, R. A. (1990) "From influence to relevance to knowledge". In: Influence Diagrams, Belief Nets and Decision Analysis (eds. R. M. Oliver, J. Q. Smith). Wiley, Chichester, pp. 3–23.

Howard, R. A. and Matheson, J. E. (1984) "Readings on the Principles and Applications of Decision Analysis, Vol. 2". Strategic Decision Group, Menlo Park, CA, pp. 719–762.

Ibrahim, J. G. and Chen, M. H. (2000) "Power prior distributions for regression models". Statistical Science 15, 46–60.

Imbens, G. W. and Rubin, D. R. (1997) "Bayesian inference for causal effects in randomised experiments with noncompliance". Annals of Statistics 25, 305–327.

Jaegar, M. (2004) "Probability decision graphs – combining verification and AI techniques for probabilistic inference". International Journal of Uncertainty, Fuzziness and Knowledge Based Systems 12, 19–42.

Jensen, F. V. and Nielsen, T. D. (2007) "Bayesian Networks and Decision Graphs". 2nd edn. Springer-Verlag, New York.

Jensen, F. V., Neilsen, T. D. and Shenoy, P. P. (2004) "Sequential influence diagrams: a unified asymmetry framework". In: Proceedings of the Second European Workshop on Probabilistic Graphical Models (ed. P. Lucas). Leiden, the Netherlands, pp. 121–128.

Jorion, P. (1991) "Bayesian and CAPM estimators of the means: Implications for portflio selection". Journal of Banking and Finance 15, 717–727.

Kadane, J. B. and Chuang, D. (1978) "Stable decision problems". Annals of Statistics 6, 1095–1110.

Kadane, J. B. and Larkey, P. D. (1982) "Subjective probability and the theory of games". Management Science 28(2), 113–120.

Kadane, J. and Winkler, R. L. (1988) "Separating probability elicitation from utilities". Journal of the American Statistical Association 83, 357–363.

Kadane, J. R., Schervish, M. J. and Seidenfeld, T. (1986) "Statistical implications of finitely additive probability". In: Bayesian Inference and Decision Techniques (eds. P. K. Goel and A. Zellner). Elsevier, Amsterdam, pp. 59–76.

Kahenman, D. and Tversky, A. (1979) "Prospect theory: an analysis of decision under risk". Econometrika 47, 263–291.

Keeney, R. L. (1974) "Multiplicative utility functions". Operations Research 22, 22–34.

Keeney, R. L. (1992) "Value-focussed Thinking: A Path to Creative Decision Making". Harvard University Press.

Keeney, R. L. (2007) "Developing objectives and attributes". In: Advances in Decision Analysis: From Foundations to Applications (eds W. Edwards, R. F. Miles Jr., D. von Winterfeldt). Cambridge University Press, pp. 104–128.

Keeney, R. L. and Raiffa, H. (1976) "Decisions with Multiple Objectives: Preferences and Value Trade-offs". John Wiley and Sons, New York.

Kjaerulff, U. B. and Madsen, A. L. (2008) "Belief Networks and Influence Diagrams: A Guide to Construction and Analysis". Springer-Verlag, New York.

Kleinmauntz, B., Fennema, M. G. and Peecher, M. E. (1996) "Conditional assessment of probabilities: identifying the benefits of decomposition". Organizational Behaviour and Human Decision Processes 66, 1–15.

Koehler, D. J., White, C. M. and Grondin, R. (2003) "An evidential support accumulation model of subjective probability". Cognitive Psychology 46, 152–197.

Koeller, D. and Friedman, N. (2009) "Probabilistic Graphical Models: Principles and Techniques". MIT Press, Cambridge, MA.

Koeller, D. and Lerner, U. (1999) "Sampling in factored dynamic systems". In: Sequential Monte Carlo Methods in Practice (eds. A. Doucet, N. de Freitas and N. Gordon). Springer-Verlag, New York, pp. 445–464.

Korb, K. B. and Nicholson, A. E. (2010) "Bayesian Artificial Intelligence". Chapman and Hall, London.

Korchnoi, V. (2001), "My Best Games, Vol. 2". Olms Press.

Koster, J. T. A. (1996) "Markov properties of non-recursive causal models". Annals of Statistics 24, 2148–2177.

Kurth, T., Walker, A. M., Glynn, R. J., *et al.* (2006) "Results of multivariable logistic regression, propensity adjustment, and propensity-based weighting under conditions of nonuniform effect". American Journal of Epidemiology 162(5), 471–478.

Lad, F. (1996) "Operational Subjective Statistical Methods". John Wiley and Sons, New York.

Lancaster, T. (2004) "An Introduction to Modern Bayesian Econometrics". Blackwell, Oxford.

Lauritzen, S. L. (1996) "Graphical Models". 1st edn. Oxford Science Press, Oxford.

Lauritzen, S. L. and Speigelhalter, D. J. (1988) "Local computation with probabilities on graphical structures and their application to expert systems (with discussion)". Journal of the Royal Statistical Society B 50, 157–224.

Lauritzen, S. L. and Wermuth, N. (1989) "Graphical models for associations between variables, some of which are qualitative and some quantitative". Annals of Statistics 17, 31–57.

Laurizen, S. L., Dawid, A. P., Larsen, B. N. and Leimer, H.-G. (1990) "Independence properties of directed Markov fields". Networks 20, 491–505.

Levinas, E. (1969) "Totality and Infinity: An Essay on Exteriorarity". Duquesne University Press, Pittsburgh.

Lindley, D. V. (1988) "Reconciliation of discrete probability distributions". In: Bayesian Statistics 2 (eds. J. M. Bernado *et al.*). North-Holland, Amsterdam, pp. 375–390.

Little, R. J. A. and Rubin, D. B. (2002) "Statistical Analysis with Missing Data". 2nd edn. Wiley, Chichester.

Liverani, S., Anderson, P. E., Edwards, K. D., Millar, A. J. and Smith, J. Q. (2008) "Efficient utility-based clustering over high dimensional partition spaces". Journal of Bayesian Analysis 04(03), 539–572.

Louchard, J., Schneider, T. and French, S. (1992) "International Chernobyl Project: Summary Report of Decision Conferences held in the USSR October – November 1990". Luxemburg City, European Commission.

Lui, J. and Hodges, J. S. (2003) "Posterior bimodality in the balanced one-way random effects model". Journal of the Royal Statistical Society B 65(1), 247–256.

Lukacs, E. (1965) "A characterisation of the gamma distribution". Annals of Mathematical Statistics 26, 319–324.

Marin, J. M. and Robert, C. P. (2007) "Bayesian Core: A Practical Approach to Computational Bayesian Analysis". Springer-Verlag, New York.

Marin, J. M., Mengersen, K. and Robert, C. P. (2004) "Bayesian modelling and inference on mixtures of distributions". In: Handbook of Statistics 25 (eds. D. Dey and C. R. Rao). Elsevier-Science, Amsterdam.

Marshall, K. T. and Oliver, R. M. (1995) "Decision Making and Forecasting". McGraw-Hill, New York.

McAllister, D., Collins, M. and Periera, F. (2004) "Case factor diagrams for structured probability modelling". In: Proceedings of the 20th Annual Conference on Uncertainty in Artificial Intelligence (UAI-04), pp. 382–391.

McLish, D. K. and Powell, S. H. (1989) "How well can physicians estimate mortality in a medical intensive care unit?". Medical Decision Making 9, 125–132.

Meek, C. (1995) "Strong completeness and faithfulness in Bayesian networks". In: Uncertainty in Artificial Intelligence 11 (eds. P. Besnard and S. Hanks). Morgan Kaufmann, New York, pp. 403–418.

Meyer, R. F. (1970) "On the relationship among the utility of assets, the utility of consumption and investment strategy in an uncertain but time invariant world". In: OR 69 Proceedings of the Fifth Conference in Operations Research (ed. J. Lawrence). Tavistock Publications, London.

Mond, D. M. Q., Smith, J. Q. and Van Straten, D. (2003) "Stochastic factorisations, sandwiched simplices and the topology of the space of explanations". Proceedings of the Royal Society of London A 459, 2821–2845.

Murphy, A. H. and Winkler, R. L. (1977) "Reliability of subjective probability forecasts of precipitation and temperature: some preliminary results". Applied Statistics 26, 41–47.

Neil, M., Tailor, M., Marquez, D., Fenton, N. E. and Hearty, P. (2008) "Modelling dependable systems using hybrid Bayesian networks". Reliability Engineering and System Safety 93(7), 933–939.

Nodelman, U., Shelton, C. R. and Koller D. (2002) "Continuous time Bayesian networks". Proceedings of the Eighteenth Conference on Uncertainty in Artificial Intelligence (UAI), pp. 378–387.

O'Hagan, A. (1979) "On outlier rejection phenomena in Bayesian inference". Journal of the Royal Statistical Society B 41, 358–367.

O'Hagan, A. (1988) "Probability: Methods and Measurements". Chapman and Hall, London.

O'Hagan, A. and Forster, J. (2004) "Bayesian Inference". In: Kendall's Advanced Theory of Statistics. Arnold, London.

O'Hagan, A., Buck, C. E., Daneshkhah, A., *et al.* (2006) "Uncertain Judgements: Eliciting Experts' Probabilities". Wiley, Chichester.

Oaksford, M. and Chater, N. (2006) "Bayesian Rationality". Oxford, Oxford University Press.

Oaksford, M. and Chater, N. (eds.) (1998) "Rational Models of Cognition". Oxford University Press.

Oliver, R. M. and Smith, J. Q. eds (1990) "Influence Diagrams, Belief Nets, and Decision Analysis". Wiley, Chichester.

Olmsted, S. M. (1983) "On representing and solving decision problems". Ph.D. dissertation. Engineering-Economic Systems, Stanford University.

Papamichail, K. N. and French, S. (2003) "Explaining and justifying the advice of a decision support system: a natural language generation approach". Expert Systems with Applications 24(1), 35–48.

Papamichail, K. N. and French, S. (2005) "Design and evaluation of an intelligent decision support system for nuclear emergencies". Decision Support Systems 41(1), 84–111.

Papaspiliopoulos, O. and Roberts, G. (2009) "Stability of the Gibbs sampler for Bayesian hierarchical models". Annals of Statistics 36(1), 95–117.

Pearl, J. (1988) "Probabilistic Reasoning in Intelligent Systems". Morgan Kauffman, San Mateo.

Pearl, J. (1995) "Causal diagrams for empirical research". Biometrika 82, 669–710.

Pearl, J. (2000) "Causality, models, reasoning and inference". Cambridge University Press.

Pearl, J. (2003) "Statistics and causal inference: a review (with discussion)". Sociedad de Estadistica e Investigacion Operativa Test 12, 2, 281–345.

Peterka, V. (1981) "Bayesian system identification". In: Trends and Progress in System Identification (ed. P. Eykhoff). Pergamon Press, Oxford, pp. 239–304.

Phillips, L. D. (1984) "A theory of requisite decision models". Acta Psychologia 56, 29–48.

Phillips, L. D. (2007) "Decision conferencing". In: Advances in Decision Analysis: From Foundations to Applications (eds. W. Edwards, R. F. Miles Jr., D. von Winterfeldt). Cambridge University Press, pp. 375–399.

Pollack, R. A. (1967) "Additive von Neumann–Morgenstern utility functions". Econometrica 35, 485–494.

Poole, D. and Zhang, N. L. (2003) "Exploiting contextual independence in probabilistic inference". Journal of Artificial Intelligence Research 18, 263–313.

Puch, R. O. and Smith, J. Q. (2004) "FINDS: A Training Package to Assess Forensic Fibre Evidence". In: Advances in Artificial Intelligence. Springer-Verlag, New York, pp. 420–429.

Puch, R. O., Smith, J. Q. and Bielza, C. (2004) "Inferentially efficient propagation in non-decomposable Bayesian networks with hierarchical junction trees". In: Advances in Bayesian Networks. Physica-Verlag, New York, pp. 57–74.

Queen, C. M. and Smith, J. Q. (1992) "Symmetric dynamic graphical chain models". In: Bayesian Statistics 4 (eds. J. M. Bernardo, J. O. Berger, A. P. Dawid and A. F. M. Smith). Oxford University Press, pp. 741–751.

Queen, C. M. and Smith, J. Q. (1993) "Multi-regression dynamic models". Journal of the Royal Statistical Society B 55(4), 849–870.

Queen, C. M., Smith, J. Q. and James, D. M. (1994) "Bayesian forecasts in markets with overlapping structures". International Journal of Forecasting 10, 209–233.

Raiffa, H. (1968) "Decision Analysis". Addison-Wesley, Reading, MA.

Raiffa, H. and Schlaifer, R. (1961) "Applied Statistical Decision Theory". MIT Press, Cambridge, MA.

Ramsey, F. P. (1931) "The Foundations of Mathematics and other Essays". Routledge and Kegan Paul, London.

Ranyard, D. C. and Smith, J. Q. (1997) "Building a Bayesian model in a scientific environment: managing uncertainty after an accident". In: The Practice of Bayesian Analysis (eds. S. French, J. Q. Smith). Arnold, London, pp. 245–258.
Rasmussen, C. E. and Williams, C. K. I. (2006) "Gaussian Processes for Machine Learning". MIT Press, Cambridge, MA.
Renooij, S. (2001) "Probability elicitation for belief networks: issues to consider". Knowledge Engineering Review 16(3), 255–269.
Riccomagno, E. and Smith, J. Q. (2005) "The Causal Manipulation and Bayesian Estimation of Chain Event Graphs". CRiSM Research Report. CRiSM, University of Warwick.
Riccomagno, E. M. and Smith, J. Q. (2004) "Identifying a cause in models which are not simple Bayesian networks". Proceedings of IMPU, Perugia, July 2004, pp. 1315–1322.
Riccomagno, E. and Smith, J. Q. (2009) "The geometry of causal probability trees that are algebraically constrained". In: Optimal Design and Related Areas in Optimization and Statistics (eds. L. Pronzato and A. Zhigljavsky). Springer-Verlag, New York, pp. 131–152.
Richardson, T. S. and Spirtes, P. (2202) "Ancestral graph Markov models". Annals of Statistics 30, 962–1030.
Rigat, F. and Smith, J. Q. (2009) "Non-parametric dynamic time series modelling with applications to detecting neural dynamics". Annals of Applied Statistics 3(4), 1776–1804.
Robert, C. (2001) "The Bayesian Case". 2nd edn. Springer-Verlag, Berlin.
Robert, C. P. and Casella, G. (2004) "Monte Carlo Statistical Methods". 2nd edn. Springer-Verlag, New York.
Robins, J. M. (1986) "A new approach to causal inference in mortality studies with a sustained exposure period – application to control of the healthy worker survivor effect". [Mathematical models in medicine: diseases and epidemics, Part 2.] Mathematical Modelling 7(9-12), 1393–1512.
Robins, J. M. (1997) "Causal inference from complex longitudinal data". In: Latent Variable Modeling and Applications to Causality (ed. M. Berkane). Springer-Verlag, New York, pp. 69–117.
Robins, J. M., Scheines, R., Spirtes, P. and Wasserman, L. (2003) "Uniform consistency in causal inference". Biometrika 90(3), 491–515.
Rossi, P. E., Allenby, G. M. and McCulloch, R. (2005) "Bayesian Statistics and Marketing". Wiley, Chichester.
Rubin, D. B. (1978) "Bayesian inference for causal effects: the role of randomisation". Annals of Statistics 6, 34–58.
Salmaron, A., Cano, A. and Moral, S. (2000) "Importance sampling in Bayesian networks using probability trees". Computational Statistics and Data Analysis 24, 387–413.
Santos, A. A. F. (2002) "A dynamic Bayesian analysis in statistical models used with certain financial risk problems". PhD thesis, University of Warwick.
Schervish, M. J. (1995) "The Theory of Statistics". Springer-Verlag, New York.
Shachter, R. D. (1986) "Evaluating influence diagrams". Operations Research 34, 871–882.
Shafer, G. (1976) "A Mathematical Theory of Evidence". Princeton University Press.
Shafer, G. R. (1996) "The Art of Causal Conjecture". MIT Press, Cambridge, MA.
Shafer, G. and Vovk, V. (2001) "Probability and Finance: It's Only a Game!". Wiley, Chichester.
Shafer, G. R., Gillett, P. R. and Scherl, R. (2000) "The logic of events". Annals of Mathematics and Artificial Intelligence 28, 315–389.

Silander, T. Kontkanen, P. and Myllymaki, P. (2007) "On sensitivity of the MAP Bayesian network structure to the equivalent sample size parameter". In: Proceedings of the 23rd Conference on Uncertainty in Artificial Intelligence (eds. R. Parr and L. van der Gaag). AUAI Press, pp. 360–367.

Small, C. G. and McLeish, D. L. (1994) "Hilbert Space Methods in Probability and Statistical Inference". John Wiley and Sons, Chichester.

Smith, J. Q. (1977) "Problems in Bayesian statistics relating to discontinuous phenomena, catastrophe theory and forecasting". PhD Thesis, Warwick University.

Smith, J. Q. (1979a) "A generalisation of the Bayesian steady forecasting model". Journal of the Royal Statistical Society B 41, 375–387.

Smith, J. Q. (1979b) "Mixture catastrophes and Bayes decision theory". Mathematical Proceedings of the Cambridge Philosophical Society 86, 91–101.

Smith, J. Q. (1980a) "Bayes estimates under bounded loss". Biometrika 67(3), 629–638.

Smith, J. Q. (1980b) "The prediction of prison riots". British Journal of Mathematical and Statistical Psychology 33, 151–160.

Smith, J. Q. (1981) "Search effort and the detection of faults". British Journal of Mathematical and Statistical Psychology 34, 181–193.

Smith, J. Q. (1983) "Forecasting accident claims in an assurance company". Statistician 32, 109–115.

Smith, J. Q. (1985) "Diagnostic checks of non-standard time series models". Journal of Forecasting 4, 283–291.

Smith, J. Q. (1988a) "Decision Analysis: A Bayesian Approach". Chapman and Hall, London.

Smith, J. Q. (1988b) "Models, Optimal Decisions and Influence Diagrams". In: Bayesian Statistics 3 (eds. J. M. Bernardo, M. H. DeGroot, D. V. Lindley and A. F. M. Smith). Oxford University Press, pp. 765–776.

Smith, J. Q. (1989a) "Influence diagrams for Bayesian decision analysis". European Journal of Operational Research 40, 363–376.

Smith, J. Q. (1989b) "Influence diagrams for statistical modelling". Annals of Statistics 17, 654–672.

Smith, J. Q. (1990) "Statistical principles on graphs" (with discussion). In: Influence Diagrams, Belief Nets and Decision Analysis (eds. J. Q. Smith and R. M. Oliver). Wiley, Chichester, pp. 89–120.

Smith, J. Q. (1992) "A comparison of the characteristics of some Bayesian forecasting models", International Statistical Reviews, 60, 1, 75–87.

Smith, J. Q. (1994) "Decision influence diagrams and their uses". In: Decision Theory and Decision Analysis: Trends and Challenges (ed. S. Rios). Kluwer, Dordrecht, pp. 32–51.

Smith, J. Q. (1996) "Plausible Bayesian games". Bayesian Statistics 5 (eds. J. M. Bernardo *et al.*). Oxford University Press, pp. 551–560.

Smith, J. Q. (2007) "Local Robustness of Bayesian Parametric Inference and Observed Likelihoods". CRiSM Research Report 07–09. CRiSM, University of Warwick.

Smith, J. Q. and Allard, C. T. J. (1996) "Rationality, conditional independence and statistical models of competition". In: Computational Learning and Probabilistic Reasoning (ed. A. Gammerman). Wiley, Chichester, pp. 237–256.

Smith, J. Q. and Anderson, P. E. (2008) "Conditional independence and chain event graphs". Artificial Intelligence 172(1), 42–68.

Smith, J. Q. and Croft, J. (2003) "Bayesian networks for discrete multivariate data: an algebraic approach to inference". Journal of Multivariate Analysis 84, 387–402.

Smith, J. Q. and Daneshkhah, A. (2009) "On the robustness of Bayesian networks to learning from non-conjugate sampling". International Journal of Approximate Reasoning 51, 558–572.
Smith, J. Q. and Faria, A. E. (2000) "Bayesian Poisson models for the graphical combination of dependent expert information". Journal of the Royal Statistical Society B 62(3), 525–544.
Smith, J. Q. and Figueroa-Quiroz, L. J. (2007) "A causal algebra for dynamic flow networks". In: Advances in Probabilistic Graphical Models (eds. P. Lucas, J. A. Gamez and A. Salmeron). Springer-Verlag, New York, pp. 39–54.
Smith, J. Q. and French, A. (1993) "Bayesian updating of atmospheric dispersion models for use after an accidental release of radio-activity". The Statistician 42(5), 501–511.
Smith, J. Q. and Papamichail, K. N. (1999) "Fast Bayes and the dynamic junction forest". Artificial Intelligence 107, 99–124.
Smith, J. Q. and Queen, C. M. (1996) "Bayesian models for sparse probability tables". Annals of Statistics 24(5), 2178–2198.
Smith, J. Q. and Rigat, F. (2009) "Isoseparation and Robustness in Finite Parameter Bayesian Inference". CRISM Research Report 07–22. CRiSM, University of Warwick.
Smith, J. Q. and Thwaites, P. (2008a) "Decision trees". In: Encyclopaedia of Quantitative Risk Analysis and Assessment, 2 (eds. E. L. Melnick and B. S. Everitt). Wiley, Chichester, pp. 462–470.
Smith, J. Q. and Thwaites, P. (2008b) "Influence Diagrams". In: Encyclopedia of Quantitative Risk Analysis and Assessment 2 (eds. E. L. Melnick and B. S. Everitt). Wiley, Chichester.
Smith, J. Q., Faria, A. E., French, S. *et al.* (1997) "Probabilistic data assimilation with RODOS". Radiation Protection Dosimetry 73(1–4), 57–59.
Smith, J. Q., Harrison, P. J. and Zeeman, E. C. (1981) "The analysis of some discontinuous decision processes". European Journal of Operations Research 7(1), 30–43.
Smith, J. Q., Anderson, P. E. and Liverani, S. (2008a) "Separation measures and the geometry of Bayes factor selection for classification". Journal of the Royal Statistical Society B 70, Part 5, 957–980.
Smith, J. Q., Dodd, L. and Moffat, J. (2008b) "Devolving Command under Conflicting Military Objectives". CRiSM Research Report 08–09. CRiSM, University of Warwick.
Speigelhalter, D. and Lauritzen, L. (1990) "Sequential updating of conditional probabilities on directed graphical structures". Networks 20, 579–605.
Speigelhalter, D. J. and Knill-Jones, R. P. (1984) "Statistical knowledge-based approaches to clinical decision support systems with applications to gastroenterology". Journal of the Royal Statistical Society A 147, 35–77.
Spiegelhalter, D. J., David, A. P., Lauritzen, S. L. and Cowell, R. G. (1993) "Bayesian analysis in expert systems" (with discussion). Statistical Science 8, 219–263.
Spirtes, P., Glymour, C. and Scheines, R. (1993) "Causation, Prediction, and Search". Springer-Verlag, New York.
Steck, H. (2008) "Learning the Bayesian network structure: Dirichlet prior versus data". In: UAI, Proceedings of the 24th Conference in Uncertainty in Artificial Intelligence (eds. D. A. McAllester and P. Myllymaki). AUAI Press, pp. 511–518.
Steck, H. and Jaakkola, T. (2002) "On the Dirichlet prior and Bayesian regularization". In: Advances in Neural Information Processing Systems [Neural Information Processing Systems, NIPS] (eds. S. Becker, S. Thrun and K. Obermayer). MIT Press, Cambridge, MA, 697704 27.

Studeny, M. (1992) "Conditional independence relations have no finite complete characterization". In: Information Theory, Statistical Decision Functions and Random Processes Transactions of 11th Prague Conference, Vol B (eds. S. Kubik and J. A. Visek). Kluwer, Dordrecht, pp. 377–396.

Studeny, M. (2005) "Probabilistic Conditional Independence Structures". Springer-Verlag, New York.

Tarjan, R. and Yannakakis, M. (1984) "Simple linear time algorithms to test chordality of graphs, test acyclicity of hypergraphs and selectively reduce acyclic hypergraphs". SIAM Journal of Computing 13, 566–579.

Tatman, A. and Shachter, R. D. (1990) "Dynamic programming and influence diagrams". IEEE Transactions on Systems, Man and Cybernetics 20(2), 365–379.

Thwaites, P., Smith, J. Q. and Cowell, R. (2008) "Propagation using chain event graphs". In: Proceedings of the 24th Conference in Uncertainty in Artificial Intelligence (eds. D. McAllester and P. Myllymaki), Helsinki, July 2008. AUAI Press, pp. 546–553.

Thwaites, P., Smith, J. Q. and Riccomagno, E. (2010) "Causal analysis with chain event graphs". Artificial Intelligence 174, 889–909.

Tian, J. (2008) "Identifying dynamic sequential plans". Proceedings of the 24th Annual Conference on Uncertainty in Artificial Intelligence (UAI-08), pp. 554–561.

Verma, T. and Pearl, J. (1988) "Causal networks: semantics and expressiveness". In: Uncertainty in Artificial Intelligence IV (eds. R. D. Schachter *et al.*). North-Holland, Amsterdam, pp. 69–76.

Verma, T. and Pearl, J. (1990) "Equivalence and synthesis of causal models". In: Proceedings of the 6th Conference on Uncertainty in Artificial Intelligence (eds P. Bonisson *et al.*). North-Holland, Amsterdam, pp. 255–270.

Vomelelova, M. and Jensen, F. V. (2002) "An extension of lazy evaluation for influence diagrams avoiding redundant variables in the potentials". In: Proceedings of the First European Workshop on Probabilistic Graphical Models (eds. J. A. Gamez and A. Salmeron). Cuenca, Spain.

von Winterfeldt, D. and Edwards, W. (1986) "Decision Analysis and Behavioural Research". Cambridge University Press.

Wakefield, J. C., Zhou, C. and Self, S. F. (2003) "Modelling gene expression over time: curve clustering with informative prior distributions". In: Bayesian Statistics 7 (eds. J. M. Bernardo *et al.*). Oxford University Press, pp. 721–732.

Walley, P. (1991) "Statistical Reasoning and Imprecise Probabilities". Chapman and Hall, London.

Wasserman, L. (1992) "Invariance properties of density ratio priors". Annals of Statistics 20, 2177–2182.

Wasserman, L. (1996) "The conflict between improper priors and robustness". Journal of Statistical Planning and Inference 52(1), 1–15.

West, M. and Harrison, P. J. (1997) "Bayesian Forecasting and Dynamic Models". Springer-Verlag, New York.

Whittaker, J. (1990) "Graphical Models on Applied Multivariate Statistics". Wiley, Chichester.

Whittle, P. (1990) "Risk sensitive optimal control". Wiley, Chichester.

Wilkie, M. E. and Pollack, A. C. (1996) "An application of probability judgement accuracy measures to currency forecasting". International Journal of Forecasting 12, 25–40.

Winkler, R. L. (1986) "Expert resolution". Management Science 32, 298–303.

Wright, G., Rowe, G., Bolger, F. and Gammack, J. (1994) "Coherence, calibration and expertise in judgemental probability forecasting". Human Decision Processes 57, 1–25.

Xiang, Y. (2002) "Probabilistic Reasoning in Multiagent Systems". Cambridge University Press.

Yates, J. F. and Curley, S. P. (1985) "Conditional distribution analyses of probabilistic forecasts". Journal of Forecasting 4, 61–73.

Index